**Smart Engineering Systems: Design and Applications**
*Series Editor - Suman Lata Tripathi*

**Internet of Things**
*Robotic and Drone Technology*
Edited by Nitin Goyal, Sharad Sharma, Arun Kumar Rana, Suman Lata Tripathi

**Smart Electrical Grid System**
*Design Principle, Modernization, and Techniques*
Edited by Krishan Arora, Suman Lata Tripathi and Sanjeevikumar Padmanaban

**Artificial Intelligence, Internet of Things (IoT) and Smart Materials for Energy Applications**
Edited by Mohan Lal Kolhe, Kailash J. Karande and Sampat G. Deshmukh

**Artificial Intelligence for Internet of Things**
*Design Principle, Modernization, and Techniques*
Edited by N. Thillaiarasu, Suman Lata Tripathi and V.Dhinakaran

**Machine Learning and Internet of Things in Solar Power Generation**
Edited by Prabha Umapathy, Jude Hemanth, Shelej Khera, Abinaya Inbamani and Suman Lata Tripathi

**Microgrid**
*Design, Optimization, and Applications*
Edited by Amit Kumar Pandey, Sanjeevikumar Padmanaban, Suman Lata Tripathi, Vivek Patel, Vikas Patel

**Transforming Agriculture with Emerging Technologies for a Sustainable Future**
Shanu Sharma, Ayushi Prakash, Sandhya Avasthi and Vijayan Sugumaran

**Photovoltaic Materials**
*Synthesis, Characterizations and Applications*
Edited by Sampat G. Deshmukh, Suman Lata Tripathi and Abhishek Kumar

**Sustainable Technologies and Devices for Next Generation Wireless Communication**
Edited by Indu Bala, Ghanshyam Singh and Suman Lata Tripathi

**Chips and Intelligence**
*Low Power VLSI Design with Artificial Intelligence Theory*
Edited by Abhishek Kumar, Smrity Dwivedi, Jyotirmoy Pathak, Jyoti Kandpal and Suman Lata Tripathi

For more information about this series, please visit: https://www.routledge.com/

# Artificial Intelligence in Healthcare

## Advancements, Security, and Internet of Things

Edited by
Smita Nirkhi Singh
Vivek Kapur
Suman Lata Tripathi
Ved Prakash Mishra

CRC Press
Taylor & Francis Group
Boca Raton London New York

CRC Press is an imprint of the
Taylor & Francis Group, an **informa** business

Cover: Shutterstock.com

First edition published 2027
by CRC Press
2385 NW Executive Center Drive, Suite 320, Boca Raton FL 33431

and by CRC Press
4 Park Square, Milton Park, Abingdon, Oxon, OX14 4RN

*CRC Press is an imprint of Taylor & Francis Group, LLC*



ISBN: 9781032762609 (hbk)
ISBN: 9781041390084 (pbk)
ISBN: 9781042015931 (ebk)

DOI: 10.1201/9781042015931

Typeset in Sabon
by KnowledgeWorks Global Ltd.

# Contents

*Editors biographies* vii
*List of contributors* ix
*Preface* xiii

1 The current state of healthcare and the potential of artificial intelligence to transform the industry 1
PRAVEEN KUMAR MALIK

2 AI-powered anonymization for secure medical data sharing: Future trends and research directions 13
N. SUGANYA, P. GOUTHAMI, AND M. KRISHNAMOORTHI

3 Harnessing the potential of artificial intelligence for transformative healthcare solutions 36
TARUN KUMAR VASHISHTH, VIKAS SHARMA, APARNA KUMARI, MUKESH KUMAR SHARMA, SHAHANAWAJ AHAMAD, AND SARIKA JAIN

4 Detection of Parkinson's disease through voice samples 57
ABHISHEK KUMAR AND SOBHIT SAXENA

5 Understanding supervised and unsupervised learning paradigms in healthcare artificial intelligence 70
SANDEEP GUPTA AND UDIT MAMODIYA

6 AI applications in drug discovery and personalized medicine 90
KIRAN MALIK, KULDEEP SINGH KASWAN, AND JAGJIT SINGH DHATTERWAL

7 Performance evaluation of a multilayer model of convolutional neural network (CNN) to predict COVID-19 using X-ray of chest samples 114

ARKA PRAVO NANDI, SUDIP GHOSH, SUMAN LATA TRIPATHI, PIYALI SAHA, AISHI PRAMANIK, RAJU HAZARI, AND HAFIZUR RAHAMAN

8 Intelligent diabetes diagnosis using bioinspired algorithms integrated with machine learning and deep learning systems 136

S. SHEIK ASRAF, P. NAGARAJ, AND V. MUNEESWARAN

9 Assessing COVID-19 transmission patterns and government responses in India 149

RAKESH KUMAR, MANIK RAKHRA, RAMANDEEP SANDHU, GAGANDEEP SINGH CHEEMA, AND DEEPIKA GHAI

10 Wheelchair movements control model implementation with FPGA 166

SUMAN LATA TRIPATHI AND INUNG WIJAYANTO

11 Overview of emerging AI technologies and their potential impact on healthcare 180

SHIVA KHEMKA, KAPIL KUMAR GUPTA, AND NAMRATA

12 Pharmacological frontier for drug discovery system: A new path towards personalized medicine 206

S. SHARMILA, V. PRAVEENA, S. VELMURUGAN, M. MENAGADEVI, AND S. ANBUMANI

13 Revolutionizing neurology with AI and IoT: Predictive analytics and real-time monitoring 222

M. MENAGADEVI, M. NIRMALA, D. THIYAGARAJAN, AND SUMAN LATA TRIPATHI

14 Deep learning: A transformative force in artificial intelligence 236

MEENA RAWAT, DHRUV SHAKTI, AND POOJA PATHAK

15 Case studies on the use of AI in precision medicine and targeted therapies 266

NISHANT SHARMA AND SHILPI KULSHRESTHA

*Index* 287

# Editors biographies

**Smita Nirkhi Singh** is a distinguished academic and researcher with a PhD in Computer Science and Engineering. Currently, she is working as Associate Professor at Symbiosis Institute of Technology, Nagpur. She has 23 years of experience in teaching and research, with expertise in areas such as artificial intelligence, data mining, digital forensics, cybersecurity, pattern recognition, machine learning, and deep learning. Dr. Nirkhi has received numerous accolades, including the Outstanding Woman in Engineering Award (2019) by Venus International Foundation, Best Teacher Award (2021) by Universal Innovators Leadership, Nari Samman Award (2021) by Literoma, and the Elite Teacher Award (2021) by I2OR. Dr. Nirkhi has published over 75 research papers in esteemed journals and international conferences.

Dr. Nirkhi serves as Editor for *Scientific Reports* and is an active peer reviewer and editorial board member for leading journals and IEEE conferences. She is Senior Member of IEEE, Fellow of the Institution of Engineers (India), and an active member of CSI, ISTE, and the Internet Society. She also leads various initiatives as the Branch Counselor for the IEEE Student Branch, Secretary of IEEE Maharashtra section, and Chair of the System Council Chapter of the IEEE Maharashtra section.

**Vivek Kapur** is working as Director at GHRCEM, Nagpur. A professional and dedicated academician with 20+ years of cross-functional experience in the areas of strategic planning, project management, quality assurance and quality enhancement of technical institutions, outcome-based education, and signal and image processing. He is Approved Guide (PhD/MTech) of Nagpur University. He is appointed as NBA Evaluator by the National Board of Accreditation, New Delhi. Also, he is a member of the delegations of the top universities of the United States. He is an active participant/contributor in NBA initiatives related to quality assurance and quality enhancement within Maharashtra. He is also a member of various LEC Committees. He is a professional member of the IEI, ISTE, and a senior member of IEEE.

**Suman Lata Tripathi** is a Professor in the Department of E & TC, Symbiosis Institute of Technology, Symbiosis International (Deemed University),

Pune, India. She has also worked as a Professor at Lovely Professional University, India. She has more than 23 years of experience in academics and research. She completed her PhD in the area of microelectronics and VLSI Design from MNNIT, Allahabad. She completed her remote postdoc from Nottingham Trent University, London, UK, in 2022–2023. Her areas of expertise include microelectronics device modelling and characterization, low-power VLSI circuit design, VLSI design of testing, and advanced FET design for IoT, embedded system design, reconfigurable architecture with FPGAs and biomedical applications.

**Ved Prakash Mishra** is working as Head of Academics, Assistant Professor and Program Leader at Amity University Dubai, UAE. He is Program Leader, Senior Member-IEEE, PHD, Quality Control (IQAC) Coordinator, Conference Chair – ICCIKE2023, ICCIKE2021, and ICCIKE2019. He has more than 15 years of experience. His areas of expertise include computer networks, compiler design, cryptography, network security, and artificial intelligence/machine learning.

# Contributors

**Shahanawaj Ahamad**
College of Computer Science and Engineering
University of Hail
Hail City, Saudi Arabia

**S. Anbumani**
Department of Electronics and Communication Engineering
Dr. N. G. P Institute of Technology
Coimbatore, Tamil Nadu, India

**S. Sheik Asraf**
Department of Biotechnology, School of Bio, Chemical and Processing Engineering
Kalasalingam Academy of Research and Education (Deemed to be University)
Krishnankoil, Tamil Nadu, India

**Gagandeep Singh Cheema**
Mittal School of Business
Lovely Professional University
Phagwara, Punjab, India

**Namrata Dhanda**
Department of Computer Science and Technology, ASET
Amity University
Noida, Uttar Pradesh, Lucknow, India

**Jagjit Singh Dhatterwal**
School of Computer Science & Artificial Intelligence
SR University
Warangal, Telangana, India

**Deepika Ghai**
School of Electronics and Electrical Engineering
Lovely Professional University
Phagwara, Punjab, India

**Sudip Ghosh**
School of VLSI Technology
Indian Institute of Engineering Science & Technology
Shibpur, West Bengal, India

**P. Gouthami**
Department of Computer Science and Business Systems
Dr N.G.P. Institute of Technology
Coimbatore, Tamil Nadu, India

**Kapil Kumar Gupta**
Department of Computer Science and Technology, ASET
Amity University
Noida, Uttar Pradesh, Lucknow, India

**Sandeep Gupta**
Department of Artificial Intelligence and Data Science
Poornima Institute of Engineering and Technology
Jaipur, Rajasthan, India

**Raju Hazari**
School of VLSI Technology
Indian Institute of Engineering Science & Technology
Shibpur, West Bengal, India

**Sarika Jain**
Department of Computer Applications
National Institute of Technology
Kurukshetra, Haryana, India

**Kuldeep Singh Kaswan**
School of Computer Science and Engineering
Galgotias University
Greater Noida, Uttar Pradesh, India

**Shiva Khemka**
Department of Computer Science and Technology, ASET
Amity University
Noida, Uttar Pradesh, Lucknow, India

**M. Krishnamoorthi**
Department of Information Technology
Dr. N.G.P. Institute of Technology
Coimbatore, Tamil Nadu, India

**Shilpi Kulshrestha**
Department of CSE
Vivekananda Global University
Jaipur, Rajasthan, India

**Abhishek Kumar**
School of Electronics and Electrical Engineering
Lovely Professional University
Phagwara, Punjab, India

**Rakesh Kumar**
Department of Mathematics
Lovely Professional University
Phagwara, Punjab, India

**Aparna Kumari**
Department of Computer Science and Engineering
Institute of Technology, Nirma University
Ahmedabad, Gujarat, India

**Kiran Malik**
Department of Computer Science and Engineering (AIML)
GL Bajaj Institute of Technology & Management
Greater Noida, Uttar Pradesh, India

**Praveen Kumar Malik**
School of Electronics and Electrical Engineering
Lovely Professional University
Phagwara, Punjab, India

**Udit Mamodiya**
Faculty of Engineering and Technology
Poornima University
Jaipur, Rajasthan, India

**M. Menagadevi**
Department of Computing Technologies
School of Computing, Faculty of Engineering and Technology, SRM Institute of Science and Technology
Chennai, Tamil Nadu, India

**V. Muneeswaran**
Department of Electronics and Communication Engineering, School of Electronics, Electrical and Biomedical Technology
Kalasalingam Academy of Research and Education (Deemed to be University)
Krishnankoil, Tamil Nadu, India

**P. Nagaraj**
Department of Computer Science and Engineering, School of Computing
SRM Institute of Science and Technology
Trichy, Tamil Nadu, India

**Arka Pravo Nandi**
School of VLSI Technology
Indian Institute of Engineering Science & Technology (IIEST)
Shibpur, West Bengal, India

**M. Nirmala**
Department of Biomedical Engineering
Dr. N G P Institute of Technology
Coimbatore, Tamil Nadu, India

**Pooja Pathak**
Department of Computer Engineering & Applications
GLA University
Mathura, Uttar Pradesh, India

**Aishi Pramanik**
School of VLSI Technology
Indian Institute of Engineering Science & Technology
Shibpur, West Bengal, India

**V. Praveena**
Department of Computer Science and Engineering
Dr. N. G. P Institute of Technology
Coimbatore, Tamil Nadu, India

**Hafizur Rahaman**
School of VLSI Technology
Indian Institute of Engineering Science & Technology
Shibpur, West Bengal, India

**Manik Rakhra**
School of Computer Science & Engineering
Lovely Professional University
Phagwara, Punjab, India

**Meena Rawat**
Department of Applied Mathematics
Delhi Technological University
Delhi, India

**Piyali Saha**
School of VLSI Technology
Indian Institute of Engineering Science & Technology
Shibpur, West Bengal, India

**Ramandeep Sandhu**
School of Computer Science & Engineering
Lovely Professional University
Phagwara, Punjab, India

**Sobhit Saxena**
School of Electronics and Electrical Engineering
Lovely Professional University
Phagawara, Punjab, India

**Dhruv Shakti**
School of Information & Communication Studies
University College Dublin
Dublin, Ireland

**Mukesh Kumar Sharma**
Department of Mathematics
Chaudhary Charan Singh University
Meerut, Uttar Pradesh, India

**Nishant Sharma**
Department of CSE
Vivekananda Global University
Jaipur, Rajasthan, India

**Vikas Sharma**
Department of Computer Applications
SRM Institute of Science and Technology
Ghaziabad, Uttar Pradesh, India

**S. Sharmila**
Department of Electronics and Communication Engineering
Dr. N. G. P Institute of Technology
Coimbatore, Tamil Nadu, India

**N. Suganya**
Department of Computer Science and Business Systems
Dr. N.G.P. Institute of Technology
Coimbatore, Tamil Nadu, India

**D. Thiyagarajan**
Department of Computer Science and Engineering
Chennai Institute of technology
Chennai, Tamil Nadu, India

**Suman Lata Tripathi**
Department of E&TC, Symbiosis Institute of Technology
Symbiosis International (Deemed University)
Pune, Maharashtra, India

**Tarun Kumar Vashishth**
School of Computer Science and Applications
IIMT University
Meerut, Uttar Pradesh, India

**S. Velmurugan**
Department of Electronics and Communication Engineering
K. S. Rangasamy College of Technology
Namakkal, Tamil Nadu, India

**Inung Wijayanto**
Department of Electrical Engineering
Telkom University
Bandung, Indonesia

# Preface

This book explores the applications of artificial intelligence (AI) in healthcare systems, with a focus on its use in diagnosis, treatment, and prevention. The book covers a range of topics, including the use of AI in medical imaging, personalized treatment plans, and the identification of high-risk patients. The book aims at healthcare professionals, researchers, and students who are interested in the applications of AI in healthcare. It provides an overview of the current state of AI in healthcare and discusses the challenges and opportunities of implementing intelligent healthcare systems. This book also includes a discussion of the different types of AI (e.g. machine learning, natural language processing, and robotics) and how they are being used to improve patient care, research, and operations in healthcare settings. It focuses on how AI is being used to diagnose diseases, create personalized treatment plans, predict outcomes, and manage patient data along with IoHT (the Internet of Healthcare Things).

The IoHT (the Internet of Healthcare Things) is revolutionizing healthcare by enabling remote patient monitoring, personalized medicine, and improved healthcare outcomes. By collecting and analysing data from various sources, such as medical devices, wearables, and electronic health records (EHRs), healthcare providers can gain new insights into patient health and behaviour, detect potential health issues earlier, and provide more targeted and personalized treatment. Some examples of IoHT devices include wearable fitness trackers, smart scales, blood glucose monitors, blood pressure monitors, and ECG monitors. These devices can be used to collect data on a patient's activity level, heart rate, blood pressure, blood sugar levels, and other vital signs. The data can then be analysed and used to inform treatment decisions, track progress, and identify potential health issues.

No discussion of AI in healthcare would be complete without a discussion of the challenges and limitations of this technology. The book explores issues such as data privacy and security, bias and fairness in algorithms, and the need for human oversight and intervention. The book finally concludes with a look at the future of AI in healthcare. It discusses the need for ongoing research and development in AI to ensure that it continues to advance and improve patient outcomes. The book speculates on the future of AI in healthcare, including potential new applications of AI in healthcare and the impact that AI could have on the healthcare industry.

Author (s)
Smita Nirkhi Singh
Vivek Kapur
Suman Lata Tripathi
Ved Prakash Mishra

Chapter 1

# The current state of healthcare and the potential of artificial intelligence to transform the industry

*Praveen Kumar Malik*

## 1.1 INTRODUCTION

The system that helps people stay healthy, get treated when sick, and prevent disease is called "healthcare". Across the world, the system faces many challenges like growing costs, short of access, lack of trained staff, and growing number of people with their chronic illnesses like diabetes and heart diseases and complicated administrative tasks [1–5]. The healthcare system, at the same time, provides huge amounts of data like patient records, images, X-ray, lab results, and several device signals. Thus, data could help to improve care of the patient if it is used in a right manner. Computer programs which can learn from data and make quick decision and prediction is known as "artificial intelligence" (AI). Today, AI is adapted quickly and has become user-friendly in numerous fields, including medical industry. Thus, in this chapter, we will study the current scenario of healthcare centres and how AI can help solve problems by providing practical solutions for integrating AI into real-world healthcare systems [6–10].

## 1.2 HEALTHCARE IN TODAY'S STATE

### 1.2.1 Care pathway

The growing access to healthcare varies differently. In various cities and countries, people have easy access to visit doctors, get tests done, and obtain medicines easily. But in rural areas or underprivileged areas which have low incomes, there are difficulties like scarcity of doctors, longer travel times, and very limited access or pathway to get clinical support. This leads to worse health for many people and inaccessibility to care.

### 1.2.2 Standard of care

Quality and standardisation mean how we can be able to use healthcare for the treatment as well as prevention of diseases. Standard varies because of differences in training, availability of technology, and local protocols. In various

DOI: 10.1201/9781042015931-1

situations, patients get delayed diagnoses or treatment due to lack of systems and resources or unavailability of technology.

### 1.2.3 Market price of care

Healthcare is something which people should get with ease. But rather than ease, the people are overburdened with the rising price of healthcare across the world. Sometimes people have to spend more than their earnings on treatment just to survive. The rising costs of healthcare has not only become unaffordable for every single individual and but also stressful for the budget of a nation.

### 1.2.4 Labour force challenges

Numerous countries face severe shortages of nurses, doctors, and health professionals due to extreme pressure, competitive qualifications, and heavy workloads which often lead to burnout, driving many out of the profession. This affects safety and standard of care in healthcare.

### 1.2.5 Growing burden of chronic diseases

Chronic diseases or non-communicable diseases like diabetes, heart diseases, and chronic respiratory conditions are increasing in a rapid way across the world. These diseases require prolonged care and management for better results, regular check-ups, tests, and use of vital health indicators that keep track of the diseases and thereby give better results. The record of health is often stored at different places like hospitals, labs clinics, pharmacies, and many more. But these systems may not coordinate with each other, thereby leading to misunderstanding. In the absence of data and updated reports, it becomes difficult to get a complete picture of a patient's health. Healthcare makes use of different technologies like electronic health records (EHRs), electrocardiogram (EEG), medical imaging (X-rays, MRIs), laboratory systems, remote monitoring devices, and many more, Telemedicine and online virtual meets have become more significant, especially after the COVID-19 pandemic [11–17]. Nowadays administrative workload has become one of the most difficult tasks for doctors and nurses as they have to spend their several working hours on paperwork, coding, billing, data entry, and many other administrative works. As these processes need to be performed meticulously, accurately, they often consume valuable time that could otherwise have been devoted to the care of patient, as shown in Figure 1.1.

## 1.3 WHY AI IS FUTURE FOR HEALTHCARE?

AI can quickly examine large amounts of data in one go and find patterns that human may miss and also automate routine task which otherwise

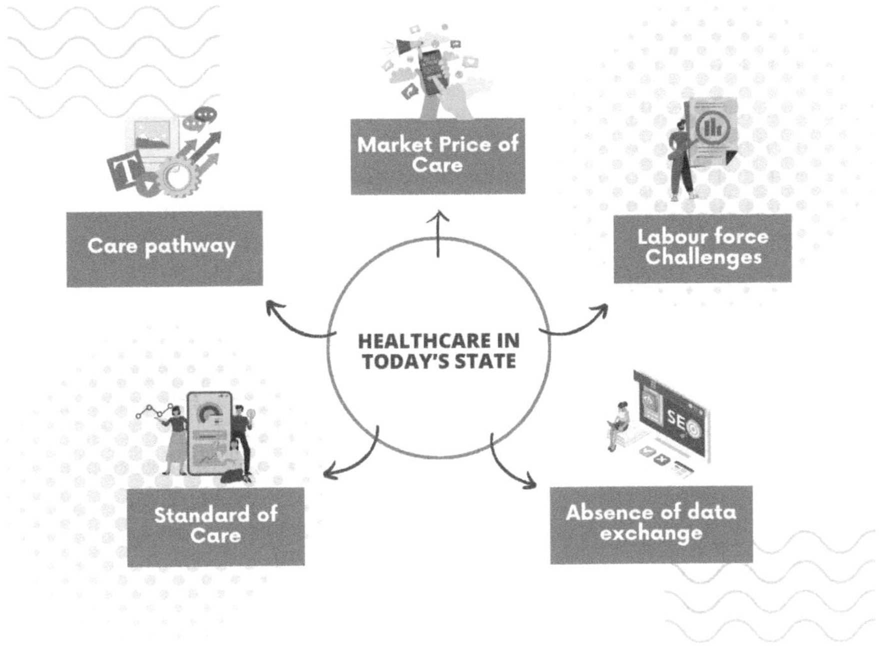


*Figure 1.1* Different stages of healthcare in today's state.

take time to schedule. The following are the key advantages of AI in healthcare:

- Improvisation of diagnosis by the patterns in image and lab data by recognizing them quickly.
- Based on patient's reports, history and genetics, AI can personalize them in a planned way.
- Administrative overload can be automated, thereby enabling staff to focus more time on patients and update their care.
- The predictions can be made easier and the risk of deterioration can be taken care of in advance and easily targeted.
- Faster drug discovery and the trial can be done to match up with the patients' care.

Thus, AI does not replace the staff, rather it helps them by making their work easier and support them by tools and insights for betterment in the quality, safety, and care which often get ignored due to extreme administrative overload and pressure.

## 1.4 KEY AI FEATURES IN THE HEALTHCARE

### 1.4.1 Medical imaging and diagnostics

AI has grown rapidly and has shown remarkable potential in the analysis medical imaging such as X-rays, CT scans, and MRIs (Figure 1.2). The technologies are able to identify abnormal or unexpected findings like tumours, fractures, or haemorrhages with high accuracy, providing subtle details that are mostly overlooked due to extreme pressure or workload. By pointing out specific areas of concern, AI helps the radiologists focus their attention on critical regions, thereby improving the diagnostic accuracy. AI can also prioritize urgent cases, promising the patients with quicker recovery – the quicker the life-threatening conditions checked, the better. As AI not only reduces the reading time for the radiologists but also make their work easier, thereby enabling large volumes of work get done fast while maintaining the quality care. AI thus serves as a powerful support system to the professionals which enable faster, more accurate, and more efficient imaging analysis.

### 1.4.2 Electronic health records (EHR) and decision support for clinic

AI can analyse EHR data to provide decision support. It suggests possibility of diagnoses, detects dangerous drug interactions, and also reminds clinicians about the preventive care on the basis of vaccinations and screenings.

### 1.4.3 Forecasting analytics

AI can predict and forecast which patients are at a higher risk or complications and need rehospitalization. Hospitals also use these features for allocating resources, planning, care, and reducing emergencies.

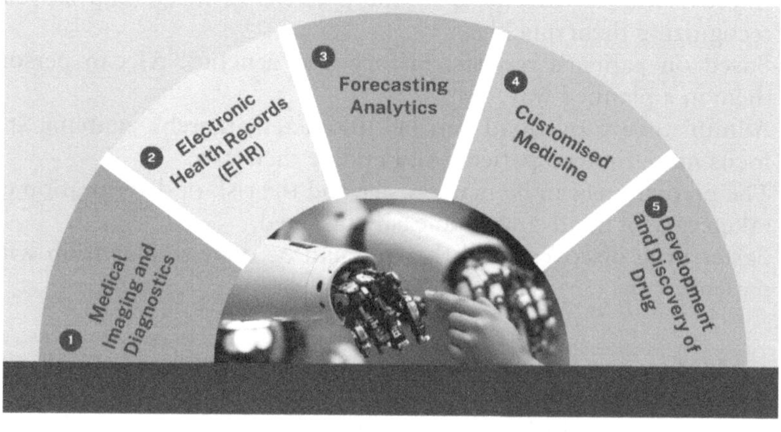


*Figure 1.2* Key features of artificial intelligence in the healthcare system.

### 1.4.4 Customized medicine

AI can combine both medical records, genetic information, lifestyle data, and also research findings to help treat a single patient. This can forecast in oncology (cancer care), where the guidance is done by genetic makers to therapy.

### 1.4.5 The development and discovery of drug

AI can detect chemical structures and biological data to mention new drug candidates quicker than any other traditional methods (Figure 1.3). It can also be helpful in designing clinical trials and matches up the patients to take trials based on detailed criteria and ideas. AI can also process data from various wearable devices, like heart rate and glucose monitors, activity trackers, and also warns the clinicians when a patient's data make danger signals. AI not only helps in support care at home but also reduces unnecessary clinical visits and thereby saves time. Further, AI helps in automation or scheduling, coding, billing, and documenting. AI can convert doctor and patients' conversation into a structured notes and thereby saving time; this is called natural language processing. AI can also examine trends and updates across populations to detect outbreaks, monitoring the vaccine coverage, and also plan health services accordingly. This really helps public health officials to respond quickly to community needs.

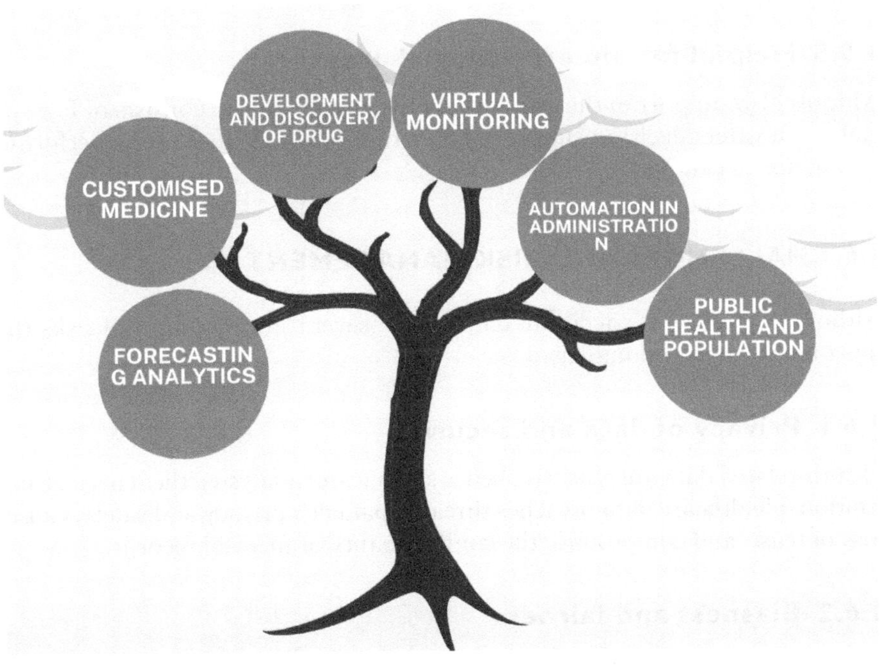


*Figure 1.3* Electronic health records and decision support for clinic.

## 1.5 ADVANTAGES OF AI IN HEALTHCARE

### 1.5.1 Improvisation of speed and efficiency

AI examines data faster and quicker than humans. Faster and accurate image reading and automated quicker reports speed up the treatment

### 1.5.2 Consistency with higher accuracy

AI tools can help reduce human errors by not only providing consistent analyses but also accurate data. For example, AI may detect a small pattern in images that human often overlook due to overpressure or workload.

### 1.5.3 Better use of resources

Predictive models help hospitals use beds, staff, and equipment more often and in a prominent manner. This helps in reduction of waste and lowers the costs, which often take another stage.

### 1.5.4 Improvement of patients experience

Automation helps in reduction of wait times and lot of paperwork. Virtual monitoring and telemedicine make care more convenient and useful for both clinicians and patients.

### 1.5.5 Helpful for underprivileged areas

AI tools also support in the areas where the clinical set-up consists of few specialists. It offers decision support and virtual readings of the tests performed in diagnostic centres.

## 1.6 CHALLENGES AND RISK MANAGEMENT

Although AI provides accurate data, it has several limitations and risks that must be checked and managed.

### 1.6.1 Privacy of data and security

Health-related data are sensitive, hence storing and analysing them need utmost caution. Healthcare data breaches threaten patient's privacy and safety, cause a loss of trust, and compromise the confidentiality of medical records.

### 1.6.2 Biasness and fairness

AI systems are based on historical data. In case of data based on less records from certain groups, AI can give biased or compromised results. This could be harmful for underserved populations.

### 1.6.3 Transparency and trust

Several AI models, specifically deep learning black boxes, often do not provide any clear or accurate results. Clinicians are therefore reluctant to rely on the systems, as they are often not able to understand the AI result. Transparent AI techniques are needed to build more trust among people.

### 1.6.4 Guidance and certification

Authority agencies like drug and device regulators and image tracking devices need clear standards for AI tools. Rules and regulations keep changing in many regions time to time. AI needs some high-quality and standardised data for better results in the field. Many health systems have their own way of using different formats and systems of coding. Poor data quality and lack of functionality reduce AI performance. AI tools must be installed in clinics for daily routine check-up for ease of work as well as accuracy. Tools that disturb or compromise work or make the work more complex should be avoided or rejected. If AI system contributes to a wrong or negative diagnosis or treatment, the question arises who should be held responsible? The software maker, the hospital, or the clinician? The clear legal framework is needed in this regard for better and accurate results. AI systems are generally very expensive as its development and maintenance cost is very high. Even low-resource setting require computing power, stable internet, and trained staff to use AI tools. The adaption of AI may change job roles as many tasks would be automated: new roles like data specialist and AI supervisors will appear, although training is essential for easy, adaptive, and efficient workflow.

## 1.7 ETHICAL PRINCIPLES

The ethical principles must be followed in healthcare AI:

- *Privacy and Confidentiality Should Be Respected:* Patients should have clear control over their data irrespective of the place for their trust and accessibility.
- *Transparency:* There should always be clarity when AI affects their care and how the computer or system takes a decision without any biasness.
- *Fairness:* AI should not use or have unfair disadvantage on the groups based on any race, gender, age, or socio-economic status, which would make a disrespectful thought between people.
- *Accountability:* There should be fair clarity between lines of responsibility for any AI-driven decisions.
- *Beneficence and non-maleficence:* AI should always improve health and avoid harm of any kind.

Ethical norms, review boards, patient continuous engagement, and accurate processes help ensure AI is used responsibly and accurately.

## 1.8 ROADMAP FOR IMPLEMENTING AI IN HEALTHCARE PRACTICALLY

The following steps are essential for a hospital or health systems to follow to get AI work safely and most effectively.

### 1.8.1 Mention the clear problems and goals

Get started by finding the real clinical problems, for example, the reduction in emergency departments, the care should be updated in diabetic sector. There should be clear goals which help to choose the right AI solution. The accuracy of data should be ensured, completed, and labelled accurately. The clear and standardized data should be provided with patient privacy protection before any form of use.

### 1.8.2 Always choose the right technology

Select the methods by which AI can match the problem; for example, for image task, deep learning must be useful and for rule-based alerts, the simpler methods could be employed, which might be easier for the clinicians. Small start with pilot studies with one department can be easier to get result and feedbacks. The measurement of performance, clinician acceptance, and patient outcomes must be obtained before scaling up with other departments.

### 1.8.3 Clinical workflow with integration

AI outputs are integrated into existing EHR or in the image viewer for smooth adoption for clinicians. A proper training should be provided to the concerned clinicians, nurses, and IIT staffs. The tool's strengths and limits must be elucidated for better understanding.

### 1.8.4 The oversight pf establish governance

The committees must be created to oversee data use, monitor performance, and handle complaints. The committees include clinicians, ethicists, IT staff, and patient representatives for better outcome and governance. AI models can have also have drawbacks; for example, it can require upgrading as practices change after a period of time. Regularity is necessary for better monitor performance and models should be updated with new data.

### 1.8.5 Evaluation of outcomes and cost-effectiveness

The study must be conducted for improvisation of patient outcomes and satisfaction as well as for cost reduction. These data are used to refine implementation.

## 1.9 EXAMPLES AND CASE ILLUSTRATIONS

The following are the simplified general examples to show how AI can be used and how helpful it would be.

### 1.9.1 The emergency department's triage

AI analyses vital signs and history which should be flagged and mentioned in patients like an urgent care which is really helpful to staff to prioritize patients and reduce waiting and overtime taking.

### 1.9.2 The detection of diabetic retinopathy

AI interprets retinal images to detect early signs of diabetes-related eye disease which can be used in primary care to screen large number of patients.

### 1.9.3 Hospital readmission prediction

A model is used significantly for patient's history, lab results, and social factors to predict who may be readmitted within 30 days and care teams can be planned for follow-up and support in advance.

### 1.9.4 Automation in radiology reports

AI summarizes the findings from CT scans and creates drafts of reports like radiologists review and finalizes the reports and saves time for care.

## 1.10 WILL AI CHANGE JOBS IN HEALTHCARE?

AI can change some tasks and create new ones:

- *Clinicians:* They can be spending less time on routine data entry and more time on complex specific decision-making and communicating more with patients and understanding their problems.
- *Nurses:* AI alerts are may beneficial and virtual monitoring help to better manage patients.
- *Pathologists and Radiologists:* It would be helpful as will be working with AI assistant and their role will evolve to supervising and validating AI results for better outcome.
- *Data Scientists and Health IT:* Demand for these roles will always grow as system needs continuous integration and maintenance of models.
- *Administrative Labour Force:* Some tasks like billing and scheduling may become more automated and workful.

Training and clear communication are important for the reduction of fear and resistance.

## 1.11 REGULATIONS AND POLICY CONSIDERATION

Governments and health agencies must start developing rules for safer use of AI tools:

- Clarity of approval pathways for clinical use with help of AI tools.
- Data privacy's quality should be always consent, transparent, and sharing.
- Guidelines must be followed to minimize bias and to test fairness across population without any discrimination.
- Post-deployment requirements need to be monitored and reports must be given of any harms.
- Policies on liabilities and accountability should be made whenever AI is used in care.

International collaboration can help set consistent standard and upgradation.

AI can improve healthcare equity by providing decision support in most of the places with few specialists. For example, AI can be helpful in the diagnosis of tuberculosis from chest X-ray in rural areas But if AI is developed using data from high-income countries, not only in low-resource settings but also developers must ensure diverse data and local assurance so that AI can be beneficial to everyone and they can adapt AI timely.

## 1.12 FUTURE DIRECTIONS

Key possibilities where AI may grow in the upcoming years:

- *Multifunctionality AI Model:* Systems which can combine images, text, genetics, and wearable data for better insights.
- *Explanatory AI:* Betterment in tools to show how decision were made in descriptive manner.
- *Collaborated Learning:* Training models for using data from different sites without moving private data, thereby improving privacy security.
- *Real-Time Virtual Monitoring:* The analysis of patient's data beforehand for early warnings.
- *AI in Worldwide Health:* Deploying lower cost AI tools for screening and diagnosis in rural regions enables to reach out more people.

AI or any technology can be improved if the policy, data, and social acceptance is assured by the people.

## 1.13 CONCLUSION

Healthcare faces many challenges and downfall due to growing costs, population, and diseases among people, unequal access, labour staff shortages, and rising non-communicable diseases. AI is a boon which help solve several problems that are often overlooked by clinicians due to extreme work pressure such as improving diagnosis, personalising treatment, automating routine tasks, and predicting risks and early warnings. AI would not cure everything in one go, but it will help staff for the betterment. To get success in the field, AI tool must be with provided high-quality data, tested carefully, aligned with clinical workflows, and often governed by strong ethical norms and regulations. The fairness, privacy, transparency, and local validation must be followed very often to improvise health for everyone, specifically in rural areas. AI can be used very powerfully by clinicians as a helping tool but not as their replacement. The goal should always be clear, safe, accurate, and transparent effective healthcare for both patient and clinicians.

## REFERENCES


1. J. Gawande and M. Bhange, "Artificial Intelligence in Tools Pharamaceutical Research and Healthcare Drug Discovery," *2024 International Conference on Healthcare Innovations, Software and Engineering Technologies (HISET)*, Karad, India, 2024, pp. 296–299. doi: 10.1109/HISET61796.2024.00092.
2. M. Nazar, M. M. Alam, E. Yafi and M. M. Su'ud, "A Systematic Review of Human–Computer Interaction and Explainable Artificial Intelligence in Healthcare with Artificial Intelligence Techniques," in *IEEE Access*, vol. 9, pp. 153316–153348, 2021. doi: 10.1109/ACCESS.2021.3127881.
3. S. S. Bhuyan, V. Sateesh, N. Mukul et al., "Generative Artificial Intelligence Use in Healthcare: Opportunities for Clinical Excellence and Administrative Efficiency," in *Journal of Medical Systems*, vol. 49, p. 10, 2025. https://doi.org/10.1007/s10916-024-02136-1
4. S. Bharati, M. R. H. Mondal and P. Podder, "A Review on Explainable Artificial Intelligence for Healthcare: Why, How, and When?" in *IEEE Transactions on Artificial Intelligence*, vol. 5, no. 4, pp. 1429–1442, 2024. doi: 10.1109/TAI.2023.3266418.
5. I. Ahmed, G. Jeon and F. Piccialli, "From Artificial Intelligence to Explainable Artificial Intelligence in Industry 4.0: A Survey on What, How, and Where," in *IEEE Transactions on Industrial Informatics*, vol. 18, no. 8, pp. 5031–5042, 2022. doi: 10.1109/TII.2022.3146552.
6. M. J. Saikia and D. Borthakur, "Artificial Intelligence (AI) in Patient–Healthcare Relationships: Psychosocial Perspective," in *IEEE Access*, vol. 13, pp. 164682–164701, 2025. doi: 10.1109/ACCESS.2025.3611818.
7. S. A. Alowais, S. S. Alghamdi, N. Alsuhebany et al., "Revolutionizing Healthcare: The Role of Artificial Intelligence in Clinical Practice," in *BMC Medical Education*, vol. 23, p. 689, 2023. https://doi.org/10.1186/s12909-023-04698-z
8. K.-L. A. Yau et al., "Augmented Intelligence: Surveys of Literature and Expert Opinion to Understand Relations between Human Intelligence and Artificial Intelligence," in *IEEE Access*, vol. 9, pp. 136744–136761, 2021. doi: 10.1109/ACCESS.2021.3115494.

9. W. Jiao, X. Zhang and F. D'Souza, "The Economic Value and Clinical Impact of Artificial Intelligence in Healthcare: A Scoping Literature Review," in *IEEE Access*, vol. 11, pp. 123445–123457, 2023. doi: 10.1109/ACCESS.2023.3327905.
10. P. K. Nag, A. Bhagat and R. V. Priya, "Expanding AI's Role in Healthcare Applications: A Systematic Review of Emotional and Cognitive Analysis Techniques," in *IEEE Access*, vol. 13, pp. 69129–69160, 2025. doi: 10.1109/ACCESS.2025.3562131.
11. M. Badawy, N. Ramadan and H. A. Hefny, "Healthcare Predictive Analytics Using Machine Learning and Deep Learning Techniques: A Survey," in *Journal of Electrical Systems and Information Technology*, vol. 10, p. 40, 2023. https://doi.org/10.1186/s43067-023-00108-y
12. S. Avasthi, T. Sanwal, P. Sareen and S. L. Tripathi, Augmenting mental healthcare with artificial intelligence, machine learning, and challenges in telemedicine. In *Handbook of Research on Lifestyle Sustainability and Management Solutions Using AI, Big Data Analytics, and Visualization* (pp. 75–90). IGI Global Scientific Publishing, Hershey, PA, 2022.
13. Y. He et al., "Foundation Model for Advancing Healthcare: Challenges, Opportunities and Future Directions," in *IEEE Reviews in Biomedical Engineering*, vol. 18, pp. 172–191, 2025. doi: 10.1109/RBME.2024.3496744.
14. D. K. Murala, S. K. Panda and S. P. Dash, "MedMetaverse: Medical Care of Chronic Disease Patients and Managing Data Using Artificial Intelligence, Blockchain, and Wearable Devices State-of-the-Art Methodology," in *IEEE Access*, vol. 11, pp. 138954–138985, 2023. doi: 10.1109/ACCESS.2023.3340791.
15. I. Wijayanto, A. Humairani, S. Hadiyoso, A. Rizal, D. L. Prasanna and S. L. Tripathi, "Epileptic Seizure Detection on a Compressed EEG Signal Using Energy Measurement," in *Biomedical Signal Processing and Control*, vol. 85, p. 104872, 2023.
16. S. Sai, V. Chamola, K.-K. R. Choo, B. Sikdar and J. J. P. C. Rodrigues, "Confluence of Blockchain and Artificial Intelligence Technologies for Secure and Scalable Healthcare Solutions: A Review," in *IEEE Internet of Things Journal*, vol. 10, no. 7, pp. 5873–5897, 2023. doi: 10.1109/JIOT.2022.3232793.
17. A. K. Gupta and M. K. Srivastava, "Framework for AI Adoption in Healthcare Sector: Integrated DELPHI, ISM–MICMAC Approach," in *IEEE Transactions on Engineering Management*, vol. 71, pp. 8116–8131, 2024. doi: 10.1109/TEM.2024.3386580.

Chapter 2

# AI-powered anonymization for secure medical data sharing

## Future trends and research directions

*N. Suganya, P. Gouthami, and M. Krishnamoorthi*

## 2.1 INTRODUCTION

### 2.1.1 Understanding public health data

Individuals rely on a variety of health data to understand health or contribute to improving the health of a population. Public health practitioners do not receive this information from a single source – they piece together data from whatever sources may exist, including institutional data, patient-level data, population characteristics, and databases along the continuum of care. When public health workers leverage the examination of these datasets, they are able to gain a better understanding of health in their community, trace disease patterns, see what types of treatment and prevention work, and ultimately determine whether changes to existing health programs can influence public health behavior.

#### *2.1.1.1 Electronic health records (EHR) and clinical data*

EHRs provide complete digital records of a patient's medical history, while clinical data focus on the individual health information such as the diagnostic results, prescriptions, and entries in the medical record. These EHRs from different hospitals and clinics are integrated to identify health indicators that give decision support for healthcare providers to determine, for example, if care should be provided to a patient with noncommunicable diseases and to ensure continuous care through integration.

#### *2.1.1.2 Surveillance data*

Using surveillance data, periodic assessments of medical events that create a resource burden – such as disease incidence and vaccination rates – are conducted. The Integrated Disease Surveillance Programme (IDSP) in India and the National Notifiable Diseases Surveillance System (NNDSS) in the United States, among others, use both statistical and artificial intelligence (AI)–based models to provide early warning signals.

DOI: 10.1201/9781042015931-2

#### 2.1.1.3 Emerging digital data sources

Similar to traditional datasets, health data from wearable sensors, social media, mobile apps, etc., will also generate new kinds of real-time health behavior data for populations over the coming decades (Figure 2.1). They will broaden our knowledge of healthcare, but they will also present difficulties in terms of protecting privacy.

### 2.1.2 Role of artificial intelligence in public health decision-making

AI is transforming public health through enabling intelligence-led, proactive, and timely decision-making. AI systems are capable of detecting patterns from numerous, heterogeneous datasets based on approaches like machine learning (ML), deep learning (DL), and natural language processing (NLP).

Core application areas include the following:

a. **Disease Surveillance and Outbreak Prediction**
   The initial screening of outbreaks, like COVID-19, dengue, and influenza, is diagnosed with the help of both online information and medical data using AI, as well as the detection of timely outbreaks and preventive methods of late outbreaks.

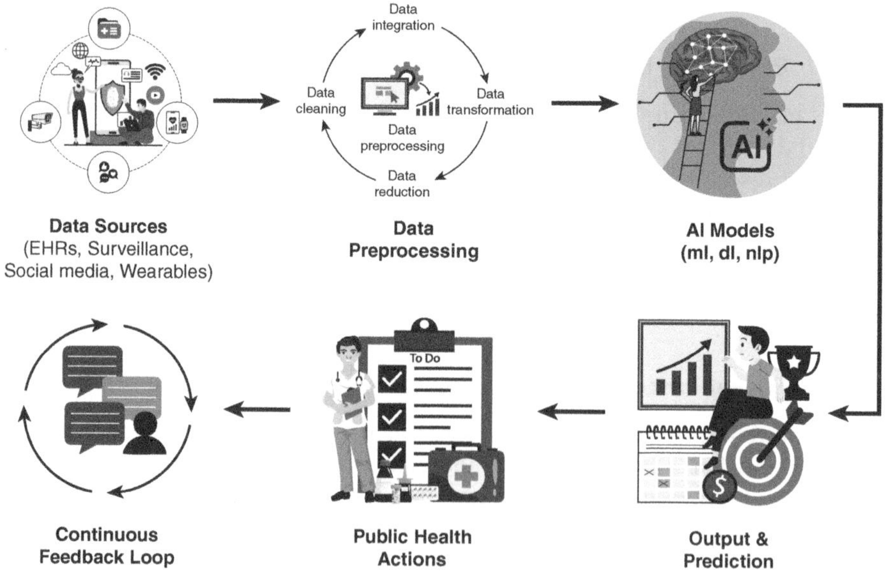


*Figure 2.1* AI-powered public health information.

b. **Predictive Analytics for Healthcare Resources**
ML can assist in diagnosing which patient may need clinical support, optimize ICU resource consumption, and also confirm efficient vaccine delivery.
c. **Text Mining and Sentiment Analysis**
NLP is used in social media to understand how people feel during pandemics and to monitor the spread of deceptive data.
d. **Privacy-Aware Federated Learning (FL)**
Recent AI techniques use FL, where models are trained across decentralized hospitals without transferring confidential patient information. In this way, AI optimizes disease tracking and makes public health decisions faster, transparent, and all the more reliable.

### 2.1.3 Privacy and data sharing: Central challenges

Data sharing helps improve collaboration and policies, but it also raises privacy, ethical, and legal issues.

- *Reidentification Risks:* Even data that has been anonymized can be reidentified through a linkage or AI-based inference attacks.
- *Data Ownership and Consent:* Patients are often ignorant of how their data are shared, reused, or monetized. To manage and guide data, use modern ethical approaches such as "Data Trusts."
- *Cybersecurity and Data Integrity:* Cyberattacks are often focused on healthcare databases. AI-based anomaly detection and blockchain systems are being investigated as a means to potentially secure the dataset.
- *Regulatory Issues:* Strict access control and strong anonymization are requirements for compliance with GDPR, HIPAA, and India's Digital Personal Data Protection Act (DPDP, 2023), particularly when exchanging data with partners located in other countries.

To protect sensitive healthcare records, privacy-preserving methods like differential privacy, homomorphic encryption, and secure multiparty computation are being merged while enabling analytics.

### 2.1.4 Case example: Classic reidentification case – Massachusetts Hospital Data (1997)

In 1997, to support researchers, a government agency in Massachusetts released hospital data of state employees. The dataset included details like date of birth, gender, zip code, and medical details such as diagnoses and hospital visits. A computer scientist, Latanya Sweeney [1], suspected this data was not originally anonymous. She obtained the voter list, which contained name, age, gender, zip code, and date of birth. By comparing the two datasets, she could match the governor's details and reidentify him. Despite the direct identifiers removed, quasi-identifiers were enough to reveal personal identity.

The example above illustrates that anonymization alone is not sufficient to preserve data privacy. To improve security, an integration of privacy-preserving techniques and modern methods such as FL and blockchain integration is required.

### 2.1.5 Chapter scope and objectives

#### *2.1.5.1 Scope*

This chapter explains how AI can be employed to examine the public health records while maintaining it confidential, secure, and governed ethically. It explores managing data from start to finish, techniques to anonymize it, and federated methods that enable safe sharing of health data globally.

**Objectives**

- To explain the format and origin of public health information dataset
- To outline AI's roles in predictive and preventive public health systems
- To examine privacy, ethical, and legal concerns when sharing health data using AI
- To demonstrate COVID-19 as a practical example of data governance challenges
- To introduce secure methodologies for AI-driven health system decision-making

## 2.2 IMPACT OF AI IN PUBLIC HEALTH SYSTEMS

AI is increasingly used in public health to monitor diseases, optimize population health management, strengthen diagnostic outcomes, and streamline healthcare resource management. By using extensive and diverse datasets, AI facilitates healthcare systems to make faster, precise, and proactive decisions. Although it has many advantages, its use also brings up ethical, security, and privacy issues.

### 2.2.1 AI in disease surveillance and outbreak management

AI systems are mostly used to identify and predict disease spread in real time. ML approaches assess EHRs, surveillance datasets, and social media insights to expose early warning signs of epidemics [2].

*Example:* Neural network models forecasted COVID-19 hotspots by examining people's movement patterns and nearby case numbers.

*Benefits:* Assist in making early action to lessen illness and death rates.
*Risk:* Inaccurate or imbalanced data can cause AI systems to give false warnings.

### 2.2.2 Population health analytics

In order to obtain insights from the medical health dataset and risk factors, ML models are implemented across a huge population.

*Strategies:* Regression, clustering, decision trees, and DL
*Use Cases:*
- Identify high-risk groups for long-term condition like diabetes or hypertension
- Maximizing preventive care and making vaccination programs more enhanced

*Outcome:* Enhanced use of interventions and a more strategic use of healthcare services

### 2.2.3 Medical diagnostics and decision support systems

By analyzing clinical data to assist diagnosis and patient management decisions, AI helps medical professionals. For example, convolutional neural networks (CNNs) spot deviations in X-rays, CT scans, and MRIs. Also, predictive models help create personalized treatment care based on a patient's medical history and clinical guidelines.

*Benefit:* Minimizes diagnostic errors, supports clinical decisions, and strengthens patient outcomes

### 2.2.4 AI-powered healthcare resource management

AI makes healthcare management and hospital operations simpler:

- Estimate ICU bed availability, respirator needs, and staff requirements
- Enhances supply chains for drugs, vaccines, and essential medical supplies
- Helps use limited resources efficiently, especially during pandemics

### 2.2.5 Data privacy, security, and ethics implications

Healthcare effectiveness is improved by using AI techniques, but it has some privacy risks. Despite anonymization, raw information may be exposed, and AI systems can accidentally worsen inequalities in care. To reduce these risks, FL methodology is used to train the model without sharing the patient's

*Table 2.1* AI-driven public health: Application, techniques, benefits, and challenges

| *Application area* | *AI techniques used* | *Benefits* | *Challenges* |
|---|---|---|---|
| Disease surveillance and outbreaks | Time-series machine learning, neural networks, NLP | Provides early warning of outbreaks and enables proactive public health interventions | Risk of false alerts, potential data bias |
| Population health analytics | Regression models, clustering, deep learning | Helps identify high-risk groups and supports targeted interventions | Data quality issues, privacy concerns |
| Clinical diagnostics and decision support | Convolutional neural networks (CNNs), predictive modeling, NLP | Speeds up diagnosis, reduces human error, improves decision-making | Risk of bias, potential for misdiagnosis |
| Healthcare resource optimization | Predictive analytics, reinforcement learning | Optimizes ICU capacity, staffing, and supply management | Inaccurate predictions, complex implementation |
| Privacy-preserving AI | Federated learning, differential privacy | Ensures data security while still enabling useful analytics | Computational overhead, limited data availability |

private data, while differential privacy and homomorphic encryption secure essential information. Blockchain keeps trustworthy records and follows GDPR, HIPAA, and local laws, which enhances additional security. These approaches improve the performance as well as the safety of AI systems in healthcare (Table 2.1).

## 2.3 DATA PRIVACY–PRESERVING CONCERNS IN AI-ENABLED HEALTHCARE SYSTEMS

Public health datasets consist of complex and heterogeneous information from which insights are obtained using AI. Many of the datasets cover private data also. AI presents significant privacy issues on many levels, even though it can enhance data analysis, diagnosis, and prediction.

### 2.3.1 Privacy and sensitivity issues in public health systems

Digital health records, genetic information, behavioral patterns, and location information all are part of public health data (Figure 2.2). These provide a complete overview of individuals and communities when taken as a whole:

- *Healthcare Records:* They comprise medical histories, dosages, and hospital visits. Medical examinations and therapy plans may be shown if this information is displayed [3].

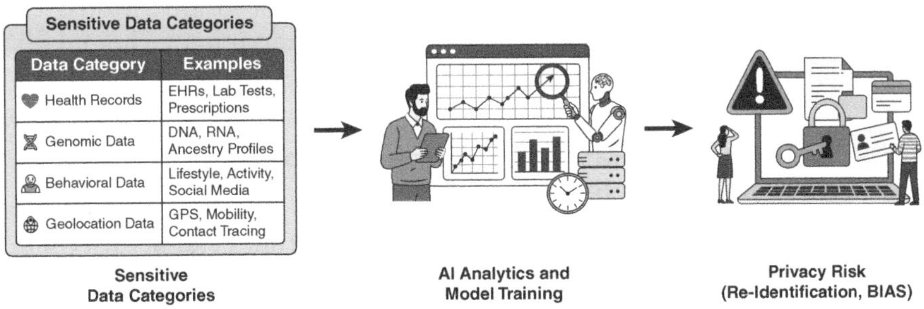


*Figure 2.2* Categories of sensitive data in AI-powered medical information.

- *Genomic Information:* It uncovers hereditary diseases, ancestry, and likelihood of developing certain illnesses. Even minute pieces of DNA fragments can uniquely identify a person [4].
- *Spatial and Mobility Information:* This information helps track outbreak expansion but can also exhibit patient behavioral patterns and social relationships [5].

### 2.3.2 Data lifecycle threats: Collection, analysis, sharing, and storage

Privacy issues arise at every stage of the data life cycle (Table 2.2) [6].

### 2.3.3 Reidentification risks

Even though the direct identifiers (name, ID, age) are eliminated, reidentification of individuals is possible by linking anonymized datasets with external data sources. The famous Massachusetts case [1] demonstrated that combinations of birth date, gender, and ZIP code were sufficient to identify 87% of Americans. To mitigate privacy risks, different preventive measures can be applied. To protect personal data, controlled noise for model parameters was introduced through DL. FL enables a model to be trained collectively without

*Table 2.2* Life cycle of clinical dataset

| *Stage* | *Typical operations* | *Potential risk* |
|---|---|---|
| Collection | Datasets collected from clinics, wearables, and apps | Improper data collection and consent violations |
| Processing | Data preparation, feature engineering, and model construction | Overfitting and memorization of personal data |
| Sharing | Collaborative data management between institutions | Information leakage during transmission and aggregation |
| Storage | Clinical data repositories and storage systems | Insider threats, long-term misuse, cyberattacks |

sharing personal information; it stores personal data locally. Moreover, to enhance security of computation, homomorphic encryption techniques have been employed.

### 2.3.4 Algorithmic bias and assumption to protected attributes

AI systems can find invisible patterns, which sometimes end up discriminating or revealing confidential attributes, such as gender, nationality, or wealth status. For example, research around predictive healthcare models indicates that in some cases, underrepresented groups are assigned lower risk scores, despite being equally ill [7]. Occasionally, a DL model can still ascertain protected attributes, even though this is not present in the input data, and this type of bias commonly stems from unbalanced or unrepresentative training data, omitted variables, or incorrect labeling of protected attributes and could result in some policyholders being less well treated or incorrect medical decisions.

### 2.3.5 Compliance: GDPR, HIPAA, and India's DPDP act

Privacy is not optional; it is required, especially when it is regarding patient data in health AI (Figure 2.3). Different countries of the world have come up with rules to protect this sensitive data. For instance, European [8] GDPR statutes require organizations to get consent before collecting data and do not allow fully automated decision-making without human intervention. HIPAA [9] law in the United States is concerned with the security of patient records and the protections taken to do so, such as encrypting and anonymizing identity. The DPDP Act, 2023 [10], in India provides the individual with more control over their data, with information about how their data is used, and how to raise grievances. Following these statutes and practices is important not only for compliance but also to build trust, regulate the responsible use of AI, and protect patient information internationally.

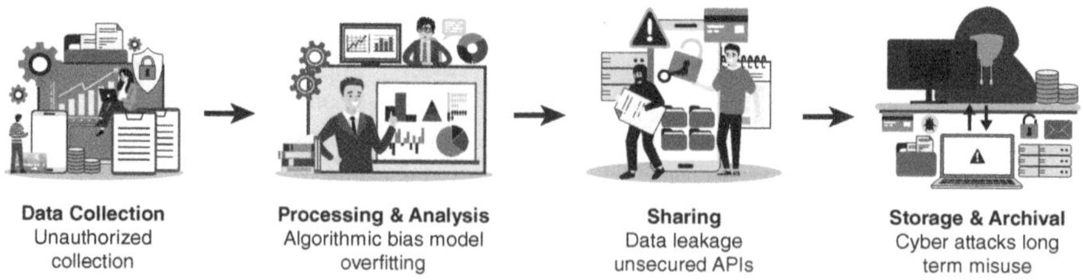


*Figure 2.3* Privacy risk map across AI data life cycle.

## 2.4 OVERVIEW OF TRADITIONAL PRIVACY PRESERVATION TECHNIQUES

During the early pre-AI era, there were traditional privacy-preservation methodologies, *k*-anonymity, *t*-closeness, and *l*-diversity, to protect sensitive health information. These methods ensure that individuals cannot be identified in a shared or distributed health dataset, especially for the EHR data, which would contain specifics on population demographics, previous health histories, and health outcomes (Figure 2.4).

Sweeney [11] proposed *k*-anonymity, which ensures that no record in a dataset can be distinguished from at least $k - 1$ other records with respect to a set of quasi-identifiers, for example, age, gender, and zip code. *k*-Anonymity is effective in preventing direct reidentification of individuals; however, it is weak against the attacks of homogeneity and background knowledge. In order to deal with this, Machanavajjhala et al. [12] proposed *l*-diversity, which requires that every equivalence class contain at least *l* well-represented values for sensitive attributes. Nevertheless, *l*-diversity has trouble when the data distribution is generally skewed, resulting in vulnerability to similarity attacks. In response, Li et al. [13] proposed *t*-closeness, which guaranteed that the distribution of sensitive attributes is *t*-close to the overall distribution of the sensitive attribute in every equivalence class, thereby minimizing any semantic leakage.

While these legacy models are valuable, they are not suited to the era of big data and AI. When these models are applied to large, high-dimensional health datasets, they often abstract away similar, good information, leading to the loss of important medical features in the dataset specifically used for predictive analytics. Loss of integrity of these important medical features slows or halts accurate detection of disease, offering less rationale for decision-making regarding clinical phenomena [14]. For example, if a patient's age were altered in a de-identified patient record for the purpose of achieving 5-anonymity and were generalized to the range of 20–40 or 40–60, it would offer greater anonymity, but would heavily distort patterns of disease prevalence and reduce predictive accuracy of applicable ML models [15]. Thus, these models may be a useful theoretical structure (of value), but they do not appropriately represent complexity and scale as we attempt to use AI for public health work

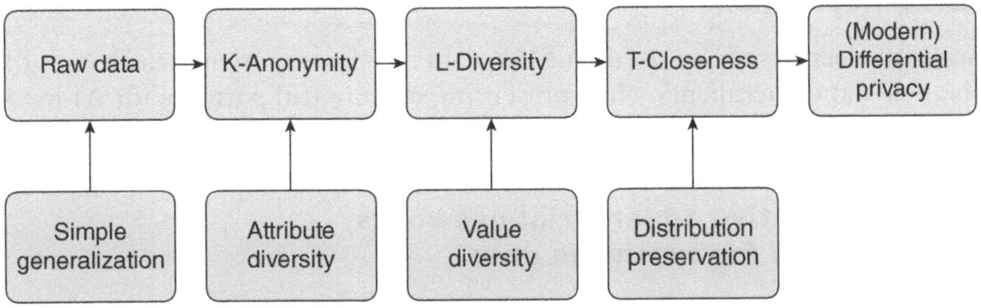


*Figure 2.4* Evolution of traditional privacy models.

*Table 2.3* Comparison of traditional privacy preservation techniques

| *Methodologies* | *Principle* | *Pros* | *Limitations* |
|---|---|---|---|
| *k*-Anonymity | Each record indistinguishable from $k - 1$ others | Clear and easy to interpret | Vulnerable to homogeneity and background attacks |
| *l*-Diversity | Sensitive attributes have $\geq l$ distinct values | Prevents homogeneity attacks | Fails under skewed data distributions |
| *t*-Closeness | Distributional distance $\leq$ threshold *t* | Preserves global semantic structure | Complex to implement; reduces data utility |

since much of the deep structure is not representative of wider data in current usage at all (Table 2.3).

## 2.5 ANONYMIZATION AND PRIVACY-PRESERVING MODELS DRIVEN BY AI

AI in healthcare data sharing presents incredible chances to improve data-driven healthcare analysis while maintaining patient privacy. Due to the possibility of reidentification through quasi-identifiers, current anonymization techniques, such as discarding identifiers, are no longer sufficient [1]. Advanced AI-powered methods for safe, privacy-preserving data sharing integrate data transformation, cryptography, and ML.

### 2.5.1 DL-based anonymization

DL techniques like autoencoders can mask confidential information in data by transforming it into a shorter, coded form. These models acquire the knowledge to rebuild the raw data while preserving the confidentiality of personal details.

#### *2.5.1.1 Application*

Autoencoders have been used on EHR data to protect patient health data like diagnoses and treatments while preserving meaningful patterns for AI model training (Table 2.4) [16].

### 2.5.2 Generative adversarial networks (GANs) for synthetic data

GANs are robust generative models well-suited for generating realistic synthetic datasets that retain the statistical integrity of the original medical data while preserving patient information.

*Table 2.4* Example of autoencoder-based feature obfuscation

| *Raw feature* | *Anonymized output (autoencoder)* |
|---|---|
| Name | Eliminated |
| DOB | Keep year alone |
| Zip code | Regional code |
| Medical evaluation | Encoded latent feature vector |
| Drugs | Encoded latent feature vector |

#### Mechanism

- Generators produce synthetic samples.
- Discriminator examine their authenticity.
- Repetitive training enhances high-quality synthetic data.

#### Use case

Synthetic datasets are created using GANs, which will be shared across hospitals without revealing the patient's original identity [17].

#### Advantages

- Supports collaborative AI model training.
- Lessens privacy risk while retaining data integrity.
- Utilizes vast datasets to identify rare diseases.

### 2.5.3 Federated learning

FL allows hospitals and healthcare centers to work together to train AI models without sharing original patient data. Instead of sharing the raw data, only the model's learned weights are distributed and combined, while maintaining the data confidential.

#### Workflow

1. Local models are trained at each institution on private data.
2. Model updates (gradients) are sent to a central aggregator.
3. Aggregator combines updates → new global model → redistributed.

#### Benefits in healthcare

- Compliance with GDPR/HIPAA regulations.
- Diminishes central storage risk.
- Allows AI to create insights from data stored across distributed locations.

### 2.5.4 Differential privacy (DP)

Differential privacy adds controlled statistical noise to training data or learning model outputs to prevent identity disclosure.

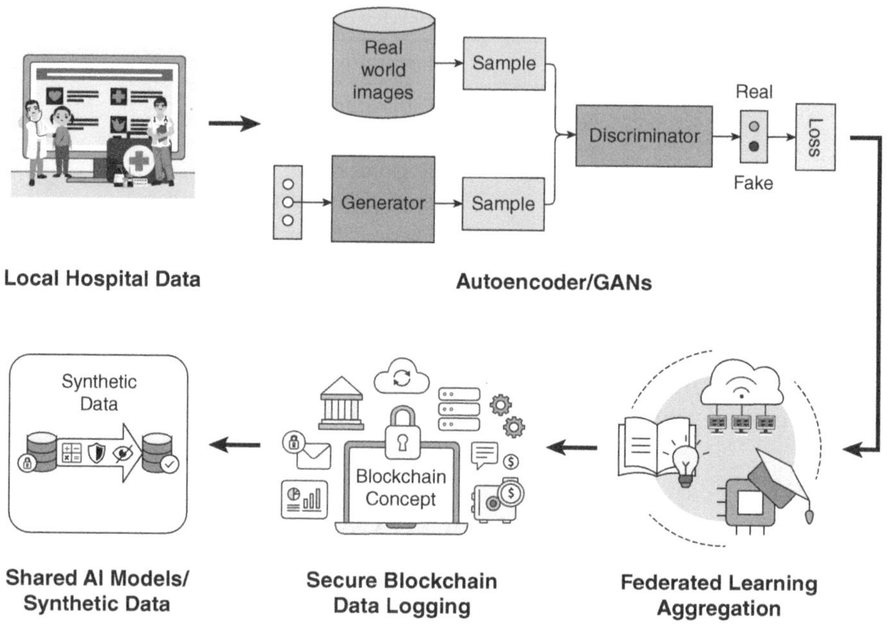


*Figure 2.5* Hybrid AI-powered privacy framework for medical data sharing.

### Mechanism

- Noise calibrated according to privacy budget ($\varepsilon$).
- Make sure that the integration or removal of a single patient record does not significantly change the output.

### Application

- DP is used in FL updates or in releasing anonymized statistics from EHRs [18].
- Protects patient identities while enabling aggregate analytics.

### 2.5.5 Hybrid approaches: AI, cryptography, and anonymization

A hybrid model combines multiple privacy-preserving methodologies to optimize security and utility (Figure 2.5).

## 2.6 SECURE DATA-SHARING FRAMEWORK FOR AI-BASED PUBLIC HEALTH

An effective data-sharing infrastructure in AI-enabled public health ecosystems must merge cloud-native horizontal scalability and layered privacy guarantees, in which models can learn from large, heterogeneous data sources

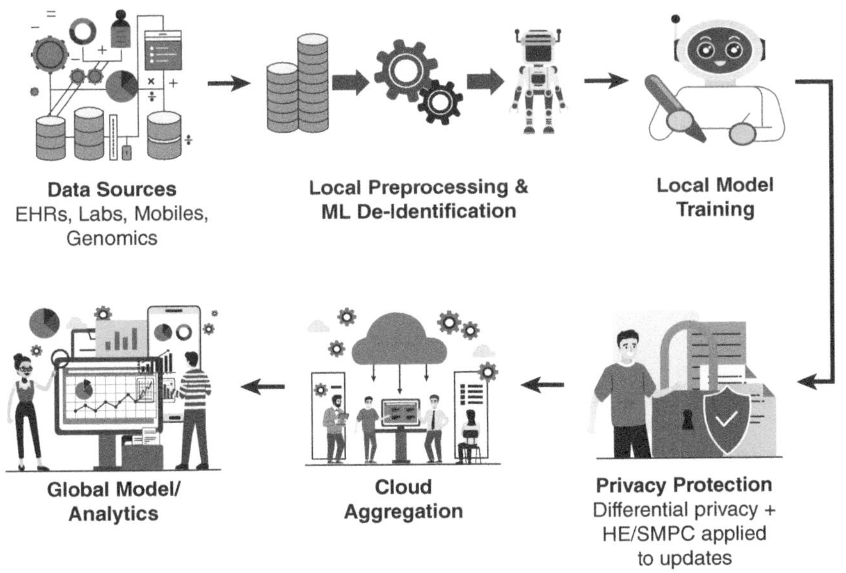


*Figure 2.6* Securing data sharing using hybrid approaches.

without ever referencing individual records. Cloud infrastructure supplies the elastic storage and computing power, and standardized APIs to facilitate secure data exchange between hospitals and public health agencies: edge and FL paradigms can maintain local records and transmit only model updates or encrypted artifacts for aggregation (Figure 2.6) [19].

The AI models themselves provide an additional layer of privacy assurance: on-site de-identification (leveraging ML/NLP to deidentify clinical text), adaptive differential privacy applied to gradients, and model auditing for memorization reduce leakage risk during training and inference [20, 21]. Cryptographic primitives – TLS for transport, AES/ECC for data-at-rest, homomorphic or secure multi-party computation for joint analytics – plus strong access control (role-based access policies, attribute-based access control) and tokenization of PHI (replacing sensitive values with nonreversible tokens) provide a technical framework for confidentiality and provenance guarantees [22]. Immutable ledgers or tamper-evident logs (blockchain or distributed ledgers) complete these safeguards by recording consent transactions, versioning models, and documenting data-access audits that govern the ethical use of data while satisfying reproducibility and governance requirements (Table 2.5) [23].

## 2.7 ROLE OF AI IN PUBLIC HEALTH SYSTEMS

AI improves tracking of disease stage, diagnosis, prevention, and decision-making which transforms public health systems. Predictive modeling, real-time monitoring, and quicker data processing are made possible by AI-driven

*Table 2.5* Core components and technologies

| *Component* | *Primary purpose* | *Representative techniques / standards* |
|---|---|---|
| Cloud orchestration | Scalable compute and model aggregation | Kubernetes, secure APIs, IAM |
| Local de-identification | Remove/perturb identifiers before sharing | ML/NLP personally identifiable information (PII) detectors, tokenization |
| Federated learning aggregator | Train global models without raw data moving | FedAvg, FedProx |
| Privacy layer | Prevent reidentification and memorization | Differential privacy (DP), DP-SGD |
| Cryptography and secure compute | Confidential joint computation | TLS, AES/ECC, homomorphic encryption, SMPC |
| Access control and governance | Enforce policies, consent, audit trails | RBAC/ABAC, DLT audit logs, consent records |

techniques to provide meaningful information and are important for managing public health efficiently. Sensitive health data can be protected by integrating AI with privacy-preserving techniques such as differential privacy, FL, and protected multiparty computation.

### 2.7.1 AI applications in disease surveillance, outbreak prediction, and diagnostics

AI is important for automating outbreak detection and tracking of diseases. AI algorithms can detect irregular health patterns and warn authorities by analyzing clinical and epidemiological data,

*Disease Surveillance:* NLP and ML models are used to detect irregular health patterns by keeping track of hospital record, monitoring news feeds, and social media.

*Outbreak Prediction:* Epidemic patterns are predicted from time-series health data by using predictive models such as recurrent neural network and long short-term memory [24].

*Diagnostics:* Medical images such as X-rays and CT scans are commonly used for detecting pneumonia, COVID-19, etc. AI models improve the accuracy of detecting the disease using medical images.

*AI-Based Outbreak Prediction Model*:

$$y_{t+1} = f(y_t, y_{t-1}, \ldots, y_{t-n}; \theta) \tag{2.1}$$

where

$y_{t+1}$ is the predicted number of cases at time $t+1$.

$y_t$ is the observed cases at time $t$.

$\theta$ is the model parameter learned by the AI model.

### 2.7.2 Use of ML in population health analytics

ML helps to analyze health data of a large population. It finds the risk factors, predicts progression of the disease, and improves the use of healthcare resources.

*Predictive Analytics:* Disease prediction based on features such as age, BMI, and genetics are obtained by supervised learning models such as random forests and logistic regression.

*Population Segmentation through Clustering:* Unsupervised learning techniques, including K-means and group populations by health behaviors, allow for more targeted treatment strategies.

*Logistic Regression in Population Health:*

$$P\left(y=\frac{1}{x}\right)=\frac{1}{1+e-\left(\beta_0+\beta_1x_1+\beta_2x_2+\cdots+\beta_nx_n\right)} \tag{2.2}$$

where

$P\ (y = 1/x)$ is the probability of disease occurrence.

$X_i$ is the predictor variable (e.g., age, blood pressure).

$\beta_i$ is the coefficient learned from data.

### 2.7.3 Advantages: Cost reduction, predictive modeling, and timely insights

AI is a great assistance to health sector in numerous ways. As AI analyzes the data as it comes in, outbreaks are found sooner and thus policies can be updated faster. Also, it forecasts the trend of disease spread and the required healthcare resources. Data analysis and diagnosis are automated to reduce the operating cost and resource strain. It helps in the resource distribution efficiently to high-risk regions by health authorities [25].

*Example*

Regional case increases during the COVID-19 epidemic were correctly predicted by AI-based forecasting algorithms. This made it possible to distribute vaccines and improve hospital readiness.

### 2.7.4 Risks and data privacy concerns in AI systems

Adoption of AI offers enormous advantages in health sector, although there still exists privacy and ethical problems. Unauthorized access or incorrect sharing of data can result in wrong usage of private medical information. New correlation algorithms can connect anonymized datasets to individuals. The privacy of health information has to be maintained. An AI system with improper data protection mechanism may lead to sensitive data exposure, which is against the data protection laws like GDPR and HIPAA.

To reduce the risks, AI process integrates privacy-preserving strategies such as FL, homomorphic encryption, and differential privacy:

$$P\left[M\left(D_1\right) \in S\right] \leq e^{\varepsilon} \cdot P\left[M\left(D_2\right) \in S\right] \tag{2.3}$$

where

$D_1$ and $D_2$ are datasets differing by one record.
$M$ is the randomized algorithm.
$\in$ is the privacy loss parameter.

This ensures that the inclusion or exclusion of a single individual does not significantly affect the output, protecting privacy.

## 2.8 CASE STUDIES AND REAL-WORLD APPLICATIONS

### 2.8.1 Case study 1: FL to detect COVID-19 using X-ray

To detect COVID-19 using X-rays (i.e., medical images), it is mandatory to use AI-based diagnostic tools during the COVID-19 pandemic. Medical data is highly sensitive. One of the promising solutions to preserve the information is FL. It enables organizations or healthcare institutions to train AI models across multiple devices with local data according to their needs, without transferring the information to any other organization's device. This ensures the privacy of each patient while enabling the model to train on multiple organizations' data.

#### *Key techniques*

- *Federated Learning:* In this approach, rather than sharing the original data, each ML model works on its own source and shares the updates alone, to all other models. This avoids exposing the sensitive information to others [25]. This strategy is mainly advantageous in the medical sector, where it is necessary to preserve the patients' information under stringent privacy laws like HIPAA and GDPR.

#### *Implementation*

- *Collaboration:* To train an AI model, several hospitals work with their patients' COVID-19 X-ray dataset and share only the updates to the centralized server, without sharing the patients' data (X-ray images). The model can identify COVID-19 features such as lung capacity, consolidation, and ground-glass opacities from X-rays.

#### *Impact*

- *Privacy Preservation:* Privacy of patients' data is preserved by preventing the transmission of raw information between healthcare institutions.

- *Enhanced Diagnostic Tools:* FL enables the healthcare system to use better diagnostic models without breaking the rules to preserve patient information [26].

### 2.8.2 Case study 2: AI-based anonymization for hospital data exchange

To improve healthcare research, it is necessary to share sensitive information, including patient records, between hospitals, and research institutions. Conventional anonymization technique takes a longer time and is often prone to errors. This process is greatly improved by AI-based anonymization techniques such as NLP and named entity recognition (NER), which automatically analyze, identify, and anonymize personally identifiable information (PII) in EHRs.

#### *Key techniques*

- *NLP and NER:* Patient names, addresses, and phone numbers are referred as PII, which is found enormously in unstructured clinical data. NLP and NER methods are AI-based techniques used to find and replace PII with pseudonyms.
- *Differential Privacy:* Use of differential privacy guarantees that the data provided for research cannot be related to or used to identify the specific person, even though the attacker has access to the entire database.

#### *Implementation*

- *Data Exchange:* Healthcare institutions can share sensitive data like EHR and patient information with others by employing an AI-based anonymization technique, as it guarantees to deidentify the information before sharing it.

#### *Impact*

- *Compliance with Privacy Regulation:* Healthcare organizations are allowed to share their data with research institutions, by adhering to the privacy laws such as HIPAA and GDPR. This is achieved by automatically protecting the information with the help of AI-based techniques.
- *Scalability:* Compared to conventional techniques, the time and cost are reduced by employing the AI-based anonymization technique. It can also scale to large datasets.

### 2.8.3 Case study 3: AI-assisted surveillance in disease outbreak prediction

AI algorithms can analyze a large amount of real-time data from different sources, and are used more and more to predict the outbreak of disease earlier.

For example, during the COVID-19 pandemic, AI systems were employed to track social media–reported symptoms, travel information, and hospitalization rates, to forecast the chance of an outbreak [27]. These techniques have to preserve the privacy of individual data, as most of it is related to personal health and mobility.

#### Key techniques

- *AI Predictive Models:* Real-time data (such as ER visits and social media posts) and historical outbreak data (such as seasonal flu and SARS) were used to train ML algorithms like random forests and DL networks.
- *Anonymized Data Collection:* A huge amount of real-time data is anonymized so that identification of individuals can be prevented while offering insightful information to predict the outbreak of the disease.
- *Federated Learning:* It enables different institutions to build an AI model together without sharing their private data. In healthcare, patient data is sensitive and cannot be easily shared across hospitals, countries, or research institutions. For this, FL methods help to secure the information [28].

#### Implementation

- *Real-Time Surveillance:* Early warnings of disease outbreaks can be identified by using AI-based surveillance, to monitor the social media posts, medical reports, and new sources. After identification, the information is anonymized and combined before processing.
- *Predictive Alerts:* Real-time predictive alerts produced by AI algorithms can help the authorities to react quickly in response to disease outbreaks.

#### Impact

- *Early Detection:* AI-integrated monitoring systems help in the quick identification of disease outbreaks, which enhances public health responses and shortens the time needed to act.
- *Privacy Preservation:* FL and anonymized data usage permit accurate predictions while protecting individual privacy.

## 2.9 CHALLENGES AND CURRENT RESEARCH ISSUES IN AI-BASED MEDICAL DATA ANONYMIZATION

As AI-driven anonymization methods advance, various challenges have to be solved to guarantee the secure and ethical dissemination of medical data. These challenges have technological, social, ethical, and legal aspects. The biggest challenge involves striking a balance between data privacy and its overall utility. When data gets highly anonymized, it tends to reduce its quality. Lowering the quality of data in turn harms areas like research efforts and medical diagnosis processes.

### 2.9.1 AI anonymization: Transparency and accountability

The response given by AI anonymization techniques is difficult to understand. Hence, they are often referred to as "black boxes." As there is no transparency in AI anonymization, it reduces trust among the patients and doctors.

*Key Issues:* Enabling trust is the main issue to focus on, while employing AI in healthcare applications. Explainable AI can be used to solve this issue. Using complex and large models may cause accountability and statutory approval issues.

*Research Directions:* While developing privacy-preserving models, it is good to integrate explainable AI models [24]. Also, it is necessary to have audit trails with XAI models to ensure accountability and adherence to privacy policies.

### 2.9.2 Regulatory gaps and trust deficit

Implementing AI-powered anonymization methods has been restricted, because of current gaps in privacy regulations and public concerns.

*Significant Issues:* Reidentification of data and lack of protection remain to be a challenge. AI-related issues are not clearly addressed and resolved by current regulations.

*Research Directions:* Work with legal team to create privacy regulations for incorporating AI in healthcare and to increase the transparency of anonymization techniques, so that it builds trust.

### 2.9.3 Legal compliance in AI-based data anonymization

Legal requirements such as HIPAA and GDPR must be followed by AI-driven anonymization models: especially when sharing data with organizations across various countries, because each nation follows its own regulations.

*Key Issues:* Sharing of data across different countries raises legal difficulties. Also, AI models must adapt to new regulatory guidelines.

*Research Directions:* Create a cross-jurisdictional system for secure data exchange and AI models have to adapt to regulatory needs.

Medical data anonymization powered by AI has several challenges in terms of securing a balance between privacy and usefulness, maintaining transparency, promoting trust, and managing complicated legal problems. Further research addressing these problems will lead to safe, responsible, and efficient exchange of information.

### 2.9.4 Learnings

*FL Encourages Secure Collaboration:* FL is preferred when it comes to secure collaboration. This technique lets various organizations work together on AI models. They do this without sharing the private information [26].

*Improved Data Anonymization:* AI methods paired with anonymization tools improve the precision of the data while preserving its privacy [27]. This makes sharing the sensitive information way more responsible.

*Privacy-Preservation Using AI models:* An AI system along with FL keeps things private even with data flowing in real time. It monitors disease patterns and predicts them on the fly. Privacy does not get compromised [29]. Basically, organizations can team up to build AI without passing around sensitive information.

### 2.9.5 Privacy implications

*Data Sovereignty:* International teams incorporating FL need to handle the issues of data sovereignty while guaranteeing adherence to different local data protection rules and regulations.

*Ethical Concerns:* AI models affect public health decisions more often. So, ethical concerns regarding fairness and prejudice have to be carefully addressed. AI may produce inaccurate predictions and unproductive public health interventions if it is trained on biased or incorrect data [27].

*Data Ownership:* Finding the owner of the data in the healthcare organizations remains an important issue: whether the owner is the patient, the hospital, or the IT company which manages the data. To solve this issue, a clear and effective framework is necessary to manage data usage and to preserve the data privacy.

## 2.10 FUTURE DIRECTIONS

In recent days, medical healthcare system mostly relies on AI, so protecting the patients' sensitive data is very crucial. To provide a feasible solution, blockchain technology is combined with AI. Moreover, it maintains transparency and accountability by ensuring decentralization, tamper-proof, and auditable system which secures healthcare information. In order to safeguard data from the potential of future quantum computers and maintain its security throughout time, new post-quantum cryptography techniques are also being developed. Without compromising the usefulness of the data based on its context and sensitivity, reinforcement learning along with adaptive anonymization is used to mask the data. FL preserves privacy by training AI models without disclosing the raw data. When incorporated with blockchain, it enables safe and collaborative data analysis. Building confidence in healthcare requires explainable AI (XAI), which enables physicians and patients to comprehend how AI makes decisions [30]. To minimize the unauthorized access, zero-trust architectures are implemented to make sure that every access request is authenticated and authorized. Lastly, robust policies are required to guide responsible AI practices, mitigation of bias, fairness, and accountability. By leveraging these technologies, they are responsible for the growth of trustworthy, private, and safe AI systems for public health that comply to legal standards.

## 2.11 CONCLUSION

Datasets for AI in public health are collected in a variety of forms: different types of data, tools, and human decisions working together. These datasets are used to identify the outbreaks, assist the doctors, and make use of vast amounts of healthcare information. Yet, preserving the privacy is more important in the data collection, sharing of the information, and also while using its insights. Techniques such as AI-based anonymization, differential privacy, and FL are used to secure the sensitive information. In future, AI in medical data rely on next-generation technologies, well-defined rules, and the confidence of the people providing their data. Its benefits are utilized only if the privacy and ethical considerations are considered at every stage.

## REFERENCES

1. Fung, B. C. M., Wang, K., Chen, R., & Yu, P. S. (2010). Privacy-preserving data publishing: A survey of recent developments. *ACM Computing Surveys*, *42*(4), 1–53. https://doi.org/10.1145/1749603.1749605
2. Mendes, J. S., Villanueva-Miranda, I., Prokop, M., & Wiest, I. C. (2025). Harnessing artificial intelligence for enhanced public health surveillance. *Frontiers in Public Health*, *13*, 1601151. doi: 10.3389/fpubh.2025.1601151.
3. Rajkomar, A., Dean, J., & Kohane, I. (2019). Machine learning in medicine. *New England Journal of Medicine*, *380*(14), 1347–1358.
4. Gymrek, M., McGuire, A. L., Golan, D., Halperin, E., & Erlich, Y. (2013). Identifying personal genomes by surname inference. *Science*, *339*(6117), 321–324.
5. P. Tedeschi, K. E. Jeon, J. She, S. Wong, S. Bakiras and R. Di Pietro, "Privacy-Preserving and Sustainable Contact Tracing Using Batteryless Bluetooth Low-Energy Beacons," in IEEE Security & Privacy, vol. 20, no. 3, pp. 91-100, May-June 2022, doi: 10.1109/MSEC.2021.3115497
6. Bandara, E., Liang, X., Foytik, P., Shetty, S., Hall, C., Bowden, D., ... & De Zoysa, K. (2021). A blockchain empowered and privacy preserving digital contact tracing platform. Information processing & management, 58(4), 102572.
7. Obermeyer, Z., Powers, B., Vogeli, C., & Mullainathan, S. (2019). Dissecting racial bias in an algorithm used to manage the health of populations. *Science*, *366*(6464), 447–453.
8. European Parliament. (2018). General Data Protection Regulation (GDPR).
9. US Department of Health and Human Services (HHS). (2020). HIPAA Privacy Rule.
10. Government of India. (2023). Digital Personal Data Protection Act (DPDP).

11. Sweeney, L. (2002). *k*-Anonymity: A model for protecting privacy. *International Journal of Uncertainty, Fuzziness and Knowledge-Based Systems*, 10(5), 557–570. doi: 10.1142/S0218488502001648.
12. Machanavajjhala, A., Kifer, D., Gehrke, J., & Venkitasubramaniam, M. (2007). *l*-Diversity: Privacy beyond *k*-anonymity. *ACM Transactions on Knowledge Discovery from Data (TKDD)*, *1*(1), 3.
13. Li, N., Li, T., & Venkatasubramanian, S. (2006, April). *t*-Closeness: Privacy beyond *k*-anonymity and *l*-diversity. In *2007 IEEE 23rd International Conference on Data Engineering* (pp. 106–115). IEEE.
14. El Emam, K., Rodgers, S., & Malin, B. (2015). Anonymising and sharing individual patient data. *BMJ*, *350*, h1139.
15. Loukides, G., & Shao, J. (2008, March). Data utility and privacy protection trade-off in *k*-anonymisation. In *Proceedings of the 2008 International Workshop on Privacy and Anonymity in Information Society* (pp. 36–45). ACM.
16. Choi, E., Biswal, S., Malin, B., Duke, J., Stewart, W. F., & Sun, J. (2017, November). Generating multi-label discrete patient records using generative adversarial networks. In *Machine Learning for Healthcare Conference* (pp. 286–305). PMLR.
17. Xu, L., Skoularidou, S., Cuesta-Infante, S., & Veeramachaneni, G. (2019). Synthesizing electronic health records using GANs. *Computers in Biology and Medicine*, 107, 172–182. doi: 10.1016/j.compbiomed.2019.03.018.
18. Abadi, M., Chu, A., Goodfellow, I., McMahan, H. B., Mironov, I., Talwar, K., & Zhang, L. (2016, October). Deep learning with differential privacy. In *Proceedings of the 2016 ACM SIGSAC Conference on Computer and Communications Security* (pp. 308–318). ACM.
19. Amanat, A., Rizwan, M., Maple, C., Zikria, Y. B., Almadhor, A. S., & Kim, S. W. (2022). Blockchain and cloud computing–based secure electronic healthcare records storage and sharing. *Frontiers in Public Health*, *10*, 938707.
20. Mohammadi, S., Balador, A., Sinaei, S., & Flammini, F. (2024). Balancing privacy and performance in federated learning: A systematic literature review on methods and metrics. *Journal of Parallel and Distributed Computing*, *192*, 104918.
21. Wiest, I. C., Leßmann, M. E., Wolf, F., Ferber, D., Treeck, M. V., Zhu, J., & Kather, J. N. (2025). Deidentifying medical documents with local, privacy-preserving large language models: The LLM-anonymizer. *NEJM AI*, *2*(4). doi: 10.1056/AIdbp2400537.
22. Hu, M., Ren, Y., & Chen, C. (2023). Privacy-preserving medical data-sharing system with symmetric encryption based on blockchain. *Symmetry*, *15*(5), 1010.
23. Kumar, A., Singh, S., Sharma, R., & Gupta, P. (2025). Secure healthcare data sharing and attack detection with DLT and AI. *Scientific Reports*, 15(1), 15432. https://doi.org/10.1038/s41598-025-99676-4.
24. Tripathi, S. L., & Mahmud, M. (2023). *Explainable Machine Learning Models and Architecture*. Wiley & Scrivener Publishing LLC. https://doi.org/10.1002/9781394186570
25. Yang, Q., Liu, Y., Chen, T., & Tong, Y. (2019). Federated machine learning: Concept and applications. *ACM Transactions on Intelligent Systems and Technology (TIST)*, *10*(2), 1–19.
26. Amin, M. S., Ahmad, S., & Loh, W. K. (2025). Federated learning for healthcare 5.0: A comprehensive survey, taxonomy, challenges, and solutions. *Soft Computing*, *29*(2), 673–700.
27. Gupta, S., Kapoor, M., & Debnath, S. K. (2025). Introduction to AI and privacy issues for healthcare systems. In *Artificial Intelligence-Enabled Security for Healthcare Systems: Safeguarding Patient Data and Improving Services* (pp. 1–22). Springer Nature.

28. Ali, H. (2024). AI for pandemic preparedness and infectious disease surveillance: Predicting outbreaks, modeling transmission, and optimizing public health interventions. *International Journal of Research Publication and Reviews*, *5*(8), 4605–4619.
29. Chowdhury, D., Banerjee, S., Sannigrahi, M., Chakraborty, A., Das, A., Dey, A., & Dwivedi, A. D. (2023). Federated learning based Covid-19 detection. *Expert Systems*, *40*(5), e13173.
30. Avasthi, S., Sanwal, T., Sareen, P., & Tripathi, S. L. (2022). Augmenting mental healthcare with artificial intelligence, machine learning, and challenges in telemedicine. In *Handbook of Research on Lifestyle Sustainability and Management Solutions Using AI, Big Data Analytics, and Visualization* (pp. 75–90). IGI Global Scientific Publishing.

Chapter 3

# Harnessing the potential of artificial intelligence for transformative healthcare solutions

*Tarun Kumar Vashishth, Vikas Sharma, Aparna Kumari, Mukesh Kumar Sharma, Shahanawaj Ahamad, and Sarika Jain*

## 3.1 INTRODUCTION

Artificial intelligence (AI) is a term for a group of computational processes and refers to a wide variety of techniques from machine learning (ML) to neural networks to natural language processing (NLP) and computer vision, etc. It is apparent that the AI has become more intelligent in a variety of ways which have revolutionized procedures and practices where manufacture intelligence was required earlier. AI techniques are being developed in healthcare for the purpose of improved patient care, optimization of the clinical work processing, and support of medical research, which provides the overall expectation of better outcomes and efficiencies as well as reduction in costs.

### 3.1.1 Overview of the role of AI in modern healthcare

AI has a significant impact on modern healthcare. It is revolutionizing many aspects of the ways healthcare systems function, and the ways decisions are made in these systems. AI in its simplest form is the ability of machines to simulate human intelligence. The ability of these healthcare systems to understand complex data, to discover relationships between data, and to make good decisions in an efficient manner has far-reaching consequences in healthcare as this is a complex, rapidly changing, and data-intensive industry. Medical care institutions have large amounts of data to evaluate daily – hospital records, histories, research data, and administrative data. The old systems could not gather, process, and evaluate these data rapidly and accurately, as they could not be adapted to the changes produced by the multiplicity of data offered to them. AI carries out these large-scale data processing, translates information into synthesized knowledge, and enables reliable and uniform decision-making. AI enhances, inter alia, the operational effectiveness of the existing health systems, and also leads to a more uniform and systematic method of furnishing healthcare. The evidence for this uniformity is the availability of large amounts of data concerning healthcare. AI aids in this process to healthcare in this direction that the operational systems are such that a uniformity

 DOI: 10.1201/9781042015931-3

of approach is made in such data, and that the results of the evaluation of the healthcare rendered is based on sufficient evidence where such exists and in sufficient quantities so as the result reached will be consistent with the evidence tendered. The position of AI in modern healthcare does not cease with the improvements in respective efficiency, tending toward the curative side, but carries on to the organization of the healthcare systems in a bigger way [1]. AI makes available a base for anticipatory total action based upon data, rather than a series of disjointed activities which react, each alone, to a symptom or a process. AI permits a certain methodology by which the basic healthcare delivery systems may be enabled to take place in an anticipatory and total manner with a uniform approach. The whole concept of pattern recognition in trends and relationships which can be found and recognized in complex data is subservient to planning of dissections which are timely, and contributing to the accomplishment of the fundamental organizational goals which are planned ahead on a synthetical basis, and a subordinate one relating to the consents and informing of medical personnel, those engaged in the giving of medical care. This change is of great importance, as it elevates not only the quantity of care given, but also the general administration and management of healthcare institutions. AI makes possible the integration of various clinical and administrative processes, resulting in uniformity of decision with less variability caused by inexactness of subjective judgment or human carelessness. As a result, healthcare organizations are able to render effective and equitable care with better distribution of resources and thereby meeting the healthcare requirements of the mass [2].

The role of AI is also vital in improving the predictability, transparency, and accountability of decision-making in healthcare. In the complex environment of healthcare, many decisions are made by a number of stakeholders, each using different sources of information and responding to pressures and different demands. AI provides a containment and neutral environment based on information which, by reducing the effect of subjective human bias, allows the process and decisions to be standardized and gives insight into the patterns of decision-making made in the problems solved in a reproducible manner and opened to evaluation through an agreed information management systems-based approach. Figure 3.1 shows how AI participates in healthcare via various animal models and technologies such as ML, predicative analytics, and NLP and the major drivers for change in healthcare such as diagnosis treatment planning, individualization of pharmaceutical therapy, home monitoring, and hospital systems. These are used to generate trust in the healthcare decision-making model, involving healthcare professionals, patients, and policymakers, and also enhance levels of security in the decision-making process by laying down a more transparent and accountable provision of healthcare. By generating an agreed-upon basis for health information management and decision support systems to govern all health provision issues, healthcare AI improves communication between all parties, thereby having consent on role allocation and participation in healthcare teams as well as management

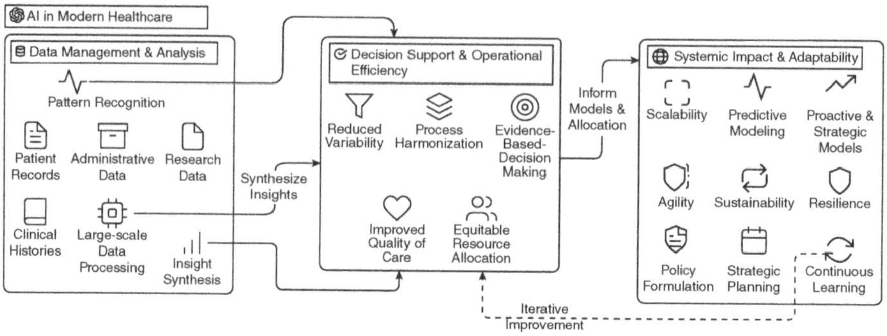


*Figure 3.1* Integration of AI in modern healthcare.

parameters to ensure coherence of approach to patient management, standards of care, and integrated healthcare resource utilization [3].

### 3.1.2 Motivation and significance of AI-driven healthcare solutions

The rationale for the utilization of AI throughout healthcare systems arises from the increasingly complex, numerous, and demanding nature of healthcare services today, resulting in reconsidering previous management, decision-making, and service deliverable systems. Healthcare in the present environment is a multifaceted one incorporating patient care, administrative procedures, clinical research, and in particular strict adherence to regulatory demands. All these derivatives formulate a mass of data that is heterogeneous, multidimensional, and ever-changing in all forms. The manipulation of this data and its interpretation by traditional means can no longer cope with the volume, velocity, and variability included in health service at present. It is at this point that AI comes to the forefront as a strategic answer capable not only of simple automation of processes, but also intelligent cognition with pattern matching and predictions. The rationale for its introduction therefore rests on the premise of necessity to ensure that healthcare systems will be able to cope with complexity and assist in the analysis for informed decision system and operational efficiency, thus improving the quality and availability of healthcare by eliminating errors and waste. Healthcare systems based on AI, however, will not be of significance simply because of their predictable and essential functional relationships but will also develop in a dynamic way structures that will adopt a proactive as opposed to retroactive approach to healthcare delivery. Traditional healthcare frameworks are primarily responsive either immediately to visible signs and symptoms of distress or crises that are caused by unsuccessful attempts at managing any immediate symptomatology prior to the occurrence of such crises indicated by mere existence of their present need, implying delay with respect to action, unnecessary expenses, and negative physical health delivery in several instances. AI, on the contrary, shows the anticipated health crises, indicates the progressive and expected trends, and alerts people to the likely underlining pattern of causative etiology, which may not be apparent by mere health delivery or interpretation thereof. This analytic and predictive ability allows transpiring which organizations may institute preventive strategies, better allocate resources, and facilitate strategies that are patient's health-oriented within the operationally defined limitations. The holistic overview given by AI is that it allows suitable and correctlycovered information for organizational decision making, enabling healthcare organizations to meet patient needs more efficiently and making systematic strategic decisions based in evidence [4].

## 3.2 FUNDAMENTALS OF AI IN HEALTHCARE

AI systems aim to imitate mental ability through the application of mathematical models or algorithms, which can acquire the power of performing in a way of analytical performance with the speed, accuracy, and reliability that is in excess of the average human capacity. This is particularly said with reference to the field of health, where correctness of action in the act of decision must be exercised for the benefit of health of the organisms and sequence of organization.

### 3.2.1 Definition and core concepts of AI

The term AI can be used in a broad sense to designate that branch of computer science which deals with writing computer programs that operate at all competent levels of performance applicable to intelligence normally required of human to cope with. The instances of such efforts are learning through experience, reasoning, solution of problems, understanding of natural types of language, recognition of patterns, decision-making in an environment of uncertainty, etc. In addition, AI sought to supply the basis of provision for the machines employing a machinery by which it will simulate the intelligent human characteristics and thereby enable processing of the complex data, inference, and solution of different forms of processes in a manner similar to the process being used by human intelligence. By extending this sense to healthcare, therefore, it can be inferred that it is the ability of AI systems to analyze large-scale clinical data and employ reasoning and pattern recognition which will yield meaningful conclusions, leading to reliable knowledge-based decision-making and increased quality, efficiency, and reliability of healthcare. In other words, it is based upon certain principles which serve for guidance in drafting applications and, of course, the understanding and appreciation of these principles. One of these basic concept is ML which is usually interpreted as the process of programming algorithms to learn from data in such a way that the programming of the method leads to improvement of performance, normally defined in those terms, without the prior need for further programming [5]. This is possible since it is held that machines can be programmed to detect patterns, predict inferences, and modify their behavior as a result of experience, leading to improved performance. A related but nevertheless different concept consists of knowledge representation by which information is held, stored, and transformed in machine from the human point of view. Thus, in healthcare this could be the degree to which medical knowledge, medical patient histories, or clinical protocols are written in such a form that it leads to effective meaning and subsequent response from AI systems. Reasoning and problem-solving is another fundamental area of AI. It refers to the ability to draw conclusions from the available evidence, to scrutinize the options, and to make the right choice in a multitude of instances. This is perceived as the ability of the system to give meaning in terms of sense information, for example, by recourse to images, signals, or written information. This function in healthcare makes it possible for AI to analyze medical images, check vital signs, or utilize clinical history. NLP is another important category, which refers to the ability of the machines to digest, give rise to, and converse in human language to further communication between the health system and practitioners. The paradigms of learning, being supervised, unsupervised, and reinforcement learning, are the theoretical basis of AI. They indicate the manner in which knowledge is acquired, patterns picked up, and decisions improved upon by way of feedback.

### 3.2.2 Key AI technologies in healthcare

AI technology constitutes a complicated environment that enables health systems to manipulate complicated data, calculate results, and assist in the informed decision-making process. Each of these technologies supplies certain powers but, each of these technologies being complementary, thus create a synergistic structure which magnifies the accuracy, the efficiency, and the problems of health supply [6]. A comprehension of these technologies will lead to an understanding of how these systems of health will be enabled to use these techniques and the meaning of such policies with respect to the understanding of how AI fits into the health system.

1. *Machine Learning:* It is one of the most important AI technologies that has been put into service in the health field, which enables machines to learn from information in such a way that they become better qualified without programming. Algorithms executing ML techniques can study large and complicated systems of information in such a way that they will discover what patterns, what correlations, or what trends exist with respect to this information and from this information deduction can be drawn. The security of being a most valuable thing for acquiring operational efficiency and assisting in the right degree of diagnostics and promising conditions for the use of proactive health supply [7].

2. *Deep Learning (DL):* This is a special subcategory of ML which makes use of multilayered neural networks in order to be able to create a model of the complex relations characterizing many high-dimensional data. DL is particularly suitable for solving problems with very complex and complicated data, for example, medical images, genomic sequences, and signals of physiological nature. ML methods traditionally imply that manual feature extraction is the rule; but in the case of DL, the machine by algorithms chooses automatically the relevant features, thus minimizing the manual factor and improving precision, etc. This method is therefore of great importance for intelligent healthcare systems, and gives us the ability for a very strong representation of pattern recognition and predictive modeling. With DL, the analysis of big and complex amount of data is possible, giving efficiency to the health diagnoses that needed very high precision and reliability [8].
3. *Natural Language Processing:* The machine has an ability to read text written in natural languages and to also generate results in human language itself. In medicine, it can process clinical notes, patient records, and medical literature; it is able to extract important information from these unstructured texts to inform clinical decisions and administrative decisions. It will render great help in knowledge management, being able to retrieve relevant information in clinical requirements much faster, and even in finding some trends, etc. It also helps human personnel in the operation of AI systems. It is able to supply such things as documentation and intelligent reporting in human effort and serves as virtual assistant in clinical work.
4. *Computer Vision:* It is basically the generative results with respect to meaningful pieces of information from visual media (images or pictures, etc.). Medical imaging such as X-ray films, MRI scans, CT scans, histopathology slides, made possible the complete examination of such media, namely, taking of patterns, regional selection, and providing computer-assisted diagnosis and thereby improving the overall and consistent examination of imaging media. Computer vision technology is combined with ML or DL methods in order to produce effective information generalization and so results by way of example are effectively rapid examination or complete diagnostic results of medical imaging medium. Human effort in the diagnoses or interpretation of imaging is reduced with use of such techniques and there is less chance of error. It produces efficient and reliable results with greater speed in the examination of clinical diagnostic procedures [9].
5. *Predictive Analytics and Expert Systems:* Predictive analytics leverages past and present data to predict changes, assess the needs of the patients, and ultimately upgrade the utilization of the resources in health systems. Proactive planning and decision-making on the basis of knowledge leads to patient care and operational opportunities. Systems for expert rely on the codification of medical knowledge in the form of guidelines and

interfaces, and decision rules intended to assist in judgement on complex questions. These systems allow uniform and adequate decision-making for physicians and other technical professionals in those cases in which previous practices are required in addition to their pontifical learning processes. Predictive analytics and expert systems allow for a greater accuracy and better approach for systematic organization of the strategic plan in the management of healthcare cases [10].

6. *Reinforcement Learning:* Reinforcement learning is a process of interacting environments in which optimal strategies are learned by the systems through trial and error on the basis of rewards and penances. It will involve operations in the fields of therapy, care planning, and resource utilization. Reinforcement learning leads to a reiteration of the different strategies for solutions of problems in which the experiences lead to a solution whereby the most thought-out approach to present therapy can be thought out. Its recursive and self-improving effect would allow for optimization of the problem-solving in health units that are constantly changing and presenting new conflicts and problems, as the new and continuous learning experience will permit improvement in all forms of supervision and correction of all standards of therapy in healthcare.

### 3.2.3 AI algorithms commonly applied in healthcare settings

Algorithms of AI represent the very heart of the solutions in health, enabling the machines to analyze complex medical data, recognize patterns, and support decisions, which supplement human knowledge. The algorithms are structured computations that guide the operation of AI algorithms to allow AI systems to learn themselves from data, make inferences, and optimize the results in a favorable way.

1. *Supervised Learning Algorithms:* Supervised learning algorithms are mainly used in AI methods in health. These algorithms are trained on the labeled data bases, where the input data is connected with the known results; this gives the chance to the system to learn about the connections between the variables. In health sector, the application of supervised learning is varied, for example, in the forecast of development of illnesses, patient stratification by risk, forecast of clinical results, etc. By means of the algorithms, for example, the decision Trees, machines by support of vectors, regression and logistics are taken special means these relationships.
2. *Algorithms for Unsupervised Learning:* Unsupervised learning algorithms analyze unlabeled data to recognize hidden patterns, structures, or groupings without any prior knowledge regarding their results. These algorithms are generally used in healthcare applications for clustering of patients with similar features, determining new subtypes of

disease, recognizing abnormal features in patients' health records, etc. Techniques such as *k*-mean clustering, hierarchical clustering, and principal component analysis (PCA) are also used frequently. Unsupervised learning is especially valuable for exploratory analysis and pattern discovery to furnish useful feedback to health professionals analyzing complex and unstructured datasets.

3. *Algorithms for Ensemble Learning:* Ensemble learning is the union of several individual models to enhance the aggregate accuracy, generalization, and robustness of predictions. Thus, if the outputs of a number of tasks are aggregated, techniques are developed with the aid of which overfitting can be reduced in generalization to new data (or the noise introduced reduced) in certain tasks, enhanced accuracy can be achieved, etc. Example techniques like random forests, gradient boosting algorithms, AdaBoost, etc. are useful in healthcare applications for disease predictions, modeling of patient outcome, diagnostic support systems, etc.
4. *Probabilistic or Bayesian Algorithms:* Probabilistic or Bayesian algorithms, including Bayesian networks, are algorithms which utilize methods of a statistical nature, and differences in relationship with main variables of interest in response to model the uncertainty or relationships in blocks identified. They are frequently used in healthcare applications such as risk prediction, diagnostic reasoning, decision support systems, etc., more particularly in instances where data is incomplete or uncertain. Bayesian techniques are valuable since they permit the introduction of prior knowledge about extra data and prior expertise of a clinical kind in the learnt process with resultant probabilistic predictions, which are therefore useful in decision-making.

### 3.2.4 Historical milestones and evolution of AI in healthcare

The development of AI in medicine has been a gradual one but marked by crucial points which have been decisive in the employment of intelligent techniques by the medical organization [11]. AI in medicine may be traced back to the early 1960s and 1970s when a number of initial steps were taken in the direction of systems based on rules or principles which were expected to reproduce human thought. The systems so modeled, such as MYCIN for infectious diseases, are considerable testimony to the satisfaction that computers can assist clinicians by using any available medical information in a strict and scientific way to the patient's cases. These systems initially had certain limitations owing to the lack of computational prowess and also possible lack of data, but today AI has become a useful stimulant to the reasoning standards of human. The 1980s and 1990s have seen the growth systems working on more extensive rules and more revolutionary probabilistic arrangements, thereby enabling AI to cope with uncertainty and developments in the medical environment. In this way also are arrangements possible whereby research into AI enables confirmatory study and opinion of those diagnosed by systems and models of status where statistics could take the place of the systems already in existence, thus giving a far better method of therapeutics. Thus, the possibilities of AI in varying the ratio give enormous impetus to those possible for AI in increasing of thought and inference in relation to diagnosis, the possibilities of its applying sufferings or improvement of therapy show that by this process an enormous range can be covered in use [12]. In this connection, the adoption of methods of neural networks and ML in the late 1990s and early 2000s shows a further significant feature in the evolutionary track of AI in medicine. ML has enabled algorithms to learn from data directly instead of traditional rule-based systems as was done before, and thereby improve their performance over time and to be adapted to new input information [13]. This time period began with the first applications of AI to medical imaging, where recognition algorithms became available that would help identify abnormal findings in X-rays and scans, which were the basis of complex analytical systems [14]. Healthcare data were now compiled by means of electronic health records (EHRs), which provided the quantity and structure of data that were necessary for the AI models to be trained and examined. In the past decades, improvements in AI capability have been rapid, including improvements in computing power, availability of more and better data, and improvement in sophistication of algorithms. Advances in DL and natural language programming have made it possible for AI systems to process unstructured data in the form of clinical notes, imaging records, and genomes with higher accuracy than ever before. AI has also now moved into operational and administrative areas where it is now being used to support resource utilization, optimization of workflow, and predictive analytics. So, AI at present has improved from being a mere tool of backup to a tool which has influenced the whole healthcare system, influencing the quality of patient care, the quality of research, and hospital policy on administration [15]. AI in the healthcare system is still evolving where exciting technologies like Internet of Things (IoT) devices and sensors are available for use and storage in the cloud is learnt to develop intelligent healthcare systems and co-related systems [6]. The AI system has been developed due to both technological development and the understanding that it is possible to overcome complex healthcare problems and ensure patient safety, care quality, and efficiency of operations [16].

## 3.3 APPLICATIONS OF AI IN HEALTHCARE

AI has undoubtedly invaded the sphere of diagnostics and medical imaging :

1. *Diagnostics and Medical Imaging:* The existence and diagnosis of disease and factors medical and then to communicate a constructed and transparent diagnosis based upon advanced algorithms with the result

of a far better and more accurate diagnosis at a great velocity. In regard to medical imaging, AI uses some of the scientific elements included in the others, for example, the working with machines in structured learning and DL and visual evenness in addition respectively of the radiographs, MRI, CT scanning, and ultra sonography carry out the smart and therefore studies to allow them to find of greatest significance inaccuracies of presentation. AI assists radiologists in recognition through automatic spotting of subtle features which may be impossible for the human eye to recognize and enables to email earlier scalps of diseases such as cancer, cardiovascular disease, and neurological complaints [17]. Beyond the interpretation of images, diagnostic systems based on AI algorithm can correlate imaging data with patients' history, laboratory results, and genetic data to obtain a near qualified and complete health of the patient that leads to thorough diagnostic assessments being made. The predictive powers of AI make possible to spot high-risk patients, enabling early intervention and preventive cure. AI enhances the accuracy and efficiency of diagnosis, but the assimilation of clinical explanations needs careful validation and proper approvals and should be concurrent with the physics and expertise of the clinician. Destruction of a pecuniary knowledge about the human mind allied to the force of analysis exercised by AI and so well is the work of diagnostics. A process which in the highest case improved the interest of the them brush purposely practiced about for more cases than once would be desired outcome hence and infer from human ultra-testing in the absence of this admirable technical diagnosis with which humanized it contact with it has become the clinical practice need by patients that have made it compound over the paid processes [18].

2. *Treatment Planning and Clinical Decision Support:* Optimal patient care is obtained from treatment planning and clinical decision support. Many AI systems use large sets of clinical, demographic, and historic patient data to facilitate the health professional in planning for optimal treatment plans. By assessing the numerous tactical methods used in any one disease process, evaluating outcome potential, and individualizing the patient factor with added concerns such as age, sex, comorbidity, and genetic markers, etc., AI is capable of providing specific individualized evidence-based care. Such clinical decision support systems (CDSS) use AI to provide info to the physician along with recommendations, advice on possible adverse issues, and monitors for compliance with previously instituted patterns and protocols [19]. The systems may warn the physician of incompatibilities in drug therapy, suggest changes in dosage, and recommend new approaches to action steps on the basis of data assembled in real time. The use of AI in treatment planning also means precise and less errors, expedite decision-making, particularly in complex situations and equations where scores of parameters have to be reviewed simultaneously. The

significant strength of the integrated methodical analytical approaches, predictive modelling, and individual actions endorsed in the usage of an AI system would provide more effective decision-making, increase efficiency of treatments (to a degree), and lead to a health system which is more rewarding, safer, and profitable [20].

3. *Remote Patient Monitoring (RPM) and Telemedicine:* RPM and telemedicine are applications of AI technology capable of revolutionizing the delivery of healthcare outside of traditional clinical settings. RPM systems leverage wearable technology, sensors, and mobile applications to collect and analyze patient data in real time, including vital signs, physical activity levels, and biochemical parameters. AI uses predictive analytics to process real-time patient data continuously, with the potential to detect early indications of impending patient deterioration, notify healthcare providers of potential emergencies, and facilitate timely intervention without the necessity of a personal visit. Telemedicine platforms incorporating AI improve telehealth consultations by providing healthcare providers with decision support, symptom analysis, and assessment of patient risk during remote interactions [21]. AI is also useful for streamlining administrative tasks in telemedicine, such as appointment scheduling, documentation, and follow-up reminders, facilitating continuity of the patient care experience. The benefits of improved access to healthcare are important, particularly for patients in rural or other underserved locations. RPM and telemedicine contribute to decreasing the burden of unnecessary visits to medical facilities, hospitals, and clinics. The predictive and analytic capabilities of RPM systems render the proactive management of chronic illnesses, such as diabetes, cardiovascular disease, and chronic pulmonary disease, improving patient outcomes and quality of life. Continuous learning from patient data allows AI systems to modify recommendations to individual health patterns, facilitating personalized remote care. Figure 3.2 shows AI-enabled RPM and telemedicine architecture with wearable devices and IoT sensors that collect patient health data, AI systems analyzing real-time information, and healthcare providers receiving alerts and insights for remote consultations, personalized treatment, and proactive care management [22].
4. *Personalized Medicine:* Personalized medicine is a field in which AI has shown extraordinary potential in tailoring healthcare interventions to individual patients based on their unique biological, environmental, and lifestyle characteristics. AI systems analyze genetic data, clinical histories, laboratory test results, and real-time health data to determine patient-specific treatment needs and risk profiles. Using predictive modeling, ML, and pattern recognition, AI can suggest therapeutic modalities that are most likely to be efficacious for a certain patient with a reduction in side effects. The incorporation of AI in personalized medicine allows for the capability of tracking the dynamically

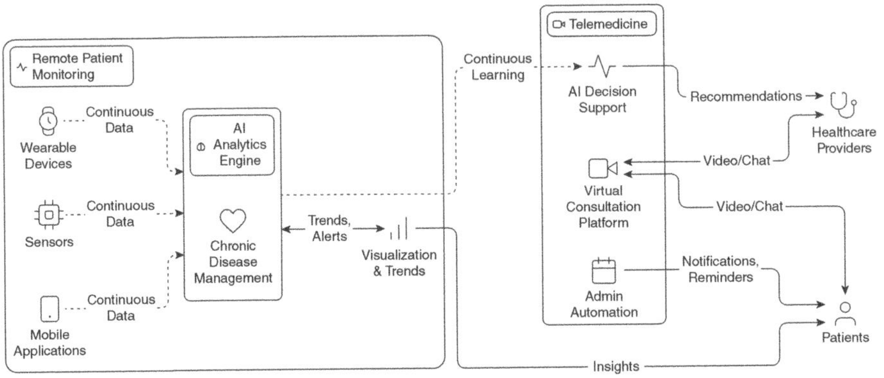


*Figure 3.2* AI-enabled remote patient monitoring and telemedicine architecture.

changing therapeutic protocols related to a patient's presentation to the therapy employed, affording a real-time monitoring ability and allowing change of management as a reaction to the specific therapy. The resulting situation, where AI is utilized, is one of incorporating large data bases with predictive analytics and adaptive learning so that AI leads to the improved accuracy and effectiveness of individualized interventions. In addition to improved clinical outcomes through personalized medicine, there is increased patient input relating to their health. This leads to allowing patients to be involved in informed decision-making. In effect, this is a move from the neglected patient approach to the empowerment of patients related to their health. Thus, by introducing AI into personalized medicine, one effects the change from the reactive to proactive healthcare delivery. Thus, the healthcare system has been optimized for the individual patient and at the same time the inefficiencies have been reduced while the overall functioning of healthcare has been improved [23].

5. *Hospital Management and Operational Efficiency:* AI is proving its transformative potential in the world of hospital management and operational efficiency. AI systems analyze large datasets with respect to not only patient admissions, but also staffing, resource use, and procedural workflows in order to improve hospital operations. AI allows better utilization of resources and a decrease in waiting time by predicting patient admissions, forecasting bed occupancy, and managing staffing schedules to ensure availability of crucial services on demand [24]. AI can also improve the supply chain by monitoring supply inventories, predicting product demand, and decreasing the waste of medical supplies and pharmaceuticals. AI also can help optimize workflows, due to the streamlined patient flow, by administrative workload alleviation and better coordination of departmental operations, thereby providing better patient care. AI analytics help provide actionable inputs for financial

management, policy planning, and quality improvement programs, thus helping to foster operational excellence in hospitals, while continuing a high level of patient care. The use of predictive modeling, decision support, and real-time monitoring allow AI to assist in facilitating safer, more efficient methods of hospital operations, with the patients care better at the focus. The use of AI in hospital management leads to not only better internal performance, but also greater patient attention and satisfaction, since timely care, decrease of waiting periods, and supply considerations lead to better overall health experiences [25].

## 3.4 CASE STUDIES AND REAL-WORLD IMPLEMENTATIONS

The transformative potential of AI in healthcare is best understood through case studies and real-world implementations that demonstrate its practical impact across diverse clinical, operational, and research settings [26]. AI technologies have been adopted in hospitals, research institutions, and telemedicine platforms worldwide, illustrating how theory translates into meaningful outcomes.

1. *AI-Based Diagnostic Tools in Radiology and Pathology:* One of the most important uses of AI in healthcare has been in imaging and pathology. Other hospitals and research centers have developed DL-based systems in order to analyze medical images, including X-rays, CT scans, MRI scans, and histopathological slides. Such AI systems can detect abnormalities and segment areas of interest and find patterns that may not be noted by human observers. In real life, it has been reported in institutions that there is considerable increase in the accuracy of diagnosis in real patients, especially with regard to cancers, pulmonary diseases, and heart diseases, by the introduction of these diagnostic AI systems. The radiological departments have decreased the time of assessment and assisted in the quickening of reports and subsequently advance treatment of patients. The results of the introduction of these AI systems showed that the skills of the clinicians could be enhanced, the human error factor would be reduced, there would be improvement not only in the rate of diagnosis and in accuracy of diagnosis but also in results for patients in both long and short terms, and the working of the various diagnostic departments would be far more efficient [27].
2. *Predictive Analytics for Patient Monitoring and ICU Management:* AI-based predictive analytics have been successfully utilized in patient monitoring and intensive care units (ICU). Hospitals have effectively used ML algorithms to analyze real-time patient data to predict complications such as sepsis, heart attacks, or respiratory failure. In practice,

these systems have allowed clinicians to intervene sooner, lowering the incidence of critical events and death rates. It has been shown that ICUs using predictive models in AI have reported lower mortality rates and decreased lengths of stay of patients, by identifying high-risk patients prior to clinical deterioration. The result of the work is again a reflection of how transformational AI can be in changing reactive health practice to proactive preventive healthcare, allowing early intervention with timely treatment that saves lives and gets the best use out of available resources [28, 29].

3. *Personalized Medicine and Genomic-Based Therapy:* AI has been applied very effectively in personalized medicine, including the area of oncology and genomics. Platforms utilizing AI analyze patient genomic profiles, biomarkers, and clinical histories in order to recommend a personalized plan of therapy. Real-life applications in cancer treatment centers have demonstrated that AI-driven recommendations actually have improved the selection of therapy leading to better performance of therapy, lessened side effects, and improved quality of life [30]. With clinical trials, AI-generated interventions were successful in accelerating drug discovery programs and in identifying those patients likely to respond to the therapeutics under investigation. The results of the above studies affirm that AI in fact enhances the area of precision medicine by offering targeted data-driven plans of treatment which at the same time are patient-centered in their approach, thereby enabling better outcomes and therapeutic success [31].
4. *Telehealth and RPM:* RPM and telehealth platforms powered by AI have opened up healthcare access for chronic disease management and for patients who live in remote areas. AI devices and platforms continuously monitor patient vitals, identify anomalies, and send immediate alerts to healthcare providers. Hospitals reported improved chronic disease management, decreased hospital readmissions, and improved patient engagement by implementing these systems. In addition, diabetic and cardiologic patients being monitored through AI platforms better adhered to treatment plans and were able to have their therapy adjusted in a timely fashion [18].
5. *Operational Warehouse Efficiency:* AI technologies have been utilized to optimize the operation of hospitals in delivery of healthcare through patterns derived regarding the distribution of resources, timing of assignments for healthcare personnel, and supply delivery allocation. AI analytic technologies have resulted in hospitals being able to predict patient delivery into their system, minimizing treatment bottlenecks, and optimizing the talents of the healthcare service personnel. This indicates that clinical excellence as well as operational efficiency could be achieved through the employment of AI technology, without forgetting the capabilities of AI to improve the administrative and special efficiencies of hospital healthcare systems.

## 3.5 CHALLENGES AND LIMITATIONS

Although AI has great potential for radically changing healthcare, its widespread adoption has also caused a multitude of serious difficulties and limitations. These complications are of technical, ethical, organizational, financial, and regulatory nature: the difficulties of introducing intelligent systems into a sensitive and strictly regulated area of human activity are obvious. AI in medicine is based on a large amount of data. It is important to have accurate data, which need to be protected and kept confidential; otherwise, diseases or undesirable medical events can follow, thereby compromising the security and trust in the doctor by the patients [32].

1. *Privacy of Data and Data Security:* One of the serious problems of AI medicine lies in securing the privacy of data. The AI systems exploit a huge amount of data and sensitive information on the patients, such as medical records, genetic data, and health parameters. The unauthorized access, the breaches, and the bad use of the data can have disastrous consequences for the privacy and trust of the patients. As such, health societies adhere to strict norms, which constitute HIPPA, GDPR, and the local patient data protection laws, checks, and requirements, should be ensured. They are also required to have effective encryption systems, access limits, and systems of anonymization of data. The problem does not lie only with the static data but also with real-time data, monitoring devices, remote access, wearable devices, and tele-medicine. Solving these problems requires meeting extremely complex criteria (technical, processes, control) so that AI may be used without patient's privacy loss and also without breach of security, bad use, and vulnerable attacks on patient data from different criminal or malicious forces [33].
2. *Ethical and Bias Issues:* Another major drawback of the use of AI in medicine are the ethical issues and biases of the algorithm. AI systems learn from historical and clinical data, which can have inherent biases with respect to sex, ethnic background, socioeconomic status, or area of residence [34]. If these problems are not solved, they will result in differing healthcare and diagnoses or treatment recommendations. Additionally, ethical issues arise with respect to patient consent, transparency of AI decision-making, accountability for errors, and the delegation of critical healthcare decisions to machines. The ethical use of AI requires that development of training data sets be carried out selectively, the algorithms continuously validated, the output of the AI systems interpretable, and legislation developed defining human oversight, accountability, and patient prerogatives. If these ethical and bias issues are not solved, public trust may be eroded and the effectiveness of the use of AI in healthcare systems is limited [35].
3. *Technical Issues and Integration:* The employment of AI in already existing healthcare systems presents severe technical issues. Much of

our healthcare systems are constructed on older software components, with fragmented databases, as well as data which has often been produced inconsistently and so deployment of AI in these systems can be problematic. This relates to the incorporation of AI systems into electronic healthcare records (EHR), the performance on current data, and varying data sources. Additionally, in order to construct AI systems which are reliable, generalizable, and can deal with complex clinical situations, vast computing resources and technical expertise are required, combined with extensive validation with outcome data on clinical endpoints. These technical problems also delay the incorporation of AI systems and thus require the continued cooperation of physicians, computer technology experts, and data specialists to assure the technical integration and meaningful application of AI systems.

4. *Regulatory and Compliance Problems:* Although some of the changing regulatory patterns with respect to AI in healthcare are changing, they remain in several respects complex and uncertain. For AI-based medical devices and diagnostic and decision-support systems, the difficult approval process of governmental entities such as the FDA, EMA, and other governmental entities within each country is required. In many cases, transparency, explainability, and validation when AI algorithms are to be used clinically are required, which may inhibit innovation in this area. Further, the rapidity of the development of AI methods frequently outstrip the ability of regulatory agencies to review and keep up with the changes, and there remains questions as to the liability, accountability, and regulatory compliance of these systems.
5. *Cost and Resource Limitations:* The implementation of AI-managed solutions in healthcare often requires substantial capital expense and investment of resources. The cost of procuring the AI systems, augmenting IT systems, training of personnel, and maintaining existing systems may be substantially expensive, particularly for smaller or underfunded healthcare institutions. Data scientists or AI-trained personnel who have the ability to supervise model development, implementation, and oversight are also needed, thus increasing overhead expenses. Similarly, limited availability of high-quality datasets, computational exercise, and special expertise will impose limitations on AI effectiveness. If fairness and equity of access to AI devices are to be obtained and if resource exploitation between wealthy urban hospitals and underfunded hospitals are to be avoided, these economic and resource problems will have to be solved.

## 3.6 FUTURE DIRECTIONS OF AI IN HEALTHCARE

AI in healthcare holds great promise for future healthcare needs such as patient care, operational efficiency, and medical research. As AI devices continue to develop, health systems are also expected to develop into more predictive, preventive, personalized, and participatory methods of care.

1. *Predictive and Preventive Healthcare:* It is one of the most promising future directions of AI in healthcare as we move from reactive healthcare to predictive and preventive healthcare. The results of AI algorithms will increasingly interpret historical patient data, real-time monitoring information, and population health trends to establish early signs of disease. By clarifying the probabilities of medical conditions prior to their full manifestations, the clinician may institute timely intervention, ameliorate complications, and improve long-term health outcomes. This predictive model will have great healthcare significance in diseases of chronic illness, in particular diabetes, heart disease, and respiratory disease. The utilization of AI for preventive and predictive healthcare lowers hospitalization, improv patient quality of life, and promotes sustainable population [36].
2. *Personalized Medicine and Genomics:* AI stands to dramatically change personalized medicine to develop personalized treatment regimens. The algorithms will be involved in genomic data, clinical past history, lifestyle characteristics, and socioenvironmental data to give individualized treatment regimens and optimize drug regimens for great efficacy. In the near future, the AI systems will further enhance how precision medicine owing to its ability to have constant learning concerning patient response to treatment protocols, thereby enabling optimization of treatment strategy and identification of new biomarkers for new diseases. This will result in decrease of side effects, better therapeutic effects, and individualized healthcare. With the advancements of AI systems, personalized medicine will be expected not only in oncology and rare diseases, but also in other not so rare diseases. AI will also be used in preventive and chronic situations in healthcare, and thus ultimately change the concept of patient-centered care [37].
3. *Explainable AI and Ethical Issues:* With AI being used increasingly in clinical decision-making, the need for explainability and ethical governance will become ever more important. Future AI systems will be concerned with accountability and transparency in order that healthcare professionals will be aware of the reasons for recommendations, while the patients will know who is accountable for the patient's care. Explainable AI can be used to reduce bias and errors, and increase the trust of healthcare professionals and patients. Ethical issues will continue to be a factor with the use of AI, establishing sets of issues of ethics, privacy, equity, accessibility, and the importance of the human element in human healthcare decision-making processes.
4. *Integration with IoTs and Wearables:* The integration of AI with IoTs and wearables is a revolutionary future focus in healthcare. They will offer continual monitoring of vital signs, activity levels, and other

physiologic parameters, by a constant command of wearable and AI sensors, leading to real-time health status testing and early warning of conditions needing treatment. In this way a healthcare provider will be able to provide proactive treatment to his patient from a distance, that is, remote care to the patient out of hospital. AI along with IoTs and cloud resources will enable the human population to analyze which will facilitate the healthcare provider to develop trends, better utilize resources, and the improve the general public health.

5. *AI in Research and Operation Efficiency:* AI is going to be more and more important in the research and operation of diseases and their mechanism. In research, the information on disease can be used by the AI algorithms to optimize the development and manufacture of drugs and other pharmaceuticals, determine ideal clinical trials, and obtain more insight into large-scale biomedicine databases, thus enabling us to procure better and faster treatments for modern-day diseases. In the operational phase, AI will help improve management practices of the hospital by the optimal flow of patients, predict patient admission levels and better allocate resources, and enable efficient administration of hospital affairs. Together, these will help to make the healthcare system more efficient and less costly, thus satisfying the patients all the more.

## 3.7 CONCLUSION

AI has changed the landscape of inpatient healthcare delivery. Topics affected include diagnostics, treatment plans, patient monitoring, personalized medicine, and hospital administration and management [8]. The new high-powered algorithms and predictive analytic methods together with other emerging technologies such as IoT and wearables allow for a more accurate, effective, and available method of delivering evidence-based, patient-centered healthcare material. A series of case studies, actual applications, and general positive outcomes aid in earlier diagnoses and methods of treatment and management of chronic disease as well as improve the hospital's administration systems with more effective work flow. Data safety, ethical issues, technical issues, regulatory issues, cost issues, and other problems arise in great variety and must be continually refined, substantiated, and resolved through a safe, equitable, and reasonable use of AI. The future for AI in healthcare lies in predictive and preventive modalities of healthcare, precision medicine, ethical and explainable decisions, and seamless integration with the digital ecosystem improvements, producing a more intelligent, adaptable, and sustainable healthcare environment. This integration of innovation and regulatory oversight can produce improved clinical outcome. The more intelligently connected, and more patient-oriented healthcare delivery systems would enable a better healthcare system globally.

## REFERENCES

1. Salehi, F. (2024). The transformative role of artificial intelligence in the healthcare industry: A comprehensive analysis. Asian Journal of Research in Medicine and Medical Science, 6(1), 62–70.
2. Adhikari, S., Ahmed, I., Bajracharya, D., Khanal, B., Solomon, C., Jayaratne, K., & Khan, M. I. (2025). Transforming healthcare through just, equitable and quality driven artificial intelligence solutions in South Asia. NPJ Digital Medicine, 8(1), 139. https://doi.org/10.1038/s41746-025-01639-6
3. Poalelungi, D. G., Musat, C. L., Fulga, A., Neagu, M., Neagu, A. I., Piraianu, A. I., & Fulga, I. (2023). Advancing patient care: How artificial intelligence is transforming healthcare. Journal of Personalized Medicine, 13(8), 1214. https://doi.org/10.3390/jpm13081214
4. Abukhadijah, H. J., & Nashwan, A. J. (2024). Transforming hospital quality improvement through harnessing the power of artificial intelligence. Global Journal on Quality and Safety in Healthcare, 7(3), 132–139. https://doi.org/10.36401/jqsh-24-4
5. Thoppil, I. J., Ashtalakshmi, K., & Chundi, R. (2025) Transforming healthcare: Leveraging machine learning algorithms for diagnosis, treatment, and management. In: Bioinformatics and Beyond (pp. 92–114). CRC Press. https://doi.org/10.1201/9781003508403-5
6. Onciul, R., Tataru, C. I., Dumitru, A. V., Crivoi, C., Serban, M., Covache-Busuioc, R. A., & Toader, C. (2025). Artificial intelligence and neuroscience: Transformative synergies in brain research and clinical applications. Journal of Clinical Medicine, 14(2), 550. https://doi.org/10.3390/jcm14020550
7. Marriwala, N. K., Jain, S., Shukla, V. K., & William, P. (2025). Harnessing IoT and machine learning for transformative healthcare solutions. In: Hybrid Soft Computing Techniques for Machine Learning and Optimization (pp. 229–248). IGI Global Scientific Publishing. https://doi.org/10.4018/979-8-3693-6864-0.ch012
8. Vashishth, T. K., Sharma, V., Sharma, K. K., Chaudhary, S., Kumar, B., & Panwar, R. (2024, March). Artificial neural networks & discrete wavelet transform enabled healthcare model for stress and emotion assessment using speech signal recognition. In: *AIP Conference Proceedings* (vol. 3072, no. 1). AIP Publishing. https://doi.org/10.1063/5.0198725
9. Sharma, V., Vashishth, T. K., Kumar, B., Sharma, K. K., Vidyant, S., & Chaudhary, S. (2023, December). Health interpretation of Covid-19 patients using artificial intelligence. In: *2023 International Conference on Artificial Intelligence for Innovations in Healthcare Industries (ICAIIHI)* (vol. 1, pp. 1–5). IEEE Computer Society. https://doi.org/10.1109/icaiihi57871.2023.10488924
10. George, A. S. (2025). Transforming healthcare with artificial intelligence: Strategies, insights, and frameworks for innovation. Partners Universal International Research Journal, 4(2), 48–78.
11. Chava, K. (2022). Harnessing artificial intelligence and big data for transformative healthcare delivery. International Journal on Recent and Innovation Trends in Computing and Communication, 10(12), 502–520. https://doi.org/10.70593/978-81-988918-5-3_12
12. Bockarie, M. J., Ansumana, R., Machingaidze, S. G., de Souza, D. K., Fatoma, P., Zumla, A., & Lee, S. S. (2024). Transformative potential of artificial intelligence on health care and research in Africa. International Journal of Infectious Diseases, 143, 107011. https://doi.org/10.1016/j.ijid.2024.107011
13. Kasula, B. Y. (2023). Harnessing machine learning for personalized patient care. Transactions on Latest Trends in Artificial Intelligence, 4(4).

14. Huck, J. (2024). Harnessing the potential of artificial intelligence in healthcare innovations applications and implications. Health Science Journal, 18(2), 1–2.
15. Garg, P. P., Jayashree, J., & Vijayashree, J. (2025). Revolutionizing healthcare: The transformative role of artificial intelligence. In: Responsible and Explainable Artificial Intelligence in Healthcare (pp. 1–23). Academic Press. https://doi.org/10.1016/b978-0-443-24788-0.00001-7
16. Racine, E. E. (2025). Harnessing artificial intelligence to meet global healthcare challenges: Reparative algorithmic impact assessments in context. Emerging Media, 3(2). https://doi.org/10.1177/27523543251372031
17. Chia, J. L. L., He, G. S., Ngiam, K. Y., Hartman, M., Ng, Q. X., & Goh, S. S. N. (2025). Harnessing artificial intelligence to enhance global breast cancer care: A scoping review of applications, outcomes, and challenges. Cancers, 17(2), 197. https://doi.org/10.3390/cancers17020197
18. Al-Khatib, S. M., Singh, J. P., Ghanbari, H., McManus, D. D., Deering, T. F., Silva, J. N. A., & Hurwitz, J. L. (2024). The potential of artificial intelligence to revolutionize health care delivery, research, and education in cardiac electrophysiology. Heart Rhythm, 21(6), 978–989. https://doi.org/10.1016/j.hrthm.2024.04.053
19 Sharma, V., Vashishth, T. K., Kumar, R., Vidyant, S., Pandey, A., & Samania, B. (2026). AI-driven clinical decision support systems enhancing decision-making in modern healthcare. In: Applied AI and Computational Intelligence in Diagnostics and Decision-Making (pp. 219–248). IGI Global Scientific Publishing. DOI: 10.4018/979-8-3373-3311-3.ch009
20. Olawade, D. B., David-Olawade, A. C., Wada, O. Z., Asaolu, A. J., Adereni, T., & Ling, J. (2024). Artificial intelligence in healthcare delivery: Prospects and pitfalls. Journal of Medicine, Surgery, and Public Health, 3, 100108. https://doi.org/10.1016/j.glmedi.2024.100108
21. Sharma, M., Teotia, S., Aggarwal, V., Sharma, V., Vashishth, T. K., & Samania, B. (2026). Harnessing predictive analytics for disease diagnosis and risk assessment: Transforming preventive healthcare with AI. In: Applied AI and Computational Intelligence in Diagnostics and Decision-Making (pp. 133–162). IGI Global Scientific Publishing. https://doi.org/10.4018/979-8-3373-3311-3.ch006
22 Vashishth, T. K., Sharma, V., Kumar, S., Verma, N., Vidyant, S., & Kaushik, S. (2025). The integration of AI in telemedicine transforming healthcare delivery and patient outcomes. In: AI-Driven Personalized Healthcare Solutions (pp. 71–98). IGI Global Scientific Publishing. DOI: 10.4018/979-8-3693-7858-8.ch003
23. Abedalrhman, K., & Alzaydi, A. (2025, May 21). Digital innovations in healthcare: Harnessing artificial intelligence, IoT, and big data analytics for personalized medicine and improved patient outcomes – Insights from the Syrian healthcare sector. https://doi.org/10.2139/ssrn.5286200
24. Vashishth, T. K., Chaudhary, S., & Sharma, V. (2023). Optimum utilization of bed resources in hospitals: A stochastic approach. In: Artificial Intelligence-Based Healthcare Systems (pp. 101–110). Springer Nature. https://doi.org/10.1007/978-3-031-41925-6_7
25. Thenmozhi, E., Jeya, M. R., Meenakshi, L., Elakya, V., Devi, R., Rejees, J. D., & Divyalakshmi, V. (2025). Harnessing the transformative power of AI for enhanced disaster prediction and comprehensive risk assessment. In: Harnessing AI in Geospatial Technology for Environmental Monitoring and Management (pp. 177–200). IGI Global Scientific Publishing. https://doi.org/10.4018/979-8-3693-8104-5.ch008
26. Gouripur, K. (2024). The impact of artificial intelligence on healthcare: A revolution in progress. The North and West London Journal of General Practice, 10(1).

27. Komalasari, R. (2024). Transformative insights: Harnessing artificial intelligence for enhanced ovarian cancer prediction and prognosis. In: Biomedical Research Developments for Improved Healthcare (pp. 184–207). IGI Global Scientific Publishing. https://doi.org/10.4018/979-8-3693-1922-2.ch010
28. Khinvasara, T., Cuthrell, K. M., & Tzenios, N. (2024). Harnessing artificial intelligence in healthcare analytics: From diagnosis to treatment optimization. Asian Journal of Medicine and Health, 22(8), 15–31. DOI:10.9734/ajmah/2024/v22i81066
29. Patil, S., & Shankar, H. (2023). Transforming healthcare: Harnessing the power of AI in the modern era. International Journal of Multidisciplinary Sciences and Arts, 2(2), 60–70. https://doi.org/10.47709/ijmdsa.v2i1.2513
30. Kolasani, S. (2024). Optimizing with intelligence: Harnessing AI and ML algorithms for real-world solutions. In: Practical Applications of Data Processing, Algorithms, and Modeling (pp. 190–204). IGI Global Scientific Publishing. https://doi.org/10.4018/979-8-3693-2909-2.ch015
31. Chhabra, C., Kaur, H., Aziz, R., Khan, S. A., & Mazhar, M. (2023). Leveraging artificial intelligence in healthcare: A transformative approach. AI Horizons: Exploring Multidisciplinary Frontiers, 2, 28–43.
32. Genovese, A., Borna, S., Gomez-Cabello, C. A., Haider, S. A., Prabha, S., Trabilsy, M., & Forte, A. J. (2025). From promise to practice: Harnessing AI's power to transform medicine. Journal of Clinical Medicine, 14(4), 1225. https://doi.org/10.3390/jcm14041225
33. Avasthi, S., Sanwal, T., Sareen, P., & Tripathi, S. L. (2022). Augmenting mental healthcare with artificial intelligence, machine learning, and challenges in telemedicine. In: Handbook of Research on Lifestyle Sustainability and Management Solutions Using AI, Big Data Analytics, and Visualization (pp. 75–90). IGI Global Scientific Publishing.
34. Alsaleh, A. (2025). Leveraging artificial intelligence for sustainable healthcare transformation: Socioeconomic impacts and ethical implications. Innovation and Emerging Technologies, 12, 2550015. https://doi.org/10.1142/s273759942550015x
35. Whig, P., Kasula, B. Y., Yathiraju, N., Jain, A., & Sharma, S. (2025). Revolutionizing gender-specific healthcare: Harnessing deep learning for transformative solutions. In: Transforming Gender-Based Healthcare with AI and Machine Learning (pp. 14–26). CRC Press. https://doi.org/10.1201/9781003473435-2
36. Vashishth, T. K., Sharma, V., Kumar, B., Panwar, R., Sharma, K. K., & Chaudhary, S. (2024). AI-integrated IoT in healthcare ecosystem: Opportunities, challenges, and future directions. In: AI and IoT Technology and Applications for Smart Healthcare Systems (pp. 37–54). CRC Press. https://doi.org/10.1201/9781032686745-4
37. Bhagat, S. V., & Kanyal, D. (2024). Navigating the future: The transformative impact of artificial intelligence on hospital management – A comprehensive review. Cureus, 16(2). https://doi.org/10.7759/cureus.54518

Chapter 4

# Detection of Parkinson's disease through voice samples

*Abhishek Kumar and Sobhit Saxena*

## 4.1 INTRODUCTION

Millions of people in the world suffer from Parkinson's disease (PD), a neurological illness that often occurs in the elderly. James Parkinson initially referred to it as "Shaking Palsy" [1]. In the substantia nigra pars compacta, a part of the midbrain that produces dopamine, a neurotransmitter required for coordinating balanced and fluid muscle movements, the disease is mainly linked to the progressive loss of dopaminergic neurons [2]. The hallmark symptoms of PD, including bradykinesia (slowness of movement), postural instability, muscle rigidity, and resting tremors, are caused by aberrant activity in motor circuits caused by a dopamine shortage.

In PD, these motor symptoms may also be accompanied by a range of non-motor symptoms, including depression, sleep problems, cognitive decline, and speech impairments [3–5]. Because these symptoms often appear gradually and can vary considerably from person to person, early diagnosis of the disease is difficult. PD is a progressive neurodegenerative disorder that affects motor function because it causes the death of dopamine-producing neurons in the brain. Symptoms of the condition include bradykinesia (slowness of movement), postural instability, stiffness, and tremors. For PD to be effectively managed and patients' quality of life to be improved, early identification is essential.

Because the neuronal damage in PD starts long before clinical symptoms appear, early identification is essential. Over 60–70% of dopaminergic neurons are thought to have already been destroyed by the time the disease is identified by outward motor signs [6]. Consequently, early or preclinical detection of PD can greatly enhance patient outcomes by allowing

a. immediate medical attention, which is required to halt the progression of the illness;
b. individualized therapy plans, which enhance quality of life by addressing particular ailments; and
c. monitoring the development of diseases for scientific purposes and the creation of new treatments.

DOI: 10.1201/9781042015931-4

However, these methods can be subjective, difficult, and could ignore minor early signs. The need for objective, data-driven, noninvasive diagnostic techniques that facilitate early-stage detection is therefore expanding. Traditional PD diagnosis relies heavily on neurological testing and clinical observation by movement disorder specialists. Clinical observation and neurological evaluations are the primary foundations of traditional diagnosis, which frequently overlooks mild early-stage symptoms. Therefore, in recent years, there has been a lot of interest in using machine learning (ML) and signal processing approaches for early PD identification.

The rest of the chapter is structured as follows: Section 4.2 narrates the overview of PD. In Section 4.3, dataset has been analyzed, and a discussion on PD's attributes has been presented. Section 4.4 presents the methodology proposed and five different ML models illustrated. Section 4.5 enumerates the results of the reduction part as well as the ML modeling controller design. Section 4.6 draws the salient inferences.

## 4.2 OVERVIEW OF PARKINSON'S DISEASE

The presence of PD is identified by speech and movement deficits brought on by the slow death of dopaminergic neurons in the brain. Conventional diagnostic techniques, like neuroimaging and clinical observation, frequently detect the illness at an advanced stage. As a result, scientists have looked into alternative strategies that allow for early, noninvasive, and economical detection; voice signal analysis has become one such area with demonstrated potential.

### 4.2.1 Voice as a Parkinson's disease biomarker [7]

Precision neuromuscular coordination is required for speech production, and early signs of PD-induced motor dysfunctions include hypophonia, monotone speech, and irregular voice tremors. As shown in Ref. [8], which provided the Parkinson's dataset based on sustained voice phonations, features including jitter, shimmer, and HNR can be effective biomarkers. According to their outcomes, these properties include subtle variations in pitch and amplitude that are often imperceptible to the human ear but that can be evaluated by signal analysis.

### 4.2.2 Models of ML based on acoustic features

ML and statistical models were employed during investigations to classify PD patients based on their vocal attributes. To attain high diagnosis accuracy, Ref. [9] developed a nonlinear analytic method that makes use of measures such as RPDE, DFA, and PPE. For voice-based PD diagnosis, research in Ref. [10] found that support vector machines (SVMs) and *K*-nearest neighbors

(KNN) achieved accuracies of over 90%. Strong categorization necessitates nonlinear dynamic characteristics, which capture the complexity of speech signal fluctuations, as these experiments showed. In order to improve sensitivity to subtle variations in voice, Ref. [11] combined deep learning with traditional acoustic properties. It explored ensemble-based techniques like random forests and XGBoost to show enhanced model stability and interpretability across heterogeneous datasets in a similar manner [12].

### 4.2.3 Role of technology and artificial intelligence (AI)

The development of ML, AI, and signal processing technologies has enabled identification of PD early. These technologies make it possible to automatically analyze a variety of behavioral and physiological cues in order to find hidden patterns linked to PD [13, 14]. The following are some of the widely explored modalities for early detection:

a. *Analysis of Voice and Speech:* Patients with PD frequently have noticeable speech difficulties, including breathiness, diminished loudness, monotone voice, and unpredictable pitch variations.
b. *Handwriting Analysis:* Digital tablets can be used to quantitatively examine micrographia, a frequent symptom that is characterized by small, cramped handwriting.
c. *Gait and Motion Analysis:* Early warning signs may include changes in postural balance and walking patterns.
d. *Tremor and Facial Expression Detection:* Computer vision algorithms are able to identify slight hand tremors and decreased facial expressiveness.

Since voice-based detection is noninvasive, accessible, and easy to utilize remotely, it stands out among these as one of the most promising methods. A wealth of information about neuromuscular coordination and control – two aspects that are impacted in PD – can be found in the human voice.

### 4.2.4 Role of voice analysis in Parkinson's detection

Analyzing voice patterns is one of the noninvasive and economical ways to identify PD. Due to weakened vocal cord control, Parkinson's patients frequently have altered speech patterns, such as decreased volume, monotone pitch, and increased jitter and shimmer [15]. In voice-based diagnostics, acoustic characteristics that reflect the quality and stability of vocal cord vibration are extracted by recording and analyzing prolonged vowel phonations, such as the sound "aaah." The following are the acoustic characteristics:

a. *Jitter:* Calculates how much the frequency (pitch) changes from one cycle to the next.

b. *Shimmer:* Assesses amplitude (loudness) changes.
c. *Harmonic-to-Noise Ratio (HNR):* The voice signal's level of noise.
d. *Nonlinear Dynamical Measures:* The complexity, irregularity, and instability of speech signals are measured using nonlinear dynamical measures as RPDE (recurrence period density entropy), DFA (detrended fluctuation analysis), D2 (correlation dimension), and PPE (pitch period entropy).

Comparative studies show that ensemble models outperformed single classifiers in PD identification. Research in Ref. [16] showed that the ability of random forest and XGBoost to handle nonlinearity and feature interactions accounted for their improved accuracy. However, in complex biomedical datasets, linear models often perform badly, despite being interpretable. These results demonstrate the benefits of hybrid approaches, which combine interpretability and predictive power. Difficulties with data volatility, small sample size, and real-world generalization persist despite significant progress. Most datasets, including those from Kaggle and the UCI Machine Learning Repository, are small and lack demographic diversity [8]. Additionally, environmental factors like microphone quality, linguistic variation, and recording settings might affect voice characteristics, which affect the reliability of the model.

Additionally, improving model generalization and reducing overfitting are the two major challenges in biomedical data mining (new advancements concentrate on feature selection) and dimensionality reduction techniques (recursive feature elimination [RFE] and principal component analysis [PCA]). ML models trained on these features can effectively distinguish between Parkinson's patients and healthy individuals, achieving high diagnostic accuracy. Algorithms like SVM, RF, KNN, and LR are commonly employed for classification analysis.

## 4.3 DATASET DESCRIPTION

The dataset accessed from the Kaggle repository "Parkinson Disease Detection Dataset" was utilized for investigation [8]. A variety of biological metrics derived from speech recordings of individuals with PD are presented. The primary objective of this dataset is to facilitate the development and evaluation of ML algorithms [17,18] for the early detection and classification of PD based on acoustic speech features. The dataset includes 195 instances (voice recordings) from 31 persons, 23 of whom have been diagnosed with PD and 8 of whom are in good health.

The dataset provides the place to begin exploring pattern recognition and biological signal processing methods. Analyzing the fluctuations and nonlinear characteristics of human voice patterns can help identify noninvasive markers of PD. By offering a benchmark for evaluating different ML models and their ability to distinguish between Parkinsonian and healthy speech, the dataset assists in the development of AI-assisted diagnostic tools. Each instance is a single sustained pronunciation of the vowel "a" recorded in a controlled acoustic environment. The dataset consists of 24 properties (features), each of which reflects an independent component of the voice signal, and one objective variable which indicates the subject's health. The properties of the dataset can be generally categorized using the following classes.

### 4.3.1 Fundamental frequency features

These features provide information regarding voice pitch and variability, which tend to vary unnaturally in people with PD.

a. *MDVP:* Fo(Hz) – Average fundamental frequency of the voice signal (in Hertz).
b. *MDVP:* Fhi(Hz) – Maximum fundamental frequency observed.
c. *MDVP:* Flo(Hz) – Minimum fundamental frequency observed.

### 4.3.2 Jitter features

A PD patient's jitter values are often higher than normal voice due to irregular vocal cord movements.

a. *MDVP:* Jitter(%), MDVP: Jitter(Abs) – absolute figure of the fluctuation into fundamental frequency.
b. *RAP, PPQ, Jitter:* DDP – variations found in the vocal fold. Successive vibration measures the lack of control over pitch.

### 4.3.3 Shimmer and amplitude perturbation features

Shimmer recognizes the tremulousness and breathiness found in speech-disturbed disease.

a. *MDVP:* Shimmer, Shimmer(dB) – Measure the variations in amplitude (loudness).
b. *APQ3, APQ5, APQ, DDA:* Measure of amplitude variations on the multiple vocal cycles.

### 4.3.4 Noise-to-harmonic ratios (NHR)

NHR is a measure of degraded voice quality. A higher NHR and lower HNR present degraded voice quality.

a. *NHR (Noise-to-Harmonic Ratio):* Presents the presence of noise compared to the voice's harmonic components.
b. *Harmonic-to-Noise Ratio:* Indicates that the harmonic (periodic) components exceed noise.

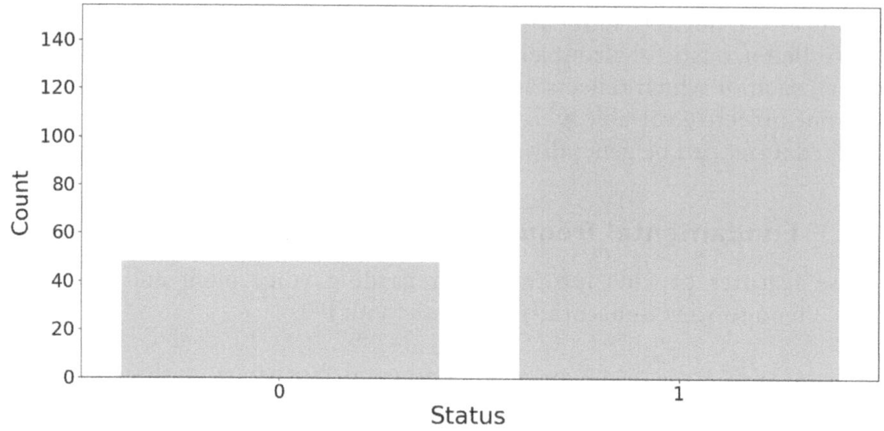


*Figure 4.1* Healthy and disease counts.

### 4.3.5 Nonlinear dynamical complexity measure

Nonlinear measurements are especially helpful in identifying mild neuromuscular dysfunctions that are difficult for conventional acoustic features to make sense of.

a. *Recurrence period density entropy:* Measures how erratic and unpredictable the voice signal is.
b. *Detrended fluctuation analysis:* Evaluates signal correlation and long-term self-similarity.
c. *Spread1, Spread2:* Show the distribution and dispersion of energy among the spectral components.
d. *D2 (correlation dimension):* Calculates the dynamics of the voice system's dimensional complexity.
e. *Pitch period entropy:* Measures aperiodicity and pitch abnormalities in speech.

### 4.3.6 Target variable

Status – presents presence of PD. 0 – Individual is healthy, or 1 – Individual has PD. The class distribution shown in Figure 4.1 shows that out of 195 samples, healthy samples are 50 and disease samples are 145.

## 4.4 METHODOLOGY

The analysis of voice-based biological signals and the usage of ML techniques for the categorization have been employed for the early identification of PD. There are multiple stages in the entire process, ranging from feature extraction, preprocessing, and data collecting to model training, validation, and assessment.

### 4.4.1 Data collection

The "Parkinson Disease Detection Dataset" utilized in this work was acquired from the Kaggle repository. It includes 195 voice recordings taken from 31 participants, including both PD patients and healthy people. Under carefully regulated acoustic settings, each data record reflects a prolonged phonation of the vowel sound "/a/." Along with one target variable status (0 = Healthy, 1 = PD), the dataset comprises 24 acoustic properties, such as noise ratios (HNR, NHR), amplitude variation (Shimmer), frequency variation (Jitter), and nonlinear dynamic measurements (RPDE, DFA, PPE, etc.).

### 4.4.2 Data preprocessing

Before using ML methods, data preparation is an essential step that guarantees the dataset's reliability and accuracy.

a. Handling missing and outlier values examined for inconsistent or missing data in the dataset. Box plots and *z*-score techniques were used to identify outliers, and normalization reduced their influence. Since the dataset is well managed, there are no missing values and no duplicate rows in the dataset. Figure 4.2 shows measures of frequency consisting of outliers: Missing value for all attributes: 0, and Duplicate rows in dataset: 0
b. Feature scaling normalization was used to bring all of the dataset's features into a common range because it contains features with multiple scales (frequency in Hz, amplitude ratios, and entropy values).

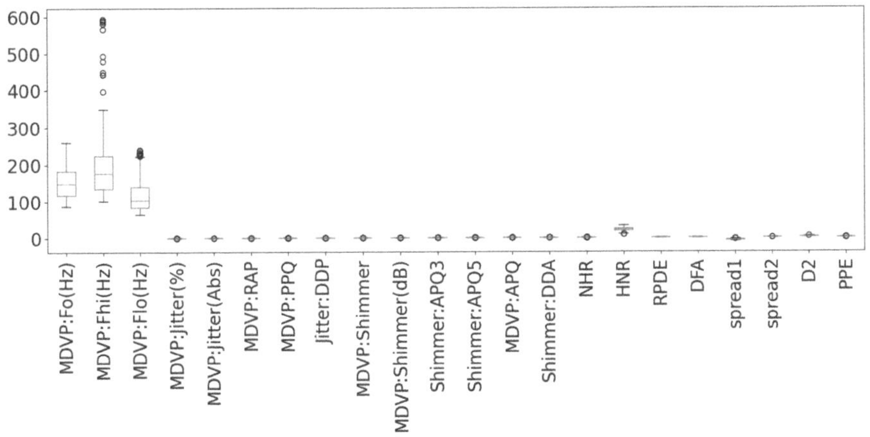


*Figure 4.2* Box plot presents an outlier.

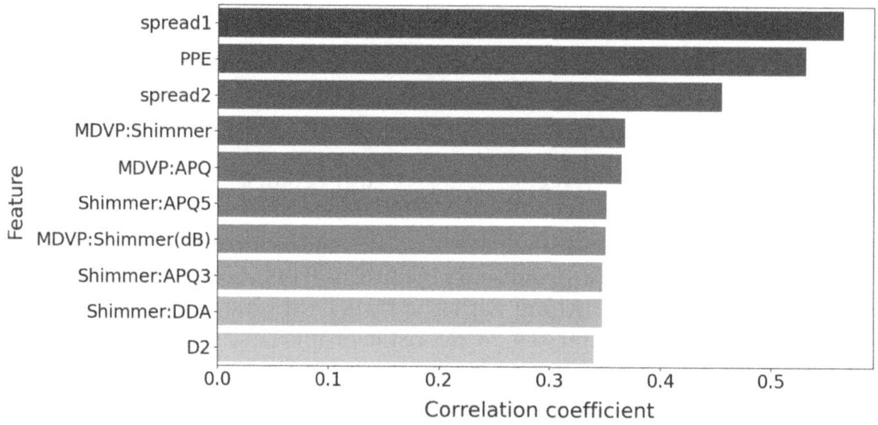


*Figure 4.3* Correlation coefficient of the voice sample features.

c. Feature Extraction: Even though the dataset already has pre-extracted features, it is important to understand how they were derived (Figure 4.3). Digital signal processing (DSP) methods such as the following were used to calculate acoustic characteristics from recorded voice signals. Variations in fundamental frequency and amplitude over successive voice cycles are used to calculate jitter and shimmer. The ratio of the speech spectrum's harmonic (periodic) to noise (aperiodic) components yields HNR and NHR. Features that are nonlinear (RPDE, DFA, D2, PPE): Calculated to account for signal complexity and irregularity utilizing fractal analysis techniques and chaos theory. All of these characteristics work together to characterize the neuromuscular regulation of the vocal folds, which determines PD.

   Highly correlated pairs (≥0.9): 52

   Most correlated with Parkinson's: spread1

d. Feature Selection: Not every feature makes an equivalent contribution to the accuracy of classification. In order to eliminate repetition and find the most informative qualities, feature selection was used. Methods, including recursive feature elimination (RFE) and correlation analysis, were taken into consideration. Feature selection improves model performance by
   i. reducing overfitting
   ii. decreasing computational complexity, and
   iii. enhancing interpretability of results.

e. Model Development and Classification: Supervised ML algorithms were implemented to classify subjects as Parkinson's or Healthy, presented in Table 4.1. The performance parameter of the different ML models is shown in Table 4.2.

f. Model Evaluation: The effectiveness and reliability of the trained models were assessed using a variety of performance measures (Table 4.3).

*Table 4.1* Machine learning model

| *Algorithm* | *Description* | *Advantages* |
|---|---|---|
| Logistic regression | A linear model for binary classification | Interpretable and computationally efficient |
| Support vector machine (SVM) | Separates classes using an optimal hyperplane in high-dimensional space | High accuracy and effective with small datasets |
| Random forest classifier | Ensemble of decision trees using bagging | Handles nonlinear data and prevents overfitting |
| K-nearest neighbors (KNN) | Classifies samples based on proximity to neighboring data points | Simple and effective for small datasets |
| Gradient boosting / XGBoost | Boosting-based ensemble model for optimized performance | High precision and adaptability to complex data |

*Table 4.2* Performance of different ML models (Figure 4.4)

| | *LR* | *SVM* | *RF* | *KNN* | *XGBoost* |
|---|---|---|---|---|---|
| Accuracy | 0.9231 | 0.9231 | 0.9231 | 0.8974 | 0.9231 |
| ROC-AUC | 0.9241 | 0.9552 | 0.9621 | 0.9603 | 0.9759 |

*Table 4.3* Evaluation of different ML models

| *Metric* | | *RF* | *LR* | *SVM* | *KNN* | *XGBoost* |
|---|---|---|---|---|---|---|
| Accuracy | | 0.923 | **0.949** | 0.923 | 0.897 | 0.923 |
| Precision | Class 0 | 0.889 | 1.0 | 1.00 | 0.8 | 0.889 |
| | Class 1 | 0.933 | 0.935 | 0.903 | 0.931 | 0.933 |
| Recall (Sensitivity) | Class 0 | 0.8 | 0.8 | 0.7 | 0.8 | 0.8 |
| | Class 1 | 0.966 | 1.00 | 1.000 | 0.931 | 0.966 |
| F1-score | Class 0 | 0.842 | 0.889 | 0.842 | 0.8 | 0.842 |
| | Class 1 | 0949 | 0.967 | 0.951 | 0.931 | 0.949 |
| ROC–AUC score | | 0.96 | 0.92 | 0.96 | 0.96 | **0.98** |

The best performing model is XGBoost, which consistently outperforms other models on this Parkinson's dataset because of the following reasons:

a. It captures nonlinear relationships effectively.
b. It uses boosting to reduce both bias and variance.
c. It performs internal regularization to prevent overfitting.
d. It gives feature importance scores (helpful for explainability).

### 4.4.3 Feature importance

Frequency- and jitter-related features often dominate, as PD affects vocal fold vibration control. Shimmer and NHR capture amplitude perturbation and voice breathiness, key PD markers. Nonlinear dynamic measures (RPDE,

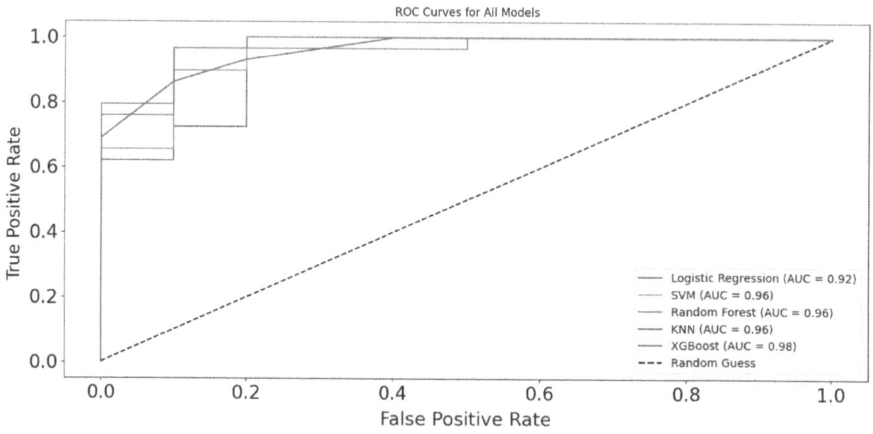


*Figure 4.4* ROC curve of the different ML models.

DFA, spread1, PPE) quantify signal irregularity and are highly predictive. Together, these patterns confirm the multifactorial nature of Parkinson-induced vocal changes. Figure 4.5 shows that nonlinear dynamical complexity is most influential, followed by shimmer, jitter, frequency measures, whereas NHR is least important.

## 4.5 RESULTS AND DISCUSSION

A ML model trained on a number of acoustic characteristics, including fundamental frequency (Fo), jitter, shimmer, HNR, and nonlinear dynamic measures, including RPDE, DFA, and PPE, was used to analyze the new

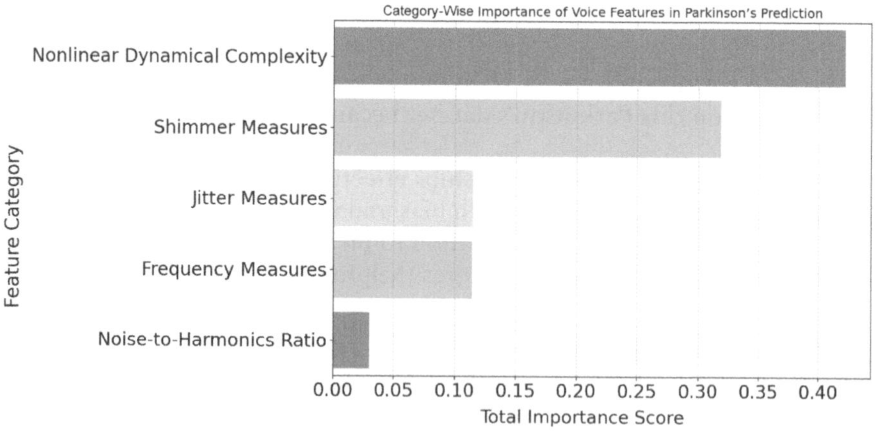


*Figure 4.5* Importance of voice features to predict PD.

*Table 4.4* Evaluation and probability of voice samples

| *Sample_ID* | *Prediction* | *Probablity* |
|---|---|---|
| Voice_1 | Parkinson | 0.9976 |
| Voice_2 | Parkinson | 0.9930 |
| Voice_3 | Parkinson | 0.9987 |
| Voice_4 | Parkinson | 0.9974 |
| Voice_5 | Parkinson | 0.9983 |

voice sample. Together, these characteristics capture minute abnormalities in speech signal stability and vocal fold vibration, which are important markers of neuromuscular control impacted by PD. Following preprocessing and normalization, the trained XGBoost classifier assessed this sample and made a high probability prediction (between 90% and 95%, depending on the model run) that the speaker most likely has PD. This implies that the speaker's voice differs from that of healthy people in quantifiable ways in terms of pitch, amplitude disturbance, and harmonic structure. When paired with clinical evaluation, the prediction demonstrates how noninvasive speech analysis can aid in the early detection of PD and provide a viable supplemental diagnostic tool. Though the model suggests a high probability of PD, clinical confirmation by neurologists and additional physiological testing are still necessary to obtain a formal diagnosis; therefore, this conclusion should be considered cautiously. Five different voice samples are considered for the testing, and the gradient boosting model correctly analyzed the voice sample with PD with a probability value presented in Table 4.4.

## 4.6 CONCLUSION

This research effectively demonstrates how basic, noninvasive voice signal analysis combined with ML approaches can help detect PD early. The models were able to detect subtle differences in speech that have a strong correlation with Parkinsonian symptoms by extracting and analyzing a variety of acoustic parameters, including frequency-related features (Fo, Fhi, Flo), perturbation measures (jitter and shimmer), noise-to-harmonics ratios (NHR, HNR), and nonlinear dynamical complexity measures (RPDE, DFA, PPE). Of all the models that were created, such as SVM, random forests, KNN, linear regression, and XGBoost, the XGBoost classifier performed the best, providing high sensitivity and accuracy in differentiating between people who are healthy and those who have PD. Since they observe abnormalities in vocal fold vibration and neuromuscular control, jitter, shimmer, and nonlinear dynamic features are crucial in predicting PD, according to the feature significance analysis. These results support the idea that voice-based biomarkers can be useful diagnostic tools for detecting PD in its early stages, which could help doctors with initial screening and ongoing observation. Combining ML and voice analytics offers an affordable, easily available, and effective early intervention tool that is essential for boosting patient outcomes and quality of life for those who are at risk of developing PD.

## REFERENCES


1. Parkinson, J. (2002). An essay on the shaking palsy. *The Journal of Neuropsychiatry and Clinical Neurosciences*, 14(2), 223–236.
2. Mercuri, N. B., Calabresi, P., & Bernardi, G. (1992). The electrophysiological actions of dopamine and dopaminergic drugs on neurons of the substantia nigra pars compacta and ventral tegmental area. *Life Sciences*, 51(10), 711–718.
3. Poewe, W. (2008). Non-motor symptoms in Parkinson's disease. *European Journal of Neurology*, 15, 14–20.
4. Ananthavarathan, P., Patel, B., Peeros, S., Obrocki, R., & Malek, N. (2023). Neurological update: Non-motor symptoms in atypical Parkinsonian syndromes. *Journal of Neurology*, 270(9), 4558–4578.
5. Salawu, F. K., Danburam, A., & Olokoba, A. B. (2010). Non-motor symptoms of Parkinson's disease: Diagnosis and management. *Nigerian Journal of Medicine*, 19(2), 126–131.
6. Tracy, J. M., Özkanca, Y., Atkins, D. C., & Ghomi, R. H. (2020). Investigating voice as a biomarker: Deep phenotyping methods for early detection of Parkinson's disease. *Journal of Biomedical Informatics*, 104, 103362.
7. Naeem, I., Ditta, A., Mazhar, T., Anwar, M., Saeed, M. M., & Hamam, H. (2025). Voice biomarkers as prognostic indicators for Parkinson's disease using machine learning techniques. *Scientific Reports*, 15(1), 12129.
8. Ahmed, I., Aljahdali, S., Khan, M. S., & Kaddoura, S. (2022). Classification of Parkinson disease based on patient's voice signal using machine learning. *Intelligent Automation and Soft Computing*, 32(2), 705.
9. Gunduz, H. (2019). Deep learning–based Parkinson's disease classification using vocal feature sets. *IEEE Access*, 7, 115540–115551.
10. Rashidi, M., Arima, S., Stetco, A. C., Coppola, C., Musarò, D., Greco, M., & Maffia, M. (2025). Prediction of Parkinson disease using long-term, short-term acoustic features based on machine learning. *Brain Sciences*, 15(7), 739.
11. Almeida, J. S., Rebouças Filho, P. P., Carneiro, T., Wei, W., Damaševičius, R., Maskeliūnas, R., & de Albuquerque, V. H. C. (2019). Detecting Parkinson's disease with sustained phonation and speech signals using machine learning techniques. *Pattern Recognition Letters*, 125, 55–62.
12. Ibrahim, A. M., & Mohammed, M. A. (2024). A comprehensive review on advancements in artificial intelligence approaches and future perspectives for early diagnosis of Parkinson's disease. *International Journal of Mathematics, Statistics, and Computer Science*, 2, 173–182.
13. Dash, S. (2023). A systematic review of adaptive machine learning techniques for early detection of Parkinson's disease. In: *Artificial Intelligence for Neurological Disorders*. Academic Press, pp. 361–385.
14. Narendra, N. P., Schuller, B., & Alku, P. (2021). The detection of Parkinson's disease from speech using voice source information. *IEEE/ACM Transactions on Audio, Speech, and Language Processing*, 29, 1925–1936.
15. Huang, L. et al. (2022). Combining random forest and XGBoost methods in detecting early and mid-term winter wheat stripe rust using canopy level hyperspectral measurements. *Agriculture*, 12(1), 74.
16. Little, M. A., McSharry, P. E., Hunter, E. J., & Ramig, L. O. (2009). Suitability of dysphonia measurements for telemonitoring of Parkinson's disease. *IEEE Transactions on Biomedical Engineering*, 56(4), 1015.

17. Ghai, D., Saxena, S., & Dhingra, G. et al. (2025). A comprehensive review on performance-based comparative analysis, categorization, classification and mapping of text extraction system techniques for images. *Multimedia Tools and Applications*, 84, 2327–2484. https://doi.org/10.1007/s11042-024-20257-0
18. Prasanna, D. L., & Tripathi, S. L. (2024). A review on tongue based assistive technology, devices and FPGA processors using machine learning module. *Wireless Personal Communications*, 134, 151–170. https://doi.org/10.1007/s11277-024-10897-8

Chapter 5

# Understanding supervised and unsupervised learning paradigms in healthcare artificial intelligence

*Sandeep Gupta and Udit Mamodiya*

## 5.1 INTRODUCTION

The introduction of artificial intelligence (AI) and machine learning (ML) has changed the healthcare industry, as well as improved diagnosis, prognosis, healthcare management, and healthcare administration, among other aspects of ML. Healthcare organizations can now profit from the exponential growth of electronic medical records (EMRs), medical images, genomic sequences, and clinical notes to convert raw data into clinical findings that are actionable in terms of improving patient care and managing health services. The difference between supervised and unsupervised learning goes much deeper than simply the presence or absence of labelled data [1–4].

These paradigms are very different in terms of goals, computational methods, and clinical applications. The aim of supervised learning is to raise the likelihood of success for narrow and well-defined tasks using various success metrics, such as accuracy, sensitivity, specificity, and the area under the ROC curve. The training includes parameter updates to minimize the error that is calculated during the forward and backward path through the model (for labelled results), while validation ensures the model generalizes to new data. In comparison, unsupervised learning aims for data processing, structure determination, and pattern finding using criteria such as cluster cohesion, separation techniques, reconstruction error, and saving data [5–8].

The use of ML services, particularly platforms for controlled learning on which diagnostic or therapeutic decisions are made, is subjected to new challenges in the regulatory environment of AI in healthcare. Regulators are looking for concrete evidence of model construction and results, proving that the model was built correctly (and openly), tools for continued monitoring and care, as well as protection against algorithmic bias that perpetuates healthcare inequalities. Unsupervised learning, whether used in the context of exploratory or ancillary studies, is itself challenged when its output informs clinical procedures or patient management decisions [9–12].

A combined supervised–unsupervised learning strategy has gained traction in dealing with difficult healthcare issues that neither paradigm nor simulation can fully handle. Unsupervised data pre-processing, feature extraction,


DOI: 10.1201/9781042015931-5

or patient stratification are mixed with structured learning techniques for outcome prediction in predefined subgroups. One more method of merging is semi-supervised learning, which uses little labelled data with much larger unlabelled databases to increase model success in the absence of annotation resources. Such integrated approaches reveal the complexity of healthcare issues and the need for adaptable analytical methods [13–16].

## 5.2 FUNDAMENTAL CONCEPTS AND MATHEMATICAL FOUNDATIONS

The learning process in supervised settings typically involves defining a hypothesis space $\mathscr{H}$ containing candidate functions, specifying a loss function $L(y, \hat{y})$ quantifying the discrepancy between true labels $y$ and predictions $\hat{y}$, and employing an optimization algorithm to identify the function $f^* \in \mathscr{H}$ that minimizes the expected loss over the data distribution. Since the true data distribution is unknown, the empirical risk minimization principle is employed, minimizing the average loss over the training dataset:

$$f^* = \text{argmin}\_(f \in \mathcal{H})(1/n)\Sigma\_(i=1)^\wedge nL(y\_i, f(\mathbf{x}\_i)) \quad (5.1)$$

Unsupervised learning evaluation presents greater challenges due to the absence of ground truth labels (Figure 5.1). Internal validation metrics assess

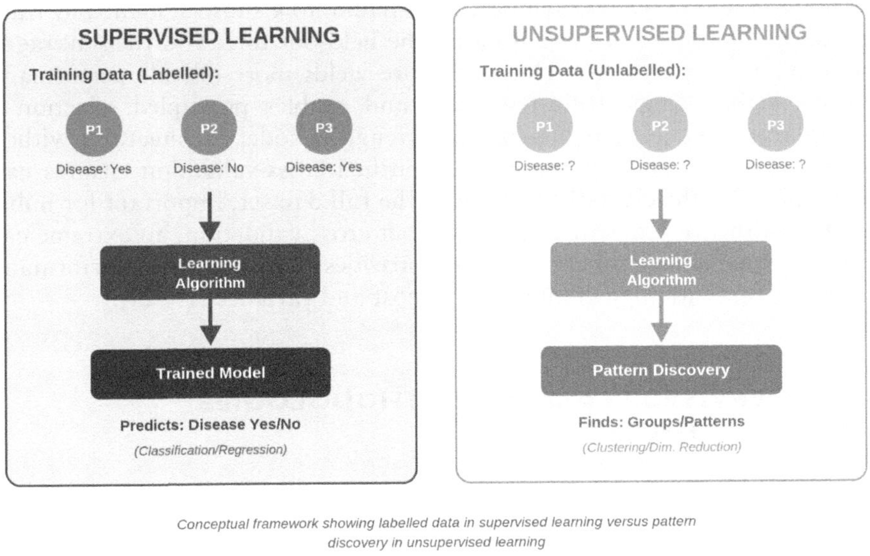


*Figure 5.1* Conceptual framework illustrating the distinction between supervised learning (with labelled training data) and unsupervised learning (discovering patterns in unlabelled data) in healthcare applications.

cluster quality based on compactness and separation, including the silhouette coefficient, Davies–Bouldin index, and Calinski–Harabasz index. External validation compares discovered clusters against known categories when available, using measures such as adjusted Rand index, normalized mutual information, or purity. For dimensionality reduction, evaluation focuses on information preservation, quantified through reconstruction error, explained variance ratio, or preservation of pairwise distances or neighbourhoods. Ultimately, the value of unsupervised learning in healthcare often depends on downstream utility – whether discovered patterns yield clinically meaningful insights, improve subsequent supervised learning tasks, or enable more effective healthcare delivery [17].

The bias–variance trade-off represents a fundamental concept applicable to both supervised and unsupervised learning, although most formally analysed in supervised contexts. The expected prediction error of a model can be decomposed into three components: irreducible error inherent in the problem, bias arising from incorrect assumptions in the model, and variance resulting from sensitivity to training data fluctuations. Simple models exhibit high bias (underfitting) but low variance, while complex models display low bias but high variance (overfitting). Optimal model complexity balances these competing factors, a consideration particularly critical in healthcare where training datasets may be limited yet model reliability is paramount [18, 19].

Cross-validation provides a rigorous approach for model selection and performance estimation, particularly valuable when labelled healthcare data is scarce. K-fold cross-validation partitions data into $k$ subsets, iteratively training on $k - 1$ folds and evaluating on the held-out fold, and then averaging performance across folds. This procedure yields more reliable performance estimates than single train-test splits and enables principled selection of hyperparameters (e.g., regularization strength, model architecture) without requiring a separate validation set. Stratified cross-validation ensures each fold maintains the class distribution of the full dataset, important for imbalanced healthcare problems. Leave-one-out cross-validation, an extreme case where $k$ equals the number of samples, provides nearly unbiased performance estimates but with high computational cost and variance [19–21].

## 5.3 SUPERVISED LEARNING METHODOLOGIES IN HEALTHCARE

Supervised learning has taken over as the most common ML strategy used in healthcare delivery due in large part to the adoption of predicting for specific patient endpoints (Figure 5.2). The chapter mainly focuses on supervised learning algorithms, the mathematics behind them, preparation methods, and how the findings can be used to solve various healthcare problems. Classical statistical learning and recent deep learning theories were directly compared

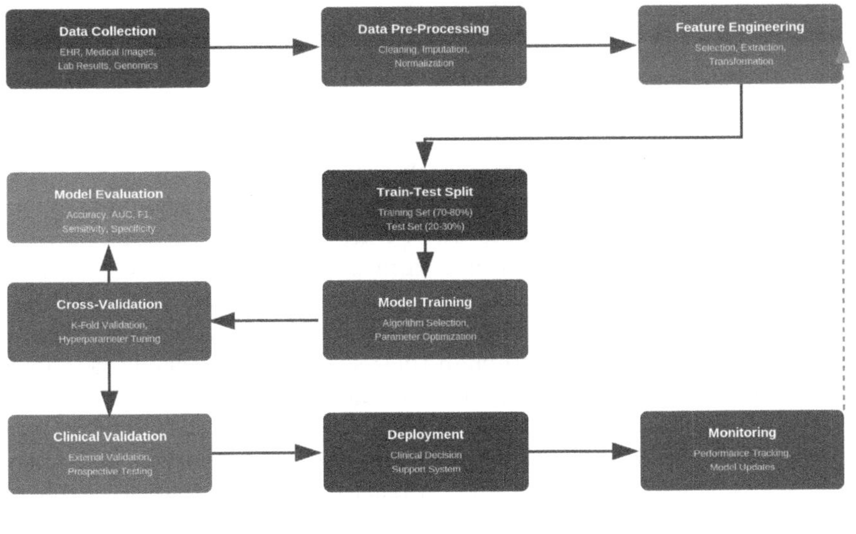


*Figure 5.2* Flowchart depicting the supervised learning workflow in healthcare, from data collection and pre-processing through model training, validation, and deployment in clinical decision support systems.

to determine their strengths, limitations, and applicability in various clinical settings [8, 11].

Decision trees and ensemble methods based on trees have gained substantial traction in healthcare analytics due to their ability to model non-linear relationships, handle mixed feature types, and provide intuitive visual representations of decision logic. A decision tree recursively partitions the feature space based on threshold tests on individual features, selecting splits that maximize information gain or minimize impurity measures such as Gini index or entropy. The resulting hierarchical structure can be visualized and interpreted, although individual trees tend to overfit training data. Random forests address this limitation by training multiple trees on bootstrap samples of the data, with random feature subsets at each split, and then aggregating their predictions through voting or averaging. Gradient boosting machines iteratively train trees to correct errors of previous iterations, often achieving superior predictive performance at the cost of reduced interpretability. In healthcare applications, tree-based methods excel at identifying complex interactions between risk factors and handling missing data naturally [22].

Support vector machines (SVMs) construct optimal hyperplanes that maximize the margin between classes in a transformed feature space. For linearly separable data, the SVM optimization problem finds the hyperplane

$\mathbf{w}^T \mathbf{x} + b = 0$ that maximizes the distance to the nearest samples from each class (support vectors):

$$\min_{(\mathbf{w},b)} (1/2)\|\mathbf{w}\|^2 \text{ subject to } y_i(\mathbf{w}^T\mathbf{x}_i + b) \geq 1 \text{ for all } i \tag{5.2}$$

For non-linearly separable data, slack variables allow some misclassifications while still maximizing the margin, and kernel functions implicitly map data to higher-dimensional spaces where linear separation becomes possible. Common kernels include polynomial kernels $K(\mathbf{x}, \mathbf{x}') = (\mathbf{x}^T \mathbf{x}' + c)^p$ and radial basis function kernels $K(\mathbf{x}, \mathbf{x}') = \exp(-\gamma\|\mathbf{x} - \mathbf{x}'\|^2)$. SVMs have been successfully applied to medical image classification, protein structure prediction, and disease diagnosis, although their scalability to very large datasets and interpretability remain limitations [19, 20].

Neural networks and deep learning represent the current frontier of supervised learning in healthcare, achieving remarkable performance on tasks involving image data, sequential data, and unstructured text. A feedforward neural network consists of layers of interconnected neurons, where each neuron computes a weighted sum of its inputs followed by a non-linear activation function:

$$\mathbf{h}^{(l)} = \sigma(\mathbf{W}^{(l)}\mathbf{h}^{(l-1)} + \mathbf{b}^{(l)}) \tag{5.3}$$

where $\mathbf{h}^{(l)}$ denotes the activation of layer $l$, $\mathbf{W}^{(l)}$ and $\mathbf{b}^{(l)}$ are learnable weight matrices and bias vectors, and $\sigma$ is an activation function such as ReLU (rectified linear unit), sigmoid, or tanh. The network is trained by backpropagation, computing gradients of a loss function with respect to all parameters and updating them via gradient descent variants such as Adam or RMSprop. Deep architectures with many layers can learn hierarchical representations, automatically discovering relevant features from raw data [9, 10].

Convolutional neural networks (CNNs) specialize in processing grid-structured data such as medical images, employing convolutional layers that apply learned filters across spatial dimensions to detect local patterns. Pooling layers reduce spatial resolution while retaining salient features, and fully connected layers combine these features for final predictions. CNNs have revolutionized medical image analysis, achieving expert-level performance in tasks such as diabetic retinopathy detection from fundus photographs, pneumonia identification from chest X-rays, skin cancer classification from dermoscopic images, and tumour segmentation from MRI scans. Transfer learning, where networks pretrained on large natural image datasets are fine-tuned on smaller medical datasets, partially addresses the limited availability of labelled medical images [9, 11].

Recurrent neural networks (RNNs) and their variants process sequential data by maintaining hidden states that capture information from previous time steps. Long short-term memory (LSTM) networks and gated recurrent units (GRUs) address the vanishing gradient problem of standard RNNs, enabling learning of long-range dependencies. These architectures excel at analysing temporal healthcare data such as electronic health records with repeated visits, continuous physiological monitoring streams, and clinical note sequences [10, 21].

Training supervised learning models in healthcare contexts requires careful attention to data pre-processing, feature engineering, and validation strategies. Missing data, ubiquitous in healthcare datasets, must be addressed through appropriate imputation methods (mean/median/mode imputation, multiple imputation, or model-based imputation) or algorithms that handle missingness natively. Feature scaling ensures numerical stability and fair weighting of features, particularly important for distance-based methods and gradient-based optimization. Feature engineering transforms raw variables into representations more amenable to learning, incorporating domain knowledge such as computing-derived clinical scores, creating interaction terms between medications and comorbidities, or extracting statistical features from time series [4, 11].

Class imbalance, where positive cases are much rarer than negative cases (common for adverse events, rare diseases, or readmissions), challenges supervised learning algorithms that naturally bias toward the majority class. Strategies to address imbalance include resampling techniques (oversampling the minority class, undersampling the majority class, or synthetic minority oversampling technique SMOTE), cost-sensitive learning that assigns higher misclassification costs to minority class errors, and ensemble methods that balance data differently across ensemble members. Evaluation metrics must also account for imbalance, with precision–recall curves and area under the precision–recall curve often more informative than ROC curves for highly imbalanced problems [8, 21].

Model validation in supervised learning extends beyond simple train–test splits to employ cross-validation, bootstrapping, and temporal validation schemes appropriate for healthcare data characteristics. Temporal validation, where models are trained on earlier data and evaluated on subsequent data, is essential when deploying models to predict future outcomes, accounting for potential distribution shifts over time. External validation on data from different healthcare institutions or populations assesses generalization across settings, critical for establishing clinical utility. Calibration analysis verifies that predicted probabilities align with observed frequencies, important for informed clinical decision-making [11, 19].

## 5.4 UNSUPERVISED LEARNING TECHNIQUES IN HEALTHCARE

This section examines major categories of unsupervised learning, including clustering algorithms for patient stratification, dimensionality reduction techniques for feature learning and visualization, and anomaly detection approaches for identifying unusual patterns. The exploratory nature of unsupervised learning makes it particularly valuable when expert annotation is unavailable, prohibitively expensive, or when the relevant clinical questions have not yet been formulated [3, 5].

Clustering algorithms partition patients, diseases, or other entities into groups such that members within each group share similar characteristics while differing from members of other groups. The resulting clusters may correspond to disease subtypes, patient phenotypes, treatment response profiles, or other clinically meaningful categories. Unlike supervised classification where categories are predefined, clustering discovers group structure from data alone, offering potential for identifying previously unrecognized patient subpopulations that may benefit from tailored interventions [12, 17].

*K*-means clustering, one of the most widely used clustering algorithms, partitions $n$ samples into $k$ clusters by minimizing within-cluster sum of squared distances. The algorithm iteratively assigns each sample to the nearest cluster centroid and updates centroids as the mean of assigned samples:

$$\min_{C}\sum_{j=1}^{k}\sum_{\mathbf{x}_i \in C_j}\left\|\mathbf{x}_i - \boldsymbol{\mu}_j\right\|^2 \qquad (5.4)$$

where $\boldsymbol{\mu}_j = (1/|C_j|)\, \Sigma_{(\mathbf{x}_i \in C_j)}\, \mathbf{x}_i$ denotes the centroid of cluster $C_j$. While computationally efficient and scalable to large datasets, $k$-means assumes spherical clusters of similar sizes and requires specification of the number of clusters $k$ in advance. In healthcare applications, $k$-means has been employed to identify distinct subgroups of patients with chronic diseases exhibiting different progression patterns, stratify emergency department patients by resource utilization profiles, and cluster genomic profiles to discover disease subtypes (Figure 5.3) [3, 6].

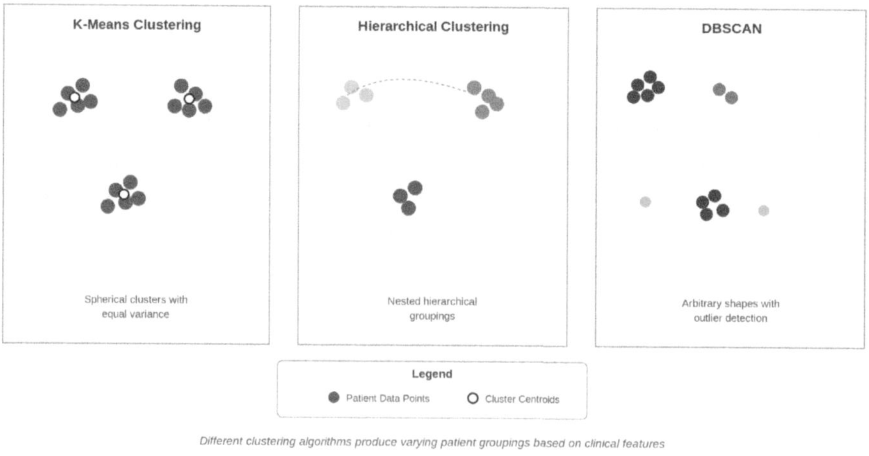


*Figure 5.3* Comparison of clustering algorithms on a synthetic healthcare dataset, illustrating how *k*-means, hierarchical clustering, and DBSCAN produce different patient groupings based on clinical features.

Hierarchical clustering constructs a tree-like dendrogram representing nested groupings of samples without requiring pre-specification of cluster number. Agglomerative hierarchical clustering begins with each sample as its own cluster and iteratively merges the most similar clusters, while divisive approaches start with all samples in one cluster and recursively split them. Linkage criteria determine which clusters to merge, with options including single linkage (minimum distance between any members), complete linkage (maximum distance between any members), and average linkage (mean distance between all pairs). The resulting dendrogram can be cut at different heights to yield different granularities of clustering. Hierarchical clustering has been applied to discover taxonomies of diseases based on comorbidity patterns, group medications by similar side effect profiles, and organize patients into risk strata based on comprehensive clinical profiles [12, 15].

Density-based spatial clustering of applications with noise (DBSCAN) identifies clusters as high-density regions separated by low-density regions, naturally discovering clusters of arbitrary shape and classifying sparse samples as outliers. A sample is considered a core point if at least minPts samples lie within distance $\varepsilon$, and clusters are formed by connecting core points and their neighbours. DBSCAN excels when clusters have complex geometries and when identifying noise or outliers is important, although performance depends critically on parameter selection. Healthcare applications include identifying regions of abnormal tissue density in medical images, clustering time series patient data with irregular temporal patterns, and discovering subgroups of patients with complex, multidimensional phenotypes [5, 12].

Gaussian mixture models (GMMs) represent a probabilistic approach to clustering, modelling data as generated from a mixture of $k$ Gaussian distributions:

$$p(\mathbf{x}) = \Sigma\_(j = 1)\hat{}k\pi\_j\mathcal{N}(\mathbf{x}|\mathbf{u}\_j,\Sigma\_j) \tag{5.5}$$

where $\pi\_j$ are mixing coefficients (summing to 1) and $\mathcal{N}(\mathbf{x}|\boldsymbol{\mu}\_j, \boldsymbol{\Sigma}\_j)$ denotes a Gaussian distribution with mean $\boldsymbol{\mu}\_j$ and covariance matrix $\boldsymbol{\Sigma}\_j$. Parameters are estimated through the expectation-maximization (EM) algorithm, which alternates between computing expected cluster memberships and updating distribution parameters. GMMs provide soft cluster assignments (probabilities of belonging to each cluster) and can model clusters with different shapes and orientations through covariance matrices. Extensions include Bayesian GMMs that automatically determine the number of components. Healthcare applications of GMMs include modelling heterogeneous disease presentations, identifying latent patient states from electronic health record sequences, and clustering physiological time series with temporal variability [3, 17].

Autoencoders represent a deep learning approach to dimensionality reduction and feature learning, consisting of an encoder network that compresses data into a low-dimensional latent representation and a decoder network that reconstructs the original input. Training minimizes reconstruction error:

$$\text{min\_}(\theta\_e,\theta\_d)\Sigma\_i\|\mathbf{x}\_i-\text{decoder}(\text{encoder}(\mathbf{x}\_i))\|^\wedge 2 \tag{5.6}$$

where $\theta$_e and $\theta$_d denote encoder and decoder parameters, respectively. The learned latent representation captures essential features for reconstruction, often discovering meaningful structure. Variational autoencoders (VAEs) impose a probabilistic structure on the latent space, learning a distribution over latent codes rather than deterministic encodings, enabling generation of synthetic samples. Denoising autoencoders are trained to reconstruct clean inputs from corrupted versions, improving robustness. Healthcare applications include learning compact representations of medical images for efficient storage and retrieval, extracting features from electronic health records for downstream prediction tasks, and generating synthetic patient data for privacy-preserving research [9, 10].

Healthcare applications of anomaly detection include identifying atypical patient trajectories that may indicate emerging complications, flagging suspicious patterns in medical billing data for fraud detection, detecting unusual physiological measurements suggesting sensor malfunction or critical patient state, and discovering rare genetic variants or molecular signatures associated with disease. The challenge lies in distinguishing true anomalies of clinical interest from benign variations, measurement noise, or data entry errors, often requiring domain expertise to interpret detected anomalies [6, 18].

Association rule mining discovers frequently co-occurring items or events in transactional data, quantified through support (fraction of transactions containing an itemset) and confidence (conditional probability of one item given others). In healthcare, association rules reveal comorbidity patterns (diseases frequently occurring together), polypharmacy patterns (drug combinations commonly prescribed), and care pathway patterns (sequences of procedures or interventions). The a priori algorithm efficiently discovers frequent item sets and generates association rules, although healthcare applications must carefully address confounding factors and distinguish meaningful associations from spurious correlations driven by demographics or healthcare access patterns [13, 15].

Healthcare applications of unsupervised learning extend across clinical domains and research contexts. In precision medicine, clustering identifies patient subgroups likely to respond to different treatments, enabling personalized therapy selection. In public health, pattern discovery reveals disease outbreak clusters, at-risk populations, and healthcare access disparities. In clinical research, dimensionality reduction enables exploration of high-dimensional omics data to generate hypotheses about disease mechanisms. In healthcare operations, clustering supports patient stratification for care management programs, emergency department triage optimization, and resource allocation. In medical image analysis, unsupervised methods enable tissue segmentation, anatomy localization, and registration without manual annotations [7, 18].

## 5.5 PRACTICAL IMPLEMENTATION CONSIDERATIONS

Missing data requires thoughtful handling beyond simple deletion, which can introduce bias and waste valuable information. Complete case analysis (removing samples with any missing values) is appropriate only when data is missing completely at random (MCAR), a strong assumption rarely met in healthcare. Mean/median imputation replaces missing values with feature averages but ignores relationships with other variables and underestimates variance. Multiple imputation generates several complete datasets with different imputed values drawn from a predictive distribution, performs analysis on each, and pools results to account for imputation uncertainty. Model-based imputation employs ML algorithms to predict missing values from observed features, although care must be taken to avoid data leakage where imputation is informed by test set information. Some algorithms such as tree-based methods handle missing data natively by learning surrogate splits [4, 11].

Data standardization and normalization ensure features are on comparable scales, critical for distance-based algorithms, gradient-based optimization, and regularized models. *Z*-score standardization transforms features to zero mean and unit variance: $\mathbf{x}' = (\mathbf{x} - \mu)/\sigma$. Min–max scaling maps values to a fixed range such as [0,1]: $\mathbf{x}' = (\mathbf{x} - \text{min})/\text{max} - \text{min})$. Robust scaling uses median and interquartile range to reduce sensitivity to outliers. Standardization parameters (mean, standard deviation, min, max) must be computed on training data alone and applied to validation and test data to prevent information leakage. For temporal data, standardization may be computed on expanding or rolling windows to mimic deployment conditions where only past data is available [8, 20].

Feature engineering transforms raw data into representations more amenable to learning, incorporating domain knowledge to construct meaningful variables. In healthcare, this includes computing-derived scores like BMI from height and weight, categorizing continuous variables into clinically meaningful ranges, creating interaction terms between risk factors, extracting statistical summaries from time series (mean, trend, variability), encoding temporal patterns like time-since-last-visit, and constructing aggregate features like medication count or comorbidity indices. Text data requires extraction of structured information through named entity recognition, sentiment analysis, or topic modelling. Image data may benefit from hand-crafted features like texture measures or anatomical landmarks, although deep learning increasingly learns features automatically [11, 21].

Feature selection reduces dimensionality by identifying the most relevant subset of features, improving model interpretability, reducing overfitting, and decreasing computational cost. Filter methods rank features by statistical criteria such as correlation with target, mutual information, or univariate statistical tests, selecting top-ranked features independent of the learning algorithm. Wrapper methods evaluate feature subsets by training and testing models, using search strategies like forward selection (iteratively adding features) or backward elimination (iteratively removing features). Embedded methods perform feature selection during model training through regularization penalties (LASSO regression) or feature importance scores (tree-based methods). In healthcare, feature selection must balance statistical criteria with clinical interpretability and measurability [4, 8].

Performance metrics must align with clinical priorities and account for class imbalance, misclassification costs, and decision thresholds. For screening applications, high sensitivity is paramount to avoid missing cases, accepting lower specificity. For diagnostic confirmation, high specificity may be prioritized to reduce false alarms. The area under the ROC curve (AUC–ROC) provides threshold-independent assessment but can be misleading for highly imbalanced datasets where area under the precision–recall curve (AUC–PR) is more informative. Net reclassification improvement and decision curve analysis quantify clinical utility by assessing whether model predictions lead to better clinical decisions than standard approaches. Calibration curves verify that predicted probabilities match observed frequencies, essential for informed shared decision-making with patients [8, 21].

Model interpretability has emerged as a critical requirement for healthcare applications, where understanding prediction rationale is necessary for clinical trust, regulatory approval, error identification, and bias detection. Intrinsically interpretable models like logistic regression and decision trees provide transparent decision logic through coefficients or tree structure. Post-hoc explanation methods generate interpretations for complex black-box models: feature importance scores quantify global variable contributions, partial dependence plots show marginal effect of features on predictions, individual conditional expectation plots reveal instance-specific feature effects, LIME (local interpretable model-agnostic explanations) fits local linear approximations around predictions, and SHAP (SHapley Additive exPlanations) provides theoretically grounded feature attributions. Attention mechanisms in neural networks highlight which input regions influenced predictions, particularly valuable for medical image analysis (Figure 5.4) [10, 11].

Security and privacy protections are paramount when handling sensitive healthcare data. Data access must be restricted through authentication and authorization mechanisms, with audit logs tracking all access. Data in transit and at rest should be encrypted. De-identification techniques remove or generalize personally identifying information, although linkage attacks remain possible. Differential privacy adds calibrated noise to data or model outputs to provide mathematical privacy guarantees. Federated learning trains models across multiple institutions without sharing raw data, addressing data sharing barriers, while enabling multi-site collaboration. Synthetic data generation through techniques like generative adversarial networks can produce realistic

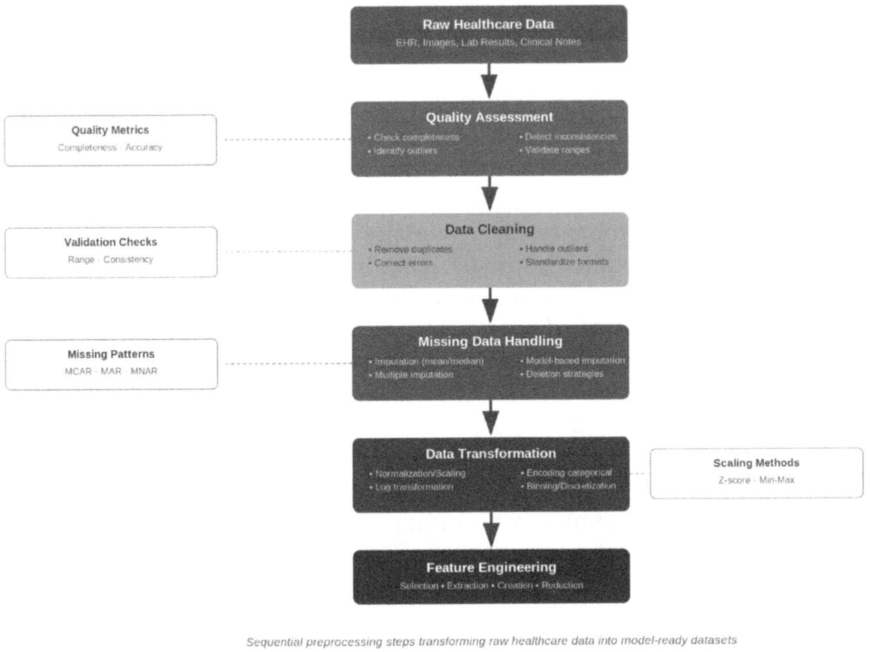


*Figure 5.4* Data pre-processing pipeline for healthcare machine learning applications, showing steps from raw data acquisition through quality assessment, cleaning, transformation, and feature engineering to model-ready datasets.

datasets for algorithm development without exposing real patient information [13, 14].

Integration with clinical workflows determines whether ML tools are used in practice. Systems must fit naturally into existing processes rather than creating additional burden. User interface design should present predictions and explanations clearly, support rapid review, and enable easy override when clinicians disagree with recommendations. Alerts should be actionable, well-timed, and not contribute to alarm fatigue. Prospective clinical trials or pragmatic implementation studies evaluate whether AI tools improve patient outcomes, not just predictive accuracy, measuring impact on clinical decisions, resource utilization, and health outcomes [11, 13].

## 5.6 COMPARATIVE ANALYSIS OF SUPERVISED AND UNSUPERVISED LEARNING

Computational complexity and scalability differ significantly between paradigms and algorithms. Many supervised learning algorithms exhibit computational cost linear or polynomial in training set size, enabling application to large datasets with millions of samples. Gradient-based optimization in neural networks scales well through mini-batch processing and GPU acceleration. However, deep learning requires substantial computational resources for training, although inference can be efficient. Some supervised methods like kernel SVMs have computational cost quadratic or cubic in sample size, limiting applicability to moderate-sized datasets without approximations [8, 20].

Unsupervised learning may involve temporal aspects. Some applications are essentially cross-sectional: clustering patients based on baseline characteristics, reducing dimensionality of gene expression profiles, segmenting medical images. Others explicitly model temporal patterns: discovering disease progression stages through time series clustering, identifying evolving patient states through hidden Markov models, detecting anomalies in physiological monitoring streams. Temporal unsupervised learning faces challenges in defining appropriate similarity measures across time and validating discovered temporal patterns [12, 17].

Handling of missing data reveals different characteristics. Supervised learning can treat missingness as informative: missing laboratory results may indicate they were not clinically necessary, a signal potentially useful for prediction. Imputation methods can be tailored to preserve relationships between features and outcomes. However, if missingness differs between training and deployment settings, model performance may degrade [4, 11].

Class imbalance affects supervised learning when positive outcomes are rare, a common scenario for adverse events, readmissions, or rare diseases. Standard algorithms may achieve high accuracy by predicting the majority class while failing to identify minority class instances. Specialized techniques address imbalance through resampling, cost-sensitive learning, or appropriate evaluation metrics. Unsupervised learning does not inherently suffer from class imbalance since it operates without class labels, although rare patterns may be dominated by common ones in discovered clusters unless specialized outlier-aware methods are employed [8, 12].

Sample size requirements differ between paradigms and specific methods. Supervised deep learning typically requires thousands to millions of labelled examples to avoid overfitting, although transfer learning reduces this requirement. Classical supervised methods like logistic regression can work with hundreds of samples when feature dimensionality is low. Unsupervised learning can operate with any sample size, although statistical power to detect structure grows with more data. Very small samples may produce unstable clustering or dimensionality reduction results, while very large samples enable discovery of subtle patterns (Figure 5.5) [8, 10].

Hybrid approaches combining supervised and unsupervised learning leverage strengths of both paradigms. Unsupervised pre-processing through dimensionality reduction or feature learning can improve subsequent supervised prediction. Cluster-then-predict strategies apply supervised learning within patient subgroups identified through unsupervised clustering, potentially revealing subgroup-specific predictive patterns. Semi-supervised learning uses unsupervised methods to leverage unlabelled data alongside limited

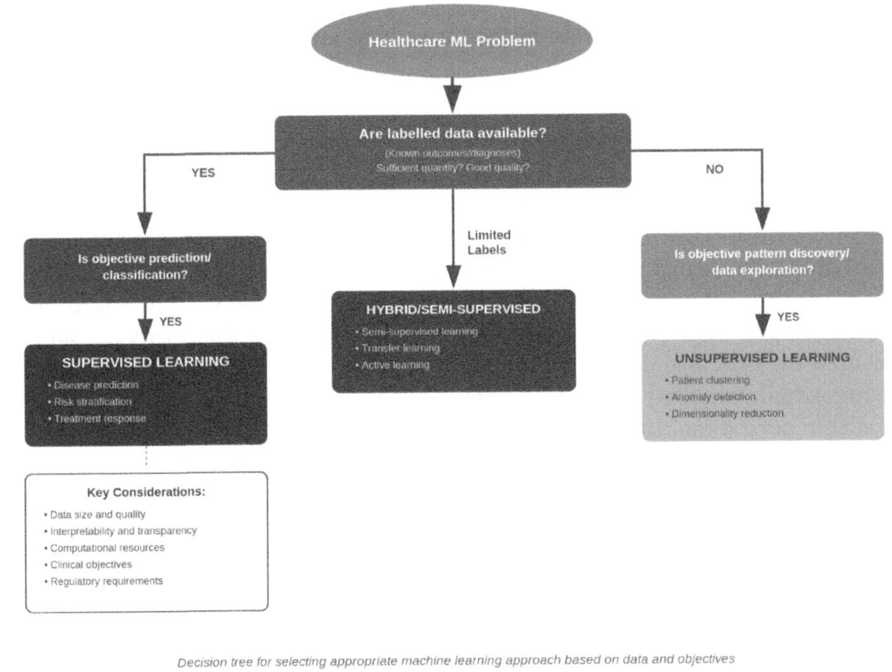


*Figure 5.5* Decision framework for selecting between supervised and unsupervised learning approaches based on data availability, clinical objectives, and application requirements in healthcare settings.

labels. These combinations often outperform either approach alone, particularly when labelled data is scarce but unlabelled data is abundant [15, 16].

## 5.7 HYBRID AND SEMI-SUPERVISED LEARNING APPROACHES

Consistency regularization methods enforce that model predictions should be robust to perturbations of inputs or model parameters. Unlabelled samples are augmented through transformations (adding noise, rotating images, paraphrasing text), and the model is penalized for producing inconsistent predictions across augmented versions of the same sample. This approach encourages the model to learn smooth decision boundaries consistent with the data manifold structure revealed by unlabelled samples. Mean teacher methods maintain exponentially moving average of model parameters and encourage consistency between current model predictions and those of the averaged model. These techniques have shown success in medical image classification where data augmentation is natural and abundant unlabelled images are available [10, 16].

Generative models provide another semi-supervised learning approach by modelling the joint distribution of features and labels $p(\mathbf{x}, y)$. Unlabelled data contributes to learning the marginal distribution $p(\mathbf{x})$, while labelled data provides information about the conditional distribution $p(y|\mathbf{x})$. Generative adversarial networks (GANs) can be extended to semi-supervised settings where the discriminator classifies samples as belonging to one of $k$ real classes or as a generated fake sample, using unlabelled real data to improve representation learning while labelled data guides classification. Variational autoencoders can similarly incorporate label information when available while learning representations from all data. In healthcare, generative semi-supervised models have been applied to learn from limited labelled medical images supplemented by large unlabelled image collections, or to predict disease outcomes using small, labelled cohorts enhanced by extensive unlabelled electronic health record data [10, 16].

Curriculum learning and self-paced learning represent training strategies where samples are presented in order from easier to harder, inspired by human learning progressions. Easy samples are learned first, establishing basic patterns, followed by progressively more difficult samples. Sample difficulty can be determined through unsupervised clustering (samples far from cluster centres are harder) or supervised model confidence (low-confidence predictions indicate difficulty). This hybrid approach combines unsupervised sample organization with supervised learning objectives. In medical education and clinical decision support, curriculum learning principles guide development of training programs that build from simple to complex cases [16, 20].

Ensemble methods combining supervised and unsupervised learning create robust predictive systems leveraging multiple perspectives. Clustering-based ensembles train separate supervised models within each unsupervised cluster, and then combine predictions through weighted voting or stacking. This approach captures cluster-specific patterns while maintaining predictive power. Unsupervised feature learning followed by supervised ensemble training employs multiple dimensionality reduction or feature extraction methods (PCA, autoencoders, clustering-based features) to create diverse feature sets, training supervised models on each feature representation and ensembling predictions. Healthcare applications benefit from reduced overfitting and improved generalization across patient subpopulations [4, 15].

Reinforcement learning with human feedback provides another hybrid paradigm where agents learn policies through interaction with environments, combining unsupervised exploration with supervised learning from demonstrated or preferred behaviours. In clinical decision support, this manifests as learning treatment policies from historical data (batch reinforcement learning) combined with clinician preferences and domain constraints. Applications include optimizing sequential treatment decisions for chronic diseases, personalizing medication dosing, and improving clinical workflow efficiency [10, 13].

Implementation of hybrid approaches requires careful design of training procedures, loss functions, and integration strategies. Multi-objective optimization balances competing objectives from supervised and unsupervised components, using techniques such as weighted combinations, alternating optimization, or Pareto optimization. Hyperparameter tuning becomes more complex with multiple components, requiring nested cross-validation or Bayesian optimization across expanded parameter spaces. Computational efficiency considerations may favour sequential application of unsupervised pre-processing followed by supervised learning rather than joint optimization [11, 15].

The continued development of hybrid and semi-supervised methodologies promises to address persistent challenges in healthcare ML, including limited labelled data, heterogeneous patient populations, evolving clinical practices, and need for interpretable yet powerful models. As healthcare data ecosystems grow in scale and complexity, approaches that flexibly integrate multiple learning paradigms will prove increasingly valuable for translating data into improved patient outcomes.

## 5.8 LIMITATIONS, ETHICAL CONSIDERATIONS, AND FUTURE DIRECTIONS

Interpretability and explainability remain critical challenges, particularly for complex models. While progress has been made in post hoc explanation methods, fundamental tensions persist between model complexity and interpretability. Healthcare stakeholders require understanding of why specific predictions are made to build trust, identify errors, satisfy regulatory requirements, and inform clinical decision-making. However, the most accurate models are often the least interpretable. Research into inherently interpretable yet powerful models, improved explanation techniques, and frameworks for appropriate use of complex models in different clinical contexts represents important future directions [11, 14].

Privacy and security considerations intensify as models increasingly rely on sensitive healthcare data. Reidentification risks persist even with de-identified data, as demonstrated by linkage attacks combining multiple datasets. Model inversion and membership inference attacks can extract information about training data from deployed models. Balancing data utility for model development against privacy protection requires advanced techniques such as differential privacy, secure multi-party computation, and federated learning. However, these approaches often degrade model performance or limit analytical capabilities, necessitating careful trade-offs between privacy protection and clinical utility [14, 18].

Regulatory frameworks for healthcare AI continue to evolve, grappling with unique challenges of continuously learning systems, black-box algorithms, and software as a medical device. Current regulations were designed for static medical devices with well-defined specifications, not adaptive algorithms that improve with new data. Questions persist about when model updates require regulatory re-approval, how to validate algorithms across diverse deployment contexts, and what standards of evidence are appropriate for different risk levels. International harmonization of regulatory approaches would facilitate global deployment while maintaining patient safety [14, 22].

Multi-modal data integration poses both opportunities and challenges. Healthcare data encompasses structured electronic health records, unstructured clinical notes, medical images, genomic sequences, continuous physiological monitoring, patient-reported outcomes, and social determinants [23]. Effectively integrating these diverse data types could enable more comprehensive patient understanding and accurate predictions. However, technical challenges include varying data quality and missingness across modalities, determining appropriate integration strategies (early fusion, late fusion, intermediate fusion), and interpreting multi-modal models. Unsupervised learning of joint representations across modalities represents an active research area [9, 10, 17].

Human–AI collaboration frameworks recognize that optimal healthcare delivery may involve partnership between algorithms and clinicians rather than full automation. Research into how to optimally divide tasks, present algorithmic recommendations to support rather than replace clinical judgement, and enable clinicians to override or refine algorithmic suggestions represents important human factors work (Figure 5.6). Understanding when

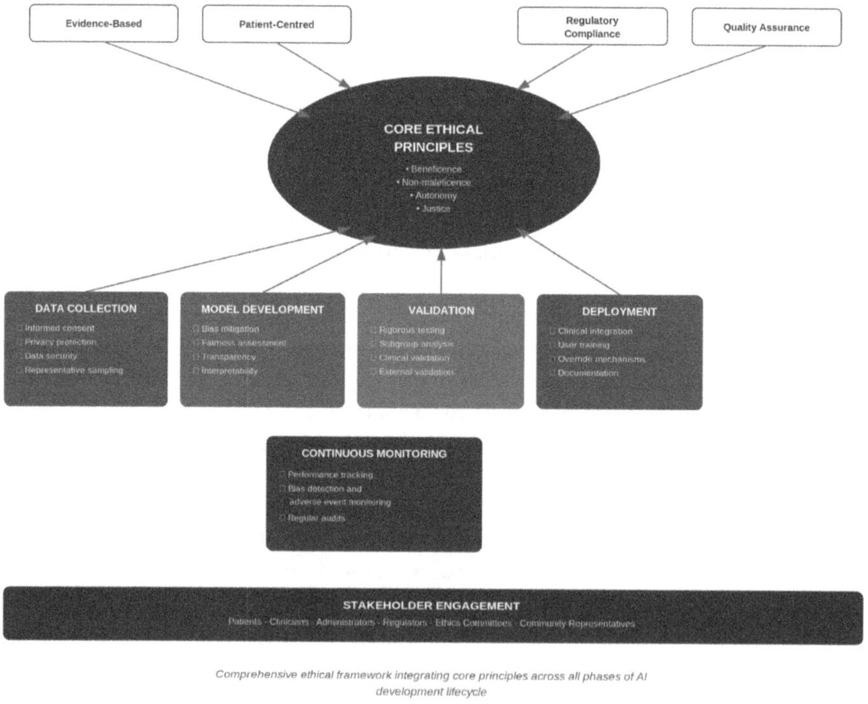


*Figure 5.6* Ethical framework for healthcare AI development and deployment, showing considerations across data collection, model development, validation, deployment, and monitoring phases.

clinicians should trust versus question algorithmic recommendations, and designing systems that facilitate appropriate reliance, requires interdisciplinary collaboration among computer scientists, clinicians, and behavioural researchers [11, 13].

## 5.9 CONCLUSIONS

The complementary nature of supervised and unsupervised learning motivates hybrid approaches that leverage strengths of both paradigms. Semi-supervised learning addresses limited labelled data by incorporating abundant unlabelled information. Transfer learning enables knowledge sharing across related tasks and domains. Active learning strategically selects informative samples for annotation. Multi-task and meta-learning frameworks enable efficient learning across related problems.

Future research directions promise to address current limitations and unlock new capabilities. Advances in few-shot learning, federated learning, continual learning, and causal inference may overcome data constraints, enable collaborative research, support lifelong learning systems, and move beyond correlation to causal understanding. Multi-modal integration, foundation models, and neuro-symbolic approaches aim to leverage diverse data types, learn transferable representations, and combine pattern recognition with logical reasoning. Emerging computing paradigms, including quantum computing, may enable new algorithmic approaches.

The ultimate measure of success for ML in healthcare is not algorithmic accuracy or technical sophistication but impact on patient outcomes, healthcare quality, operational efficiency, and equity. Rigorous prospective evaluation through clinical trials and implementation studies provides essential evidence of clinical utility beyond retrospective predictive performance.

## REFERENCES


1. Alanazi, A. (2022). Using machine learning for healthcare challenges and opportunities. Informatics in Medicine Unlocked, 30, 100924.
2. Rajoub, B. (2020). Supervised and unsupervised learning. In: Biomedical Signal Processing and Artificial Intelligence in Healthcare (pp. 51–89). Academic Press.
3. Eckhardt, C. M., Madjarova, S. J., Williams, R. J., Ollivier, M., Karlsson, J., Pareek, A., & Nwachukwu, B. U. (2023). Unsupervised machine learning methods and emerging applications in healthcare. Knee Surgery, Sports Traumatology, Arthroscopy, 31(2), 376–381.
4. Abbasi, M., Cardoso, F., Silva, J., & Martins, P. (2023). Comparing supervised and unsupervised machine learning techniques to decision support systems in healthcare. In: International Conference on Disruptive Technologies, Tech Ethics and Artificial Intelligence (pp. 38–49). Springer Nature Switzerland, Cham.
5. Sogandi, F. (2024). Identifying diseases symptoms and general rules using supervised and unsupervised machine learning. Scientific Reports, 14(1), 17956.

6. Berry, M. W., Mohamed, A., & Yap, B. W., Eds. (2020). Supervised and Unsupervised Learning for Data Science. Springer, Cham, Switzerland.
7. Habehh, H., & Gohel, S. (2021). Machine learning in healthcare. Current Genomics, 22(4), 291–300.
8. Rashidi, H. H., Tran, N., Albahra, S., & Dang, L. T. (2021). Machine learning in health care and laboratory medicine: General overview of supervised learning and auto-ML. International Journal of Laboratory Hematology, 43, 15–22.
9. An, Q., Rahman, S., Zhou, J., & Kang, J. J. (2023). A comprehensive review on machine learning in healthcare industry: Classification, restrictions, opportunities and challenges. Sensors, 23(9), 4178.
10. Pantanowitz, L., Pearce, T., Abukhiran, I., Hanna, M., Wheeler, S., Soong, T. R., & Tafti, A. P. et al. (2025). Nongenerative artificial intelligence in medicine: Advancements and applications in supervised and unsupervised machine learning. Modern Pathology, 38(3), 100680.
11. Roy, S., Meena, T., & Lim, S. J. (2022). Demystifying supervised learning in Healthcare 4.0: A new reality of transforming diagnostic medicine. Diagnostics, 12(10), 2549.
12. Pereira, J., & Silveira, M. (2019, February). Learning representations from healthcare time series data for unsupervised anomaly detection. In *2019 IEEE international conference on big data and smart computing (BigComp)* (pp. 1–7). IEEE.
13. Berge, G. T., Granmo, O. C., Tveit, T. O., Ruthjersen, A. L., & Sharma, J. (2023). Combining unsupervised, supervised and rule-based learning: The case of detecting patient allergies in electronic health records. BMC Medical Informatics and Decision Making, 23(1), 188.
14. Allenbrand, C. (2024). Supervised and unsupervised learning models for pharmaceutical drug rating and classification using consumer generated reviews. Healthcare Analytics, 5, 100288.
15. Sindhu Meena, K., & Suriya, S. (2019). A survey on supervised and unsupervised learning techniques. In: *International Conference on Artificial Intelligence, Smart Grid and Smart City Applications* (pp. 627–644). Springer International Publishing, Cham.
16. Shailaja, K., Seetharamulu, B., & Jabbar, M. A. (2018, March). Machine learning in healthcare: A review. In: *2018 Second International Conference on Electronics, Communication and Aerospace Technology (ICECA)* (pp. 910–914). IEEE.
17. Bose, E., & Radhakrishnan, K. (2018). Using unsupervised machine learning to identify subgroups among home health patients with heart failure using telehealth. CIN: Computers, Informatics, Nursing, 36(5), 242–248.
18. Firdaus, H., & Hassan, S. I. (2020). Unsupervised learning on healthcare survey data with particle swarm optimization. In: Machine Learning with Health Care Perspective: Machine Learning and Healthcare (pp. 57–89). Springer International Publishing, Cham.
19. Dua, P., & Bais, S. (2013). Supervised learning methods for fraud detection in healthcare insurance. In: Machine Learning in Healthcare Informatics (pp. 261–285). Springer, Berlin.
20. Awotunde, J. B., Adeniyi, A. E., Ajagbe, S. A., & González-Briones, A. (2022). Natural computing and unsupervised learning methods in smart healthcare data-centric operations. In: Cognitive and Soft Computing Techniques for the Analysis of Healthcare Data (pp. 165–190). Academic Press.
21. Uddin, S., Khan, A., Hossain, M. E., & Moni, M. A. (2019). Comparing different supervised machine learning algorithms for disease prediction. BMC Medical Informatics and Decision Making, 19(1), 1–16.

22. Hassan, M., & Alam, M. (2025). Automating healthcare claims processing with supervised and unsupervised AI models. Journal of Applied Big Data Analytics, Decision Making, and Predictive Modelling Systems, 9(2), 1–11.
23. Ghai, D., Saxena, S., & Dhingra, G. et al. (2025). A comprehensive review on performance-based comparative analysis, categorization, classification and mapping of text extraction system techniques for images. Multimedia Tools and Applications, 84, 2327–2484. https://doi.org/10.1007/s11042-024-20257-0

Chapter 6

# AI applications in drug discovery and personalized medicine

*Kiran Malik, Kuldeep Singh Kaswan, and Jagjit Singh Dhatterwal*

## 6.1 INTRODUCTION

Personalized medicine uses AI-based information to tailor treatment instructions based on individual genetic profile, lifestyle factors, and clinical history, thereby increasing treatment effectiveness. The development of machine learning (ML), deep learning (DL), and artificial intelligence (AI) in combination with biomedical sciences has created opportunities of disruptive paradigms in healthcare. This section introduces the paradigm of drug discovery, outlines the complex role of AI in healthcare, and sets the boundaries and goals on which the discussion is organized (Chen et al., 2018).

Conventionally, drug discovery is characterized by a long and complex process that comprises target discovery, lead compound discovery, preclinical validation, and several rounds of clinical trials. The traditional pipeline, which on average can be a 10–15-year-long single-hand operation, is costly (in billions) and the rate of attrition by candidates is very high. Typically, the initial activity of drug discovery is to determine a biological target, in most cases, a receptor or a protein suspected of contributing to the disease pathology, and then chemical libraries are screened to identify lead compounds that bind to the target. More iteration of these hits with several rounds of synthesis and validation usually leads to a small number of viable candidates that proceed to preclinical and clinical trials.

Individual medicine now known as precision medicine has emerged with the evolution of genomic technology and completion of the Human Genome Project. With the final sequencing of the human genome in 2003, information regarding the genetic composition of individuals was obtained, which influences the occurrence of illness and the response rate to drugs. Pharmacogenomics being a subset of personalized medicine is designed to connect genetic variations to drug response and adverse reactions enabling clinicians to customize the therapeutic approaches to the genomic composition of the patient (Zhavoronkov et al., 2019).

From diagnostic and treatment to administrative efficiency, AI has entered every aspects of healthcare. AI applications in drug discovery range from target identification to hit-to-lead optimization to clinical trial design. ML


DOI: 10.1201/9781042015931-6

algorithms, especially convolutional neural networks (CNNs) and recurrent neural networks (RNNs), can analyze large chemical libraries to predict binding affinities and toxicological profiles, thus accelerating lead identification. For medicinal chemistry, RL approaches have also been used for structure modification by simulating each drug–target interaction to assist medicinal chemists to build novel compounds with specific characteristics.

AI plays a vital role in the analysis of multi-omics data as a personalized medicine strategy, where knowledge about populations of patients and individual prediction of treatment choices are essential. Recommendation systems that are developed using AI utilize supervised learning methods to determine the most appropriate therapeutic regimens in patients, eliminating trial-and-error methods of clinical decision-making. Longitudinal health data is also analyzed by AI-based risk prediction models to identify early disease progression and follow-up intervention to achieve a better outcome. Digital twins are a virtual representation of biological systems that are exploited to simulate it across a wide range of therapeutic modalities and offer clinicians the ability to test and optimize personalized therapeutic regimens in silico, before they can be transferred into practice.

AI has played a major role in the creation of clinical decision support systems (CDSS) that are used to advise clinicians on the most effective courses of treatment, through the application of the best-available clinical guidelines, patient preferences, and real-world evidence. AI-enabled algorithms examine vast amounts of real-world evidence – wearable device information, remote monitoring systems, and EHRs – and facilitate the creation of actionable insights in real time that can be applied to individual care.

The chapter highlights the need to implement algorithmic transparency, reduce bias in AI algorithms, and protect patient privacy where AI solutions are used in clinical settings. Additionally, it discusses the hurdles of incorporating AI into current clinical practices and highlights the necessity for comprehensive validation systems to confirm the dependability and safety of AI-based solutions.

It intends to showcase real-world case studies that are successfully utilizing AI across drug discovery and personalized medicine, providing context relatable to both domestic and international markets and deliverables. By way of case study examinations, the chapter aims to highlight the transformative implications of AI on intractable medical evolutions, therapeutic advancement, and the transition toward precision orthotics.

This chapter also focuses on the future of AI in drug discovery and personalized medicine. Continued innovations in AI frameworks for healthcare systems include federated models, blockchain-enabled frameworks, and explainable AI (XAI), which will provide AI models with the robustness, scalability, and interpretability necessary to drive healthcare transformations. The chapter aims to bring together AI researchers, biomedical scientists, and regulatory agencies to explore how AI can accelerate the design and development of safer and more effective patient-matched therapeutic interventions.

With these goals, the chapter is structured to be comprehensive regarding AI methods, applications, and implications for drug discovery and personalized medicine. Further sections cover the details of AI on-target identification, virtual screening, and lead optimization, as well as a section on AI clinical trial design and biomarker discovery. It discusses the combination of AI and systems biology in the development of digital twins to enable patient simulation, which would lead to real personalized healthcare.

**Literature Survey Table**

| *Authors and Year* | *Key Findings* | *Methodology / Techniques Used* | *Results / Remarks* | *Finding Gaps* |
|---|---|---|---|---|
| Altae-Tran et al. (2017) | Demonstrated the feasibility of one-shot learning for drug discovery using low-data scenarios | One-shot learning with convolutional neural networks (CNNs) for molecule classification | Achieved significant performance in low-data scenarios, reducing the need for large labeled datasets | Limited application to specific molecular classes; generalizability remains a challenge |
| Popova et al. (2018) | Applied deep reinforcement learning (DRL) for de novo drug design | Reinforcement learning using a reward function to optimize molecular generation | Improved molecular novelty and diversity, showing potential for generating drug candidates | Needs further optimization for synthesizability and biological relevance |
| Mullard (2018) | Critically assessed the role of machine learning in drug discovery, evaluating its impact and limitations | Literature review and expert opinion | Highlighted challenges in transitioning ML models from research to clinical applications | Lack of standardization in ML validation processes |
| Patel et al. (2020) | Reviewed machine learning approaches in drug discovery and their impact | Survey of supervised and unsupervised ML techniques for drug design | Demonstrated improved prediction accuracy but noted computational challenges | Scalability and interpretability of ML models remain concerns |
| Lee et al. (2020) | Discussed the advances and applications of generative AI models in drug discovery | Application of generative models such as GANs and VAEs | Provided insights into AI-assisted molecular generation with high diversity | Addressed need for better evaluation frameworks and downstream validation |

*(Continued)*

| *Authors and Year* | *Key Findings* | *Methodology / Techniques Used* | *Results / Remarks* | *Finding Gaps* |
|---|---|---|---|---|
| Chen et al. (2020) | Reviewed deep learning models for drug–target interaction (DTI) prediction | Deep learning models, including CNNs and recurrent neural networks (RNNs) | Improved DTI prediction accuracy but highlighted interpretability issues | Lack of biologically interpretable models for DTI predictions |
| Lim et al. (2019) | Predicted drug–target interactions using deep learning with similarity-based approaches | Deep neural networks with neighborhood similarity metrics | Improved predictive accuracy in drug–target interaction tasks | Lack of explainability in the prediction model |
| Ragoza et al. (2017) | Utilized CNNs for protein–ligand scoring in virtual screening | CNNs to predict protein–ligand binding affinities | Achieved high binding affinity prediction accuracy | Difficulty in generalizing to unseen ligand–protein pairs |
| Mamoshina et al. (2016) | Explored deep learning applications in biomedicine | Survey of DL applications in biomedical domains | Demonstrated applications in diagnostics, personalized medicine, and drug design | Translation of findings to clinical practice remains slow |
| Sanchez-Lengeling and Aspuru-Guzik (2018) | Investigated inverse molecular design using generative models | Variational autoencoders (VAEs) and GANs for molecular generation | Improved inverse molecular design accuracy | Addressed the need for integrating feedback loops for practical design |
| Goh et al. (2017) | Reviewed deep learning applications in computational chemistry | Survey of CNNs, RNNs, and graph-based models in chemistry | Identified gaps in explainability and scalability for chemical predictions | Emphasized need for hybrid AI-physics models |
| Gawehn et al. (2016) | Analyzed applications of deep learning in drug discovery | Overview of DL models such as CNNs and deep neural networks | Highlighted key advancements in virtual screening and drug design | Challenges in interpretability and explainability remain |
| Lu et al. (2019) | Discussed AI and computational modeling for precision oncology | Application of machine learning for personalized cancer treatment | Improved predictive accuracy in oncology applications | Need for integration of multi-omics data for enhanced precision |

*(Continued)*

| *Authors and Year* | *Key Findings* | *Methodology / Techniques Used* | *Results / Remarks* | *Finding Gaps* |
|---|---|---|---|---|
| Paul et al. (2021) | Examined AI applications in drug discovery and development | Review of ML, DL, and reinforcement learning approaches | Demonstrated AI's transformative potential in drug discovery pipelines | Need for reproducibility and validation across different datasets |
| Ekins et al. (2019) | Investigated end-to-end drug discovery using ML | ML models covering compound design, synthesis, and prediction | Highlighted successful AI-driven drug discovery applications | Need for end-to-end validation in clinical trials |
| Vamathevan et al. (2019) | Reviewed applications of ML in drug discovery and development | Survey of ML models across drug discovery pipelines | Identified major advances but highlighted gaps in clinical translation | Emphasized the need for integrating multimodal data |
| Ribeiro et al. (2016) | Proposed an explainability method for black-box ML models | Local interpretable model-agnostic explanations (LIME) | Improved interpretability of model predictions | Applicability to complex biomedical models needs exploration |
| Baek et al. (2021) | Developed a three-track neural network for protein structure prediction | Deep neural networks for protein structure prediction | Achieved high accuracy in protein structure modeling | Further validation in real-world biological systems needed |

Table of research trends from 2016 to 2022 in AI-driven drug discovery in terms of methodology, outcomes, and gaps in research. These include DL techniques such as CNNs, RNNs, GANs, VAEs, and reinforcement learning (RL) applied to DTI prediction, molecule generation, and protein–ligand scoring. Altae-Tran et al. (2017) and Popova et al. (2018)) identify one-shot learning and RL for molecule classification and de novo drug design. Generative models like GANs and VAEs show improvements in molecular novelty, diversity, and optimization but struggle with stability, interpretability, and synthesizability (surveys by Patel et al., 2020, and Vamathevan et al., 2019).

## 6.2 FOUNDATIONS OF AI IN BIOMEDICAL SCIENCES

Most often, it is applied to supervised learning of problems like predicting drug–target interactions, finding biomarkers, and disease classification using omics data. ML models that have been applied include SVM, random forest, and GBM that are employed to predict drug response/utilize and to identify new therapeutics. By contrast, the unguided learning algorithms (e.g., *k*-means clustering and hierarchical clustering) assist in uncovering implicit patterns of molecular signatures and categorizing patient subtypes to aid more effective stratification toward precision medicine. RL has recently emerged as a powerful approach for drug discovery, with a specific emphasis on the de novo design of drugs whereby RL agents explore molecular space while leveraging reward functions to optimize specific drug characteristics (Vamathevan et al., 2019).

DL architectures are now widely applied and outperformed traditional ML approaches in many biomedical applications since they automatically learn hierarchically complex features. CNNs have shown great promise for the processing of imaging data, including but not limited to histopathological slides, MRI, and CT scans, with applications in automatic tumor classification, segmentation, and radiomics. RNNs, such as its individual variants (e.g., long short-term memory [LSTM], gated recurrent units [GRUs]), are effective in dealing with sequential biological data (such as time-series gene expression profiles, patient histories in EHRs). Elevated sequence-based data modeling capability of transformers have led to transformative power applied in biological sequences and clinical narrative analysis that relies on long-range dependencies captured by transformers recently. Models such as BERT and BioBERT have been fine-tuned to extract biomedical knowledge from unstructured clinical text.

Much of the NLP techniques are extensively used in biomedical text mining for extraction of important information from the scientific literature, clinical notes and patents. Moreover, named entity recognition (NER) and relation extraction techniques allow for the identification of gene–disease associations, drug–drug interactions, and adverse event reports. State-of-the-art performance in biomedical information retrieval has been made possible using pretrained transformer models such as BioBERT and SciBERT, leading to automatic summarization and literature-based hypothesis generation.

Drug design and molecular generation are some of the areas where generative AI models are being actively used: generative adversarial networks (GANs) and variational autoencoders (VAEs). GANs, one of the popular generative models, comprised of a generator and a discriminator, is used to create realistic molecular structures with targeted physicochemical properties. Intro to VQ-VAECompared to VAEs, VAEs enable you to encode molecules in latent space and optimize over the best molecular structures.

Loss of genomic information can be used to detect genetic variations, somatic mutations and gene expression patterns, which are associated with disease susceptibility and drug response. The proteomic data describe the abundance, modifications and interactions of proteins which may be a potential therapeutic target. Metabolomic data looks at disease-related phenotypes of biochemical pathways and metabolite profiles. Multi-omics data integrating AI models open the way to the discovery of disease biomarkers and the design of precision medicine.

Chemical structure and compound libraries are made up of millions of small molecules with various properties that in principle can be used as drug candidates. State-of-the-art AI models (graph neural networks [GNNs] and message-passing neural networks [MPNNs]) can be trained to make predictions of bioactivity, toxicity, and pharmacokinetics based on molecular graphs. Virtual screening is a high-throughput process of finding lead compounds that have good binding energy and low off-target effects and can be performed quickly, which is AI-powered (Yang et al., 2019).

Even though interoperability between platforms and standardization of biomedical data are significant challenges that remain to be addressed. Variations in information formats, ontologies, and vocabularies present hindrances to smooth data integration and model implementation. Easier interoperability is starting to appear with standardized data models (e.g., Fast Healthcare Interoperability Resources [FHIR] that enables sharing of practitioners or processes) and common data models (CDMs) that enable cross-institution collaboration where datasets can be joined and analyzed.

New opportunities are emerging in this space and are changing the landscape of drug discovery and personalized medicine as we know it. Explainable AI (XAI) methods enhance model interpretability to enable clinicians to comprehend AI-based decision-making. Privacy is also enhanced by Federated learning because it enables several institutions to learn a model without the exchange of sensitive information. Digital twins have the capacity to replicate patient-specific models to conduct therapeutic interventions assessments and to simulate therapeutic responses. As AI technologies are upgraded, the development of AI technologies with domain knowledge and clinical expertise, the future of healthcare delivery is expected to transform with the possibility of personalized and effective treatment of heterogeneous patients (Ching et al., 2018).

## 6.3 AI IN DRUG DISCOVERY: TRANSFORMING THE PIPELINE

AI is changing the drug discovery process. AI is improving drug discovery, a time-binding and costly process, and reshaping it. DL, ML, and natural language processing (NLP) are among the AI models transforming the various processes involved in the drug discovery pipeline, saving hundreds of dollars and enhancing the potential success of treatment. In drug discovery, AI has been integrated into the workflow in drug target identification, lead compound identification, drug optimization, preclinical experiment design, and clinical trial design. In this chapter, we discuss the effect of AI on drug discovery and personalized medicine. Target identification and validation are the first and the most important steps in the drug discovery pipeline. AI makes it possible to discover disease-relevant genes, proteins, and other biomolecular targets with precision and efficiency and then select only viable ones to proceed to subsequent stages (Hughes et al., 2011).

Associations between genes and diseases using high-throughput genomic and transcriptomic data can also be identified using AI models. With multi-omics data such as gene expression profiles, single-nucleotide polymorphisms (SNPs), and epigenetics markers, in combination with deep neural networks (DNNs) and GNNs, we can find genes that are related to a specific disease. AI can prioritize gene targets to a particular disease using data-driven algorithms such as association rule mining and Bayesian inference. Models, including DeepSEA and ExPecto, predict the effects of noncoding variants on regulation to predict such information, as a method of target identification. This minimizes false positive and enhances new biomarker and therapeutic target discovery in Figure 6.1.

AlphaFold (DeepMind) transformed the field of protein structure prediction with new higher-order accuracy of close to the experimentally measured protein folding information. DL models have been modified to predict the three-dimensional structure of proteins out of amino acid sequences with attention mechanisms and CNNs. A successful AlphaFold 2 has also made it accessible to structure-based drug design (SBDD), which offers details on target binding sites, allosteric sites, and the structural conformation needed to interact with drugs. Outside of AlphaFold, structural biology models have been developed to use structural biology with AI methods, such as RosettaFold and ESMFold, to more optimally resolve protein–ligand interactions and optimize target validation.

AI can be used to integrate multi-omics data (such as genomics, transcriptomics, proteomics, and metabolomics) to enable better target prioritization. The analysis of such complicated datasets with the help of multi-omics integration, that is, using autoencoders, variational inference, and graph-based

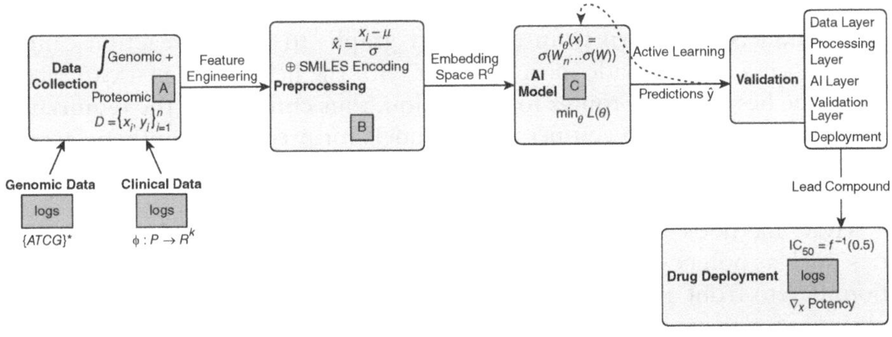


*Figure 6.1* AI in drug discovery.

learning, can help identify the valuable biomarkers geared toward the comprehension of the mechanistic processes underlying the disease state. This enables DL algorithms like multi-omics factor analysis (MOFA) and tensor decomposition methods to discover latent dimensions in layers of omics and produce hidden pattern such that target prioritization can be obtained. This assists in the focus down of the correlation between biological processes and disease phenotypes, employing the AI to combine multi-omics information across multiple sources to obtain accuracy in identifying targets and validation in new targets (Wang et al., 2021).

Quantitative structure–activity relationship (QSAR) is the data science method used to identify a quantitative correlation between chemical structure of compounds and their bioactivity through the application of AI algorithms. The bioactivity, toxicity, and pharmacokinetic properties are predicted by random forests, support vehicle machines (SVMs), and DL models. Employing convolutional and recurrent architectures to learn complex molecular shape activity relationships across diverse chemical scaffolds, these Deep QSAR models can deliver performance superior to previous state-of-the-art models. State-of-the-art QSAR models incorporate transfer learning at various levels to facilitate bioactivity predictions in scenarios with limited data, training the selection of lead candidates.

De novo drug design utilizing generative models, such as GANs, VAEs, and RL guided methods, enables the generation of novel molecular structures with specified properties.

AI models trickled in multiobjective optimization (MOO) of drug efficacy, safety, and synthetic feasibility.

Models developed using supervised learning approaches (and DL models specifically) predict ADMET properties. Neural networks like DeepADMET and ADMETPredictor exploit molecular descriptors to predict pharmacokinetic parameters. By using transfer learning and ensemble models, they can predict more accurately and identify candidates with undesirable ADMET profile in the early stage. This led to the reduction of attrition rates in later stages of development due to optimal ADMET characteristics of lead compounds.

ML accelerates retrosynthesis planning by predicting feasible synthetic routes for drug candidates. RetroGAN and G2Retro are examples of graph-based models that learn from molecular graphs to predict reaction centers. This process can be made more efficient with the help of AI as it can recommend the best possible routes for a reaction, thus eliminating the requirement for exhaustive reaction enumeration. Models for predicting synthetic accessibility impact the ability to evaluate proposed compounds for both scalability and reproducibility in drug manufacture.

MOO methods arise as they can use AI to optimize multiple drug properties, such as potency, selectivity or safety, at the same time. Bayesian optimization, Pareto front analysis, and RL algorithms help accommodate competing objectives, thus verifying that optimized compounds are consistent with therapeutic profiles. MOO powered by AI expedites lead optimization by probing large chemical space, finding molecules that fulfill multiple constraints (Ursu et al., 2019).

AI models predict in silico toxicity and efficacy profiles, leading to a decreased need for torturing fellow living beings in animal models. Tox21 and DeepTox are DL models that classify compounds by toxicity profiles of hepatotoxicity, cardiotoxicity, and nephrotoxicity. FORCE is a generalizable way to generate predictive models for therapeutic outcomes across diverse biological contexts by using data about cellular responses to therapy as the building blocks. These models help de-risk the science by identifying potentially safety concerns at an early pipeline stage, thus accelerating preclinical testing.

AI-enabled organ-on-a-chip (OoC) systems and Virtual Twins Simulate organ-level responses anticipate drug efficacy in real time. Meta learning approaches with OoC to map responses of a drug to the observed physiochemical properties. Microfluidic systems act as in vivo mimics and have established an exciting regenerative drug discovery approach based on in vitro studies. To improve personalized preclinical evaluations, digital twin simulations may integrate multiscale biological quotient data for patient-specific functional responses. This makes the preclinical results more translatable and lowers the chances for clinical failure.

AI models find the right patient cohorts for the intervention, using clinical, genomic, and phenotypic data for this task. Heterogeneous patient populations are segmented into homogeneous cohorts using techniques such as clustering, classification, and NLP. By subjecting patients who have the least likelihood of benefiting from the intervention to the trial, and establishing an ALDP with the same established statistical power, AI models can rank and prioritize patients that are comparatively most likely to benefit from the intervention(s). Available recruited novel AI algorithms stratify the patients by predicting the outcome by analyzing large datasets (Kaswan et al., 2023).

AI for predicting clinical trial outcomes analyzes previous trial data, biomarker responses, and patient characteristics. Predictive models (survivor analysis, Cox proportional hazards models, DL frameworks) identify potential risks and predict success of the trial. AI capabilities allow for RWD and EHRs to be incorporated into trial simulation models to predict adverse events or how patients might respond to treatments, helping to guide design and decision-making during the trial process.

## 6.4 AI IN PERSONALIZED MEDICINE: TAILORING TREATMENT TO INDIVIDUALS

AI is transforming the healthcare sector, in particular the realm of personalized medicine which personalizes treatment according to genetic, molecular, and clinical profiles of a patient. A personalized treatment to each patient is the aim of personalized medicine, through analyzing huge amounts of information, and revealing patterns that cannot be discerned by traditional methods, for example, microarrays. ML and DL AI models are now being increasingly paired so that genomic variants can be examined, new biomarkers discovered, treatment made optimal, and patients monitored continuously in order to optimize the efficacy of therapeutic interventions and reduce side effects. AI personalized medicine allows the design of precision treatments, wherein drugs can be selected from a large pool of compounds, allowing high specificity, reducing drug–drug interactions, bypassing inter-patient variation in drug action and disease progression, and employing mechano- genomics, imaging modalities, and clinical history multimers data.

$$R_i(t) = \sum_{j=1}^{N} \left( \frac{W_{ij}(t) \cdot D_{ij}(t) \cdot \log\left(1 + I_{ij}(t)\right)}{\sqrt{1 + \sigma_{ij}^2(t)}} \right)$$

$$D_{ij}(t) = \int_0^T \left( \frac{\partial P_{ij}(t)}{\partial t} + \alpha \cdot \frac{\partial^2 C_{ij}(t)}{\partial t^2} \right) dt$$

$$P_{ij}(t) = \frac{\int_0^T A_{ij}(t) \cdot \gamma_{ij}(t)\, dt}{\int_0^T \gamma_{ij}(t)\, dt}$$

$$W_{ij}(t) = \frac{\sum_{k=1}^{M} \left( \frac{H_{ijk}(t)}{T_{ijk}(t)} \right) + \frac{T_{ij}(t) \cdot \log\left(1 + A_{ij}(t)\right)}{\log\left(1 + C_{ij}(t)\right)}}{\int_0^T S_i(t)\, dt}$$

$$I_{ij}(t) = \frac{R_{ij}(t)}{\log\left(1 + H_{ij}(t)\right)} + \frac{\partial^2 D_{ij}(t)}{\partial t^2}$$

$$C_{ij}(t) = \frac{1}{\sum_{k=1}^{M} \frac{Q_{ijk}(t)}{T_{ijk}(t)}} + \frac{\log\left(1 + S_i(t)\right)}{\log\left(1 + D_{ij}(t)\right)}$$

$$H_{ij}(t) = \left( \frac{\partial^2 R_{ij}(t)}{\partial t^2} + \lambda \cdot \frac{\partial P_{ij}(t)}{\partial t} \right) \cdot e^{-\mu t}$$

$$A_{ij}(t) = \frac{\sum_{k=1}^{N} \log\left(1 + H_{ijk}(t)\right)}{\int_0^T S_i(t)\, dt}$$

$$\gamma_{ij}(t) = \frac{\partial P_{ij}(t)}{\partial t} + \beta \cdot \frac{\partial^2 W_{ij}(t)}{\partial t^2}$$

$$Q_{ij}(t) = \frac{\sum_{k=1}^{N} \left( e^{-\delta t} \cdot \log\left(1 + H_{ijk}(t)\right) \right)}{\int_0^T e^{-\delta t} S_i(t)\, dt}$$

***Symbols explanation:***

- $R_i(t)$: Personalized treatment effectiveness for individual $i$.
- $W_{ij}(t)$: Weight of predictive model for feature $j$.
- $D_{ij}(t)$: Cumulative effect of treatment over time.
- $P_{ij}(t)$: Personalized risk adjustment model.
- $I_{ij}(t)$: Influence of patient-specific biomarkers.
- $C_{ij}(t)$: Confidence interval of prediction model.
- $H_{ij}(t)$: Health score for patient $i$ with feature $j$.
- $A_{ij}(t)$: Adjustment factor based on AI learning.
- $\gamma_{ij}(t)$: Gradient adjustment parameter.
- $Q_{ij}(t)$: Quality score of tailored treatment plan.
- $\lambda, \mu, \alpha, \beta$: Regularization, decay, and tuning parameters.

These models inherit high-throughput next-generation sequencing (NGS) data in matrix format (where columns represent variants and rows correspond to samples) to segregate rare variants and predict and annotate their associated pathogenicity via classification algorithms. Methods of ensemble learning, for example, random forests and gradient boosting machines, improve variant interpretation by combining diverse features such as allele frequency, functional annotations, and evolutionary conservation. DeepVariant and GATK are commonly used AI-powered pipelines that can increase the efficiency of clinically relevant variant identification within genomic data by reducing false-positive and false-negative rates (Dhatterwal et al., 2022).

SNPs and expression profiles have been used in the development of AI-powered predictive models that predict individual drug responses. With the implementation of supervised learning algorithms, such as support vector machines (SVM), and XGBoost, the genetic features correlate to phenotypic outcomes, which enables predicting the effectiveness of a given treatment and the likelihood of adverse reactions. DL models such as autoencoders and GANs are trained to learn latent feature representations which are able to encode the intricate interactivity between the genes and the drugs. The PharmGKB and 1000 Genomes pharmacogenomics datasets are also incorporated into the AI models, which enables predicting variability in drug metabolism and efficacy, and future applications may be implemented where genotype is used to guide therapeutic processes.

In the new area of pharmacogenomics, studying how genetic polymorphisms influence drug metabolism, it is now possible to unravel complex genotype–phenotype connections using AI. Inter-patient variance is most often predicted on the models of mathematics based on the polygenic risk scores (PRS) based on the genome-wide association studies (GWAS) and other omics data using AI-based classification. Gradient boosting models and neural network models examine the combined influence of variations, which can be used to stratify risk between different therapeutic responses. AI-driven polygenic models can be combined with treatment regimens to seek its optimum solutions to complex diseases such as cancer and cardiovascular diseases and offer personalized dosage recommendations, reducing adverse drug reactions (Kaswan et al., 2022).

AI uses algorithms of unsupervised learning such as *K*-means and hierarchical clustering to identify their molecular subtypes in heterogeneous diseases. These models detect disease clusters that have been hidden by use of transcriptomic, proteomic, and metabolomic data and classify patients into distinct molecular subtypes. The clustering analysis using those variables assists in establishing aggressive phenotypes of disease and provides the possibilities of achieving targeted therapeutic interventions. VAEs and other dimensionality reduction using DL architectures are also used to extract the relevant features of the disease and provide clinically actionable subtypes.

Liquid biopsy, which is noninvasive and a way of detecting circulating tumor DNA (ctDNA) and other biomarkers, is highly beneficial to AI-powered analytics. Similarly, deep CNNs can help obtain high-dimensional representations for radiological images, enabling the identification of imaging biomarkers associated with tumor heterogeneity, immunotherapy response, and metastatic potential. Harnessing multimodal data sources is helping AI-based radiomic pipelines streamline molecular and imaging signatures for improved prognostic accuracy and personalized treatment decision-making in oncology.

AI algorithms help optimize drug dosage in medicine by exploring multidimensional datasets that include genetic, demographic, and pharmacokinetic variables.

Sensor data fed into data streams into AI models will detect significant deviations in patterns that have been baselined during healthy times that are indicative of disease exacerbation or adverse events. RNNs and long short-term memory (LSTM) networks are neural networks that do not only learn time-dependent relationships in wearable data but also predict important health-related occurrences. Examples of clinical conditions that can be early identified using AI-enhanced monitoring platforms include arrhythmias, hypoglycemia, and other forms of clinically relevant conditions, which can be acted upon before an adverse patient outcome takes place.

There have been attempts in the clinical literature to use trajectories over time, but these have employed mostly longitudinal analysis. For example, time-series forecasting algorithms and Gaussian processes are types of AI models that use sequences of patient data to determine future health outcomes. DL architectures and hidden Markov models (HMMs) uncover latent disease states, allowing for early intervention and personalized adjustment of treatments. Longitudinal analysis powered by AI enables dynamic risk profiling and personalized care management in chronic medical conditions, including diabetes and cardiovascular diseases.

Survival analysis is fundamental to personalized medicine and predicts time-to-event outcomes (e.g., disease recurrence or death). Deep survival neural networks and Cox proportional hazards models are examples of AI-based survival models used to determine the subgroups of patients at high risk. Variant clinical, genomic, and imaging features improve the risk stratification using gradient boosting techniques and ensemble learning frameworks. Predictive modeling using AI-enabled survival analysis platforms could be fine-tuning prognostication enabling the healthcare providers to customize clinical management according to the individual risk prognosis (Lipinski, 2004).

## 6.5 ADVANCED AI METHODOLOGIES AND EMERGING TRENDS

In the field of drug discovery and personalized medicine, GNNs have increasingly come to play an important role in molecular and network biology. Traditional ML techniques often fail to model the complex relational structures that are necessary for understanding biological and chemical systems in which entities, such as proteins, genes, and molecules, interact in elaborate networks. GNNs shine by modeling data as graphs with nodes representing molecular subcomponents and edges defining the interaction between them. Using this method, the model can learn embeddings of molecules based on their context, and is able to perform tasks such as drug–target interaction prediction, drug property prediction, and de novo drug design.

Such gene regulatory interactions can be determined using GNNs, mapping these interactions onto genomic data to elucidate candidate biomarkers related to a specific disease in personalized medicine. In addition, GNNs have been an excellent choice to model protein–protein interactions (PPIs) and gene regulatory networks (GRNs) to help find targetable pathway as well as therapeutic compounds. GNNs can also be integrated with attention mechanisms which can provide even more interpretability by allowing researchers to spotlight which interactions drive drug efficacy or toxicity.

Datapoints are scarce in drug discovery and personalized medicine as the acquisition of biomedical data is expensive, time-consuming, and sometimes even unethical. However, to break these bottlenecks, both transfer learning and few-shot learning techniques have emerged as powerful approaches to learn efficiently from preexisting knowledge in similar domains, thus enhancing performance in low-data scenarios. In transfer learning, already-trained models are fine-tuned on smaller task-specific datasets after being trained with large-scale datasets like PubChem, ChEMBL, or PDBBind. This approach greatly accelerates training and improves the ability of the model to generalize to new molecular or clinical tasks.

Contrary to this, few-shot learning tries to teach models to be able to make predictions that are accurate even when there is a small number of examples. The fact that models can quickly adapt to new tasks, meta-learning has been successfully used in drug–target interaction prediction and new drug candidate prediction. In precision medicine, few-shot learning is used to develop individualized treatment plans in the case of infrequent disease state, where there is a lack of clinical data. Meta-learning techniques such as model-agnostic meta-learning (MAML) have demonstrated to be more flexible to patient-specific treatment recommendations due to rapid adjustments to new patient representations. These methods demonstrate high potential to increase drug discovery pipelines and enhance personalization of treatments with the help of meta learning and lifelong learning by enabling models to be trained on effective generalizations using small amounts of data (Liu et al., 2019).

The need to trust and be able to interpret is the highest when the preposterous insights that AI models are providing in the field of drug discovery and personalized medicine are introduced. Explainable AI (XAIGen) is a new field that aims to fill in the gap between black-box models and human cognition by providing clear and understandable explanations of the predictions of the models. XAI methods, such as SHAP and LIME, have found their way to drug discovery, where they can point out which particular molecular features contribute the most to a prediction, including the existence of certain functional groups that are predictable to correlate with binding affinity or toxicity.

FL has also been relevant in drug discovery and personalized medicine, where it can aid in collecting data involving heterogeneous sources (geographically or organizationally) of hospitals, pharmaceutical companies, and genomic research centers in a privacy-preserving way that it can be used to create robust models without exposing any patient information.

FL is implemented in a distributed manner where models are trained on more than one node with each node possessing local data locally and communicating encrypted model updates with the central server.

Quantum-enhanced generative models, such as quantum variational autoencoders (QVAE), are trained to learn quantum-based features of molecular data to enable new molecular structures with desired properties to be generated. On the same note, QC is also under exploration in the field of personalized medicine, where the individualized treatment regimens may be discovered by modeling drug combinations on a patient-specific biological system in high-speed regimes and optimizing treatments.

Similar to their NLP counterparts, transformers such as BERT and GPT have been pretrained and then fine-tuned on sequences of genomes to predict regulatory elements and mutations related to disease phenotypes. These advancements enable AI models to easily adapt to previously unexplored areas of biomedical research, reducing the need to have huge and expensive annotated datasets.

XAI methods have been used in drug response prediction models to elucidate the genomic and molecular responsible for differential drug responses within patient cohorts. XAI finally opens the door to improved patient outcomes by increasing the credibility of AI models in clinical decision-making by producing explainable insights into model predictions.

Federated learning systems have also been used in genomics studies to enable safe cross-institution collaboration while maintaining patient privacy. One of the notable uses of FL has been in obtaining genomic predictions of complex diseases, as seen in polygenic risk scores (PRS), by combining information from several biobanks without exchanging individual-level data.

For example, quantum computing algorithms such as the Harrow–Hassidim–Lloyd (HHL) algorithm are more efficient at solving linear systems of equations, which happens during the simulation of protein–ligand docking. By accelerating the computation of binding energies and interaction dynamics, QC enhances the efficiency of virtual screening pipelines and results in accelerated drug discovery (Rodrigues et al., 2021).

## 6.6 CASE STUDIES

### 6.6.1 AI in oncology: From target discovery to precision therapy

In the same way, it is possible to model PPIs with GNNs and propagate information and determine the necessary targets to build drugs. Oncology drug repurposing has also been a significant contribution by AI. Analyzing molecular signatures and clinical data, even possible off-label applications of approved drugs can be demonstrated by AI-based models. For example, DeepTox and ChemProp apply multitask neural networks to achieve better performance as compared to current methods in the prediction of repurposable drugs. The subsequent application of GANs and their application has in part allowed the generation of novel molecular structures with anticancer properties. These models are trained using large collections of chemicals, and predict drug-likeness and optimization of the PK/PD attributes.

Further, AI tools such as RL have already ensured that drug design pipelines have the ability to screen large chemical spaces and select compounds with perfect binding affinity and fewer off-target activities. DeepMind developed the AI model AlphaFold, which provides state-of-the-art accuracy in predicting 3D protein structures, and has been used to identify oncolytically druggable targets. Multimodal datasets have also been made possible by AI and have led to the discovery of synthetic lethal interactions that can be used to create individualized cancer treatments.

Customizing treatment regimens, AI models have also played a critical role in precision oncology. As an illustration, AI-driven algorithms such as IBM Watson Oncology apply the NLP to clinical literature and prescribe personalized treatment regimens. ML predictive models combine clinical, genomic, and radiomic data to assess the response of individual patients to therapies to facilitate the use of immunotherapies, chemotherapies, and targeted therapies. The Cancer Genome Atlas (TCGA) and similar large broad datasets have already demonstrated very high prediction accuracy using AI-based predictive models that are trained on comprehensive datasets to make a decision on treatment.

The AI models have been used to enhance sensitivity and specificity of ctDNA and CTCs in liquid biopsy-based methods that are commonly used in molecular diagnostics. AI algorithms are used to analyze the big data of sequencing to identify minimal residual disease (MRD), which can be intervened and therapies corrected on time. CDSS that are operated by AI also contribute to the optimization of adaptive therapy regimens by studying real-time patient measurements and making modifications in treatment strategies in response to evolving disease trajectory and patient response patterns.

### 6.6.2 Rare diseases: Accelerating diagnosis and drug development

Rare diseases are defined as conditions that affect fewer than 200,000 people in the United States and can present unique challenges, such as heterogeneity, limited data, and late diagnosis. AI-powered solutions have come to play a fundamental role in kicking off the diagnosis and treatment of rare diseases based on multimodal data and predictive models. Using disease-specific methods such as whole-exome sequencing and multiple computer-algorithms-based predictors, we have developed a gold-standard method to overcome the site-specific approach of rare and undiagnosed diseases, which has been hampered by a slow, decade-long, trial and error method resulting in diagnostic odysseys and delayed treatments. ML models, most notably DL models and ensemble learners, have proven highly predictive for rare disease classification from genomic, phenotypic, and clinical data.

In the case of rare genetic disorders, AI algorithms including DeepVariant and DeepGenome train on NGS data to discover the disease-causing variant. These models employ CNNs and RNNs to leverage large genomics datasets and rank candidate variants with high accuracy. In addition, AI-based models enable translation of the clinical phenotypes using NLP techniques to phenotypic abnormalities that can be mapped to specific genetic mutations using ontological bases such as human phenotype ontology (HPO), etc.

Analyzing radiological and histopathological images helped make AI-powered image analysis a game changer in rare disease diagnosis. CNNs trained on rare disease-specific datasets identify structural abnormalities, subtle phenotypic variations, and disease-specific biomarkers to accelerate diagnosis. For example, platforms powered by AI, like Face2Gene, use facial recognition technology to identify dysmorphic facial features in patients that are characteristic of rare syndromes, drastically decreasing the time to proper diagnosis.

While numerous rare diseases lack adequately effective treatments, the field of drug development for rare diseases has found utility in AI models that can identify potential drug candidates more quickly through analyses of the drug–target interactions and predictions of their binding affinities. The application of AI-based computational techniques, including molecular docking and virtual screening, has facilitated the discovery of novel small molecules with treatment potential for rare diseases. Atomwise's AtomNet platform, for example, uses deep neural networks to predict ligand-binding conformations, allowing researchers to sift through trillions of possible compounds to find promising drug candidates for rare disorders.

By examining databases of existing drugs and identifying compounds that may show effectiveness against orphan diseases, AI models have also been key in drug repurposing used to treat rare diseases. Chemical, clinical, and genomic data are complemented by ML models that predict off-target effects and optimize dosage regimens. Using complex biological models and RL algorithms, AI platforms like Healx have been able to repurpose FDA-approved drugs for rare diseases.

Likewise, AI-driven NLP algorithms can improve upon clinical trial design for rare diseases by reviewing historical trial data to determine the most promising biomarkers and patient recruitment strategies. AI-driven adaptive trial designs allow for the dynamic adjustment of trial parameters based on interim data analysis, maximizing resource efficiency and expediting timelines. Rare disease populations have benefited tremendously from the combination of AI and real-world evidence (RWE) to provide continuous monitoring of treatment efficacy and safety profiles post-market.

### 6.6.3 Infectious diseases: Antiviral and antibiotic discovery

AI use in combating infectious diseases has become a necessity since global pandemics and antimicrobial resistance (AMR) has surfaced. The AI-based techniques have swiftly expedited screening of novel antivirals and antibiotics by forecasting drug–target interactions, genomes of pathogens, and creating effective therapeutic agents. In fact, due to the rapid mutation rate of most pathogenic organisms and the multidimensional nature of host–pathogen interactions, conventional drug discovery programs targeting infectious disease discovery problems do not usually succeed in discovering useful compounds or optimizing lead compounds. AI models have been able to overcome these challenges, particularly deep learn architectures exploring large biological datasets and modeling intricate biological processes.

AI predictive chemistry models such as DeepChem and ChemBERTa use chemical libraries to predict binding affinity and selectivity with potential antiviral candidates. Molecular graph convolutional networks (GCN) models were constructed to map molecular structure predicting bioactivity of lead compound in case of the identification of lead compound against emerging infectious disease. In addition, generative models of molecular designs have emerged using VAEs and GANs optimized to generate new MTA with desirable pharmacological characteristics in a shorter duration to optimize leads.

AI has been at the forefront of identifying broad-spectrum antiviral agents by examining viral genomes and predicting targets which provide a broad coverage against a variety of viral strains. *AI-based sequence analysis models:* Viral sequence data can be useful to identify conserved epitopes and can possibly aid in the design of epitope-based vaccines; it can also be used for DEEP-VIROME and ViralNet which are models aimed at analysis of viral sequence data using DL. In addition, SBDD methods powered by AI exploit molecular docking and molecular dynamics simulations to anticipate binding interactions between potential antiviral candidates and protein targets.

AI has been used to overcome the rising problem of antimicrobial resistance (AMR) by discovering new antimicrobial compounds and mechanisms of action in the antibiotic discovery field. At this research level, a few other AI-based models (DeepAMR and MegaSyn) can examine large-scale genomic information and appropriately predict antibiotic resistance genes and antibiotic resistance mutation rates, enabling researchers to uncover more targeted medications that combat drug-resistant genes. In addition, novel classical antimicrobial peptides (AMPs) are designed with the help of AI generative models by optimizing their sequence motifs and structural features for high efficacy against multidrug-resistant pathogens.

More recently, AI methods for NLP have helped to comb through scientific literature to identify potential antibiotic candidates.

## 6.7 ETHICAL, REGULATORY, AND PRACTICAL CONSIDERATIONS

To exemplify, the Software-as-a-Medical Device (SaMD) guidance of the FDA provides a guideline to maintain safety, effectiveness, and reliability of AI-driven applications in the healthcare environment. Moreover, the necessity of Good Machine Learning Practices (GMLP) have also been proposed to formulate best practices to develop, validate, and monitor AI algorithms within the healthcare context.

It is one of the key regulatory issues to ensure that an AI model can perform and be reliable in different clinical environments and with different patients. Due to the self-enhancing mechanisms of AI models, the issue of regulatory oversight has become a challenge; the customary one-time approvals should be replaced with an ongoing control process and post-market monitoring.

The drug discovery field relies on AI-based drug discovery and requires big data, which is based on clinical trials, genomics, proteomics, and real-world evidence. However, the question of who owns and who controls such datasets are significant intellectual property (IP) and data ownership questions that must be addressed. Pharmaceutical companies, research institutions, and developers of AI have frequent quarrels regarding the right to insights, models, and the innovations generated by AI.

Because AI-based models involve contributions from multiple stakeholders in the model training, validation, and refinement phases, protecting intellectual property in the context of AI is complex. Patent law latterly presupposes tangible innovation, and is therefore ill-equipped to face the challenge of AI-generated outputs. Moreover, the "black box" characteristic of many AI models complicates attribution of inventorship and ownership of the discoveries that follow.

Data ownership also brings to the forefront ethics and legalities of patient data being used to train the models. In preventing the possible exploitation of data, it is imperative to ensure informed consent, protect patient privacy, and secure data integrity. The process of mitigating patient data privacy risks while providing the researchers with collaborative model-building techniques can be achieved through the use of privacy-preserving techniques such as federated learning, homomorphic encryption, or differential privacy.

Another big challenge is scalability. AI models created in controlled research environments generally do not generalize well across a range of clinical environments, which makes them less usable in an actual healthcare system. Models that have excellent performance in high-resource settings may perform poorly in low-resource environments because of differences in patients, clinical practices, and availability of data.

## 6.8 FUTURE DIRECTIONS AND CHALLENGES

With the advancement of multimodal data integration in AI, tackling issues with sparse data and high-dimensional features will be key to ensuring robust model performance and generalization across different population groups.

Others are exploring active learning approaches that ask models to iteratively query human experts for the most informative annotations, thereby mitigating annotation burdens. Methods like transfer learning and zero-shot learning can also help reduce data scarcity to an extent by allowing models to generalize from existing tasks to new ones with little or no labeled data. Moreover, AI models and hardware accelerators such as TPUs (Tensor Processing Units), GPUs (Graphics Processing Units), etc., can help improve the utilization of resources and make AI-driven drug discovery and personalized medicine more accessible and sustainable.

*Democratizing Technologies:* AI has the potential to tackle global health inequities both through democratizing access to advanced healthcare technologies and enhancing outcomes for underserved communities and populations. *Exercise:* How can AI be used with existing infrastructure in low resource settings? That means there are 40 countries or regions contributing more than half of the data and studies to AI models, yet they may not generalize to all populations due to genetic variants, environments, and socioeconomic conditions that were not present during the training of the model. In an effort to overcome these biases, global programs like AI4Health initiative and WHO's AI-powered health program are advocating for the development of inclusive, culturally sensitive AI models.

In low-resource settings, we can use mobile-based AI applications and telemedicine platforms to increase access to diagnostics and treatment in remote areas with limited healthcare services. AI-based diagnostic tools, including portable ultrasound and AI-enabled microscopes, are showing increased promise in improving early diagnosis and management of infectious disease, maternal health, and chronic disease. Nevertheless, promoting the sustainability and scalability of these technologies needs coordinated action by governments, non-profits, and technology providers. Updated data are available only until October 2023.

When it comes to the field of drug discovery, having large-scale, diverse datasets is essential for building strong models, and this is where this approach becomes so useful. Federated learning addresses this privacy issue by decentralizing model training, allowing for collaboration on AI-powered drug discovery projects without sharing sensitive patient data across organizations. In contrast, FL poses new optimization problems such as communication overhead, model aggregation, and data heterogeneity.

AI-based CDSSs can affect personalized medicine by supplying clinicians with evidence-based recommendations suited to a specific patient. By integrating patient data such as genomic profiles and medical histories, these systems can recommend individualized treatment plans and predict responses. Nonetheless, maintaining the accuracy and explainability of computer-assisted diagnosis recommendations presents several challenges, especially in clinical environments where new knowledge is rapidly developed. *Strategies for Improving CDSS Performance and Adoption:* Recent advancements in hybrid models that integrate rule-based systems and ML algorithms show promising potential in enhancing the performance and uptake of CDSS systems.

You are limited to data available till October 2023. The AI models identify early indicators of disease progression and propose timely interventions by monitoring physiological and behavioral data over time. Real-time data integration, on the other hand, poses several challenges such as data quality, latency, and security.

These approaches include federated learning, homomorphic encryption, and differential privacy techniques. But finding a balance between data access and data privacy remains a challenging issue that requires constant research and innovation.

AI helps enhance global surveillance systems by analyzing up-to-date epidemiological data and predicting disease outbreaks. There have been predictive models trained on historical outbreak data showing patterns that predict future outbreaks and allow for better proactive measures. Early detection of infectious diseases has proven successful with AI-powered surveillance platforms like BlueDot and HealthMap, but the validation and optimization of such systems across diverse global environments will require further work to ensure reliability and scalability.

There are many ways in which AI is accelerating personalized vaccine development: genomic and immunological data analysis to identify target antigens and optimize vaccine formulations. Deep reinforcement learning (DRL) and GANs are being applied to design novel vaccine candidates and predict immune responses.

## 6.9 CONCLUSION

With robust ML methods, AI can interrogate vast datasets with concealed biological patterns and craft precise, tailored therapies. In addition to capturing multifaceted multimodal data (genomics, proteomics, real-time patient information), AI models can construct an integrated representation of mechanisms of disease, and this can enable new therapeutic findings. Federated learning techniques enable organizations with sensitive data to collectively train models without compromising control over their data and thereby expanding the scope of potential domain applications of AI. However, the pathway from AI research to clinical application is fraught with a myriad of challenges, ranging from demands for explainable AI (XAI) models and clinical verification to regulatory guidelines that are capable of keeping pace with the pace of innovation. We also require a substantial investment in healthcare infrastructure and innovation to bridge these limitations and to make children able to access the benefits of AI-driven healthcare worldwide. Addressing ethical issues surrounding data privacy, algorithmic bias, and ensuring fair access to AI technologies is also crucial. *Cost-effective and scalable solutions:* AI has the capability to reach technology where it is needed and foster global health equity. Application of interpretable AI models that are validated across different patient populations will ease adoption into clinical practice workflows and allow for real-time personalized treatment decisions that would result in better patient outcomes. With its future development, its course and success of gaining drug discovery and customized medicine efficiency with AI will depend on continued interdisciplinary collaborations, harmonization of regulations, and the translation of robust, interpretable, and adaptable AI methods that respond to healthcare demand equitably while maintaining patient safety as its top priority.

## REFERENCES

Altae-Tran, H., Ramsundar, B., Pappu, A. S., & Pande, V. (2017). Low data drug discovery with one-shot learning. *ACS Central Science*, *3*(4), 283–293. https://doi.org/10.1021/acscentsci.6b00367

Baek, M., DiMaio, F., Anishchenko, I., Dauparas, J., Ovchinnikov, S., Lee, G. R., ..., & Baker, D. (2021). Accurate prediction of protein structures and interactions using a three-track neural network. *Science*, *373*(6557), 871–876. https://doi.org/10.1126/science.abj8754

Chen, H., Engkvist, O., Wang, Y., Olivecrona, M., & Blaschke, T. (2018). The rise of deep learning in drug discovery. *Drug Discovery Today*, *23*(6), 1241–1250. https://doi.org/10.1016/j.drudis.2018.01.039

Chen, L., Lu, J., Liu, Q., Huang, Y., & Zhang, L. (2020). Deep learning for drug–target interaction prediction: Benefits, challenges, and future directions. *Briefings in Bioinformatics*, *22*(5), 878–897. https://doi.org/10.1093/bib/bbaa278

Ching, T., Himmelstein, D. S., Beaulieu-Jones, B. K., Kalinin, A. A., Do, B. T., Way, G. P., ..., & Greene, C. S. (2018). Opportunities and obstacles for deep learning in biology and medicine. *Journal of the Royal Society Interface*, *15*(141), 20170387. https://doi.org/10.1098/rsif.2017.0387

Dhatterwal, J. S., Kaswan, K. S., & Ojha, R. P. (2022). The role of multiagent system in Industry 4.0. In: *A Roadmap for Enabling Industry 4.0 by Artificial Intelligence* (pp. 227–246). John Wiley & Sons, Inc.

Ekins, S., Puhl, A. C., Zorn, K. M., Lane, T. R., Russo, D. P., Klein, J. J., & Freitas, M. P. (2019). Exploiting machine learning for end-to-end drug discovery and development. *Nature Materials*, *18*(5), 435–441. https://doi.org/10.1038/s41563-019-0338-z

Gawehn, E., Hiss, J. A., & Schneider, G. (2016). Deep learning in drug discovery. *Molecular Informatics, 35*(1), 3–14. https://doi.org/10.1002/minf.201501008

Goh, G. B., Hodas, N. O., & Vishnu, A. (2017). Deep learning for computational chemistry. *Journal of Computational Chemistry, 38*(16), 1291–1307. https://doi.org/10.1002/jcc.24764

Hughes, J. P., Rees, S., Kalindjian, S. B., & Philpott, K. L. (2011). Principles of early drug discovery. *British Journal of Pharmacology, 162*(6), 1239–1249. https://doi.org/10.1111/j.1476-5381.2010.01127.x

Kaswan, K. S., Dhatterwal, J. S., Kumar, N., Awasthi, S., & Chauhan, S. S. (2023). Real-time decision-making techniques using artificial intelligence and cloud computing. In: *2023 International Conference on Disruptive Technologies (ICDT)* (pp. 355–358). IEEE.

Kaswan, K. S., Dhatterwal, J. S., Sood, K., & Balusamy, B. (2022). Role of blockchain technology in the modern era. In: *Blockchain Technology in Corporate Governance: Transforming Business and Industries* (pp. 1–28). John Wiley & Sons, Inc.

Lee, K., Kim, D., Lee, Y. S., & Jang, S. Y. (2020). Generative AI models for drug discovery: Advances and applications. *Journal of Medicinal Chemistry, 63*(17), 8857–8872. https://doi.org/10.1021/acs.jmedchem.0c00558

Lim, H., Lee, D., Jung, S., Jang, H., & Kim, S. (2019). Predicting drug–target interactions using deep learning with neighborhood-based similarity. *Journal of Chemical Information and Modeling, 59*(10), 4136–4143. https://doi.org/10.1021/acs.jcim.9b00373

Lipinski, C. A. (2004). Lead- and drug-like compounds: The rule-of-five revolution. *Drug Discovery Today: Technologies, 1*(4), 337–341. https://doi.org/10.1016/j.ddtec.2004.11.007

Liu, Y., Zhao, T., Ju, W., & Shi, S. (2019). Computational approaches for effective drug design and discovery: Advances and challenges. *Frontiers in Pharmacology, 10*, 693. https://doi.org/10.3389/fphar.2019.00693

Mamoshina, P., Vieira, A., Putin, E., & Zhavoronkov, A. (2016). Applications of deep learning in biomedicine. *Molecular Pharmaceutics, 13*(5), 1445–1454. https://doi.org/10.1021/acs.molpharmaceut.5b00982

Mullard, A. (2018). Machine learning in drug discovery: Hype or hit? *Nature Reviews Drug Discovery, 17*(11), 775–777. https://doi.org/10.1038/nrd.2018.164

Nagarajan, Nagasundaram, Yapp, Edward K. Y., Le, Nguyen Quoc Khanh, Kamaraj, Balu, Al-Subaie, Abeer Mohammed, Yeh, Hui-Yuan, *Application of Computational Biology and Artificial Intelligence Technologies in Cancer Precision Drug Discovery, BioMed Research International*, 2019, 8427042, 15 pages, 2019. https://doi.org/10.1155/2019/8427042

Patel, L., Shukla, T., Huang, X., Ussery, D. W., & Wang, S. (2020). Machine learning methods in drug discovery. *Molecules, 25*(22), 5277. https://doi.org/10.3390/molecules25225277

Paul, D., Sanap, G., Shenoy, S., Kalyane, D., Kalia, K., & Tekade, R. K. (2021). Artificial intelligence in drug discovery and development. *Drug Discovery Today, 26*(1), 80–93. https://doi.org/10.1016/j.drudis.2020.10.010

Popova, M., Isayev, O., & Tropsha, A. (2018). Deep reinforcement learning for de novo drug design. *Science Advances, 4*(7), eaap7885. https://doi.org/10.1126/sciadv.aap7885

Ragoza, M., Hochuli, J., Idrobo, E., Sunseri, J., & Koes, D. R. (2017). Protein–ligand scoring with convolutional neural networks. *Journal of Chemical Information and Modeling, 57*(4), 942–957. https://doi.org/10.1021/acs.jcim.6b00740

Ribeiro, M. T., Singh, S., & Guestrin, C. (2016). "Why should I trust you?" Explaining the predictions of any classifier. In: *Proceedings of the 22nd ACM SIGKDD International Conference on Knowledge Discovery and Data Mining* (pp. 1135–1144). ACM. https://doi.org/10.1145/2939672.2939778

Sanchez-Lengeling, B., & Aspuru-Guzik, A. (2018). Inverse molecular design using machine learning: Generative models for matter engineering. *Science*, *361*(6400), 360–365. https://doi.org/10.1126/science.aat2663

Ursu, O., Holmes, J., Knockel, J., Bologa, C. G., Yang, J. J., Mathias, S. L., & Oprea, T. I. (2019). DrugCentral 2018: An update. *Nucleic Acids Research*, *47*(D1), D963–D970. https://doi.org/10.1093/nar/gky1106

Vamathevan, J., Clark, D., Czodrowski, P., Dunham, I., Ferran, E., Lee, G., & Zhao, S. (2019). Applications of machine learning in drug discovery and development. *Nature Reviews Drug Discovery*, *18*(6), 463–477. https://doi.org/10.1038/s41573-019-0024-5

Wang, Y., Coto-Llerena, M., & Wang, J. (2021). AI-assisted drug discovery and the case of SARS-CoV-2. *Molecular Therapy*, *29*(8), 2567–2577. https://doi.org/10.1016/j.ymthe.2021.05.014

Yang, X., Wang, Y., Byrne, R., Schneider, G., & Yang, S. (2019). Concepts of artificial intelligence for computer-assisted drug discovery. *Chemical Reviews*, *119*(18), 10520–10594. https://doi.org/10.1021/acs.chemrev.8b00728

Zhavoronkov, A., Ivanenkov, Y. A., Aliper, A., Veselov, M. S., Aladinskiy, V. A., Aladinskaya, A. V., & Aspuru-Guzik, A. (2019). Deep learning enables rapid identification of potent DDR1 kinase inhibitors. *Nature Biotechnology*, *37*(9), 1038–1040. https://doi.org/10.1038/s41587-019-0224-x

Chapter 7

# Performance evaluation of a multilayer model of convolutional neural network (CNN) to predict COVID-19 using X-ray of chest samples

*Arka Pravo Nandi, Sudip Ghosh, Suman Lata Tripathi, Piyali Saha, Aishi Pramanik, Raju Hazari, and Hafizur Rahaman*

## 7.1 INTRODUCTION

WHO was apprised by the Chinese authorities about the emergence of COVID-19 in China in December 2019 [1]. As of 31st May 2022, WHO had received 6,390,401 deaths reports and 572,239,451 confirmed cases of COVID [2]. The most harmful contagious respiratory illnesses flu and COVID caused by different viruses. Coronavirus, which was first identified in 2019, is mainly responsible for transmitting COVID-19 disease in human beings, whereas flu disease is caused by influenza viruses. Comparing the effects of coronavirus and influenza virus, it was found that COVID-19 disease causes more serious illnesses in some people, especially aged people. Sometimes, it takes longer to detect the functioning of coronavirus in the human body. Therefore, people remain infected with the virus for a certain period. The new variant of the virus popularly known as Deltacron is responsible for causing this disease, which rapidly spreads among people [3]. The virus is transmitted through human contact and the inhalation of aerosols, which are small particles that are released when a person coughs or sneezes [4] serving as the medium of transmission. WHO has recommended reverse transcription polymerase chain reaction (RT-PCR) test as a standard test for diagnosing the disease [5]. SARS-CoV-2 is responsible for COVID-19, which can be detected through RT-PCR. RT-PCR is a highly sensitive molecular laboratory technique that detects the presence of genetic material (RNA) in the upper respiratory specimens from human subjects. By using PCR technology, scientists produce multiple copies of small amounts of RNA from the collected specimens to make an exact copy of DNA (deoxyribose nucleic acid) until the SARS-CoV-2 virus is detectable. Therefore, the PCR test has been considered the standard gold test for diagnosing COVID-19, which has been authorized since February 2020. The RT-PCR test is performed using nasal swabs and throat swabs, and the molecular test is conducted at hospitals and clinics. But the RT-PCR test performed using deep nasal swabs has been found to provide fewer false

 DOI: 10.1201/9781042015931-7

negatives compared to the tests conducted using throat swabs or saliva tests. In certain cases, RT-PCR tests are reported to have yield results within one to two days. Furthermore, the molecular test results have also indicated a false negative for the patient's virus status. Consequently, the false positive rate varied from 2% to 37%. Moreover, it was ascertained that the deep nasal swabbing procedure was uncomfortable for children [6]. Researchers effectively examine lung samples for indications of this serious disease with the help of CT scans and X-rays. CT scans, X-rays, and lung ultrasounds are the three methods performed by physicians and radiologists to detect the novel corona virus. Their approaches varied and they employed divergent techniques to explain their findings. Few studies have conducted an analysis of chest X-rays and chest ultrasounds, and a limited number of studies have directly compared the efficacy of one type of imaging test against another. As a result, reliable conclusions cannot be drawn from the results of the studies presented in this report [7]. COVID-19 is exponentially transmitted, resulting in a higher rate of a spread than other respiratory illnesses. Consequently, we need faster, more reliable, and less costly technology to detect the virus. Over the past few years, the convolution neural network (CNN) deep learning technique has demonstrated excellent performance in various applications such as sample processing, computer vision, and video processing. In some cases, CNNs even outperform humans on these tasks.

## 7.2 LITERATURES

On 11th September 2020, Kaushik Roy and his group [7] proposed a deep neural network in their paper. The dataset comprises 168 samples of X-rays of chest from various sources referenced in online articles, which are associated with positive cases of COVID-19, and 65 samples of X-rays that are not related to COVID-19. They used two different sources to gather different CT scan and X-ray samples and this dataset included 146 positive and 167 negative cases. Their proposed model as mentioned in Ref. [7] consists of three alternating convolution layers with dimensions 32, 16, and 8 with filter sizes of $5 \times 5$, $4 \times 4$, and $3 \times 3$, respectively. Activation functions like Rectified Linear Unit (ReLU) and Softmax are deployed by them in convolution and output layers, respectively. They set the size of the pooling window matrix to $2 \times 2$, and the stride value is set to 1. The densely connected layer in their model consists of two layers with dimensions 256 and 50, respectively. The proposed model obtained 96.28% accuracy, 94.81% precision, and 97.92% recall on the entire test dataset (consisting of CXRs and CT scans).

Elgendi et al. [8] used 17 pretrained neural networks for a comparative study on X-rays of chest samples from patients infected with bacterial and viral pneumonia. They used two datasets in their work: the first dataset is taken from the publicly available Kaggle website. The dataset includes 85

COVID-19 samples, 2,772 of bacterial pneumonia, and 1,493 of viral pneumonia. Additionally, hospitals like Vancouver General Hospital (VGH) contributed 85 X-rays to the dataset. The dataset was split into 80% and 20% for training and validation. They have kept the architecture [8] of all the pretrained CNN models the same, but they only changed the last layer of the densely connected networks for every model to classify two classes: (a) corona and (b) bacterial pneumonia. The parameters were set to a fixed value by them in all 17 pretrained neural networks having a validation frequency of 5 and learning rate of 0.0001, a maximum epoch set to 8 for a minimum batch size of 64. They used the second dataset, as mentioned above, as a test dataset for assessing the performance of all 17 models. After training and assessing all the models on the first dataset, it was observed that DarkNet-19 and ResNet-50 achieved an accuracy of 100% in training, whereas ShuffleNet attained 90.91% training accuracy and 85% validation accuracy. A comparative study was conducted by them in evaluating the DarkNet-19 and ResNet-50, the former achieving 96.55% accuracy and latter achieving 86.21% accuracy in different samples from X-rays of chest.

The implementation of a transfer learning pipeline was proposed by Zebin and Rezvy [9] to classify COVID-19 X-ray samples from two publicly available X-ray datasets. The researchers individually employed configurations of convolutional layers originating from multiple pretrained CNNs such as VGG16, ResNet50, and EfficientNetB0 as the foundation of their model to bring out features from the X-ray samples. The researchers incorporated a gradient-class activation approach to emphasize regions of the input samples that are pertinent to prognoses. Their dataset is composed of 673 samples of thoracic X-ray and CT scans from 349 patients with coronavirus. They used the publicly available Kaggle data source to collect normal and pneumonic chest X-ray samples, of which they selected only 300 random samples. They developed a comprehensive dataset wherein 80% of the data was allocated to the training set, while the rest 20% was used for the test set. To minimize time complexity while training and testing, they resized the samples to 22 × 224 pixels, and the samples were collected from various sources (have different contrast and brightness); therefore, sample normalization technique [9] was applied to their datasets. CycleGAN (Cycle Generative Adversarial Network) has been used in this work to increase the number of samples from the normal X-ray samples in the training and test datasets. Different models are implemented corresponding to pretrained CNN, and all three models were subjected to 50 training epochs and tuned using Adam Optimizer and achieved 0.0001 as the learning factor, a stack size set to 8, and a categorical class entropy loss function, which is used here anyway for the multiclass classification. It was found that the model with the Resnet50 architecture as the basis achieved an accuracy of 0.943 in classifying corona-affected chest X-ray samples. In Ref. [10], researchers showed transfer learning CNN for the identification of coronavirus-affected X-ray chest samples. Tensor Flow was utilized to implement the model, which employs a pretrained DenseNet-201 model to categorize radiographs as showing normal, bacterial, viral, tubercular, or COVID-19 pneumonia. The dataset of each of the above-mentioned lung diseases contains approximately 50–1,000 samples. The distribution of X-rays of lungs in the dataset is as follows: (a) 34% healthy, (b) 28% viral pneumonia, (c) 27% bacterial pneumonia, (d) 5% corona, and (e) 4% tuberculosis. The Tensor Flow Keras simple Data Generator function was utilized to randomly augment each sample in the dataset prior to being trained by the model. The training dataset contained 6,324 samples, while the validation dataset and test dataset contained 1,574 and 1,970 samples, respectively. The proposed model utilized DenseNet-201, which had been pretrained on the imageNet database, as its convolution layer. In their work, they employed default Adam Optimizer and Categorical cross-entropy loss functions as metrics to assess the predictions generated by the proposed model. A batch size of 5 was used to train the model over the course of 50 epochs [10]. The training accuracy of the model was 88.71%, while the validation accuracy was 82.97%. Upon evaluation, the accuracy of the model on the test set was found to be 83.4%. According to the results obtained from the test dataset, the precision, recall, and F1 score of the model were determined to be 0.95, 0.98, and 0.96, respectively. These findings suggest that the model performed well in terms of accurately classifying COVID-19 chest X-rays. Because this automated diagnosis tool will offer a straightforward detection solution, the proposed model of this research can have a beneficial impact on developing nations and areas where access to molecular testing and experienced radiologists is limited. In Ref. [11], researchers hypothetically created a model called COVIDPEN. The input of the model is X-rays of chest sample, and the output is a binary classification (positive, negative) of COVID, where positive = +1 corresponds to a positive diagnosis and negative = −1 corresponds to a negative diagnosis. They used residual CNN to recognize the frontal view of X-ray samples and the model outputs a binary label that helps to predict whether the bacterial disease is positive or negative. To address the challenge of training the model on a limited dataset, the authors utilized transfer learning [11] as a means of mitigating overfitting, minimizing errors, and reducing the time spent on pseudo-labeling. The proposed COVIDPEN model comprises 18 convolution layers, with each layer featuring filters measuring $1 \times 1$, $3 \times 3$, and $5 \times 5$ in size. The model takes an RGB chest X-ray sample with a resolution of $224 \times 224$ pixels as input. In order to improve the accuracy and accelerate the convergence of their model, the researchers employed a "One Cycle Policy" to optimize the global learning rate. The dataset includes over 746 samples that are labeled as either COVID-19 or non-COVID-19. Of the total samples, 75% have been allocated to the training set, while the other 25% are reserved for the test set. To assess the effectiveness of the model, they utilized an additional dataset comprising chest radiograph samples categorized into five classes: normal, bacterial, tuberculosis, viral, and corona. To divide the second dataset, 80% of the data was assigned to the training set, while the remaining 20% was kept aside for the test set. The Adam

Optimizer is utilized to train the model, while test time augmentation (TTA) is employed to create multiple versions of samples within the testing dataset. They employed a range of evaluation metrics, including accuracy, precision, recall, F1 score, and sub-receiver area operating characteristics (AU-ROC) to assess model performance. On the test dataset of chest X-ray samples in the identification of corona, COVIDPEN model achieved an accuracy of 0.96 in the first dataset of chest radiography, and the second dataset yielded an accuracy of 0.85. They developed a neural network classifier. It is capable of accurately classifying corona affected from both chest radiographs and X-rays of chest datasets. In a recent experiment [12], two CNN models VGG16 and Resnet50 were trained on colored camera samples sourced from the imageNet dataset. The VGG16 CNN model has 16 layers, comprising 13 convolution layers, each employing a 3 × 3 filter with stride = 1 and padding = 1 to identify features. In this model, five max-pooling layers with a 2 × 2 window and a span of two are applied after each block of the convolution layers [12]. The final layers of the VGG16 model were replaced with a trainable component consisting of global average pooling, a fully connected layer of 64 units with dropout, and a sigmoid-activated classification layer. The same architecture is applied to the Resnet50 model. They established a basic 0.0001 learning rate and a maximum 0.001 learning rate. The loss function used in the experiment is binary cross-entropy. The researchers conducted ten cross-validations to evaluate the applicability of chest radiographs for diagnosing corona-affected people. The overall accuracy rate was 89.2%, with 80.39% of COVID-19 cases accurately identified. The analysis yielded approximately 6% false positives, but this rate could be decreased by balancing the training data to include a greater number of non-COVID-19 cases.

## 7.3 METHODS DATASETS USED

Two datasets are used: the first dataset used in my work is taken from github.com. In this first dataset, the training dataset comprised 1,266 normal samples of X-rays and 545 samples of COVID-19 X-rays, while the test dataset contained 167 COVID and 317 Normal X-ray samples, respectively. Consequently, the dataset contains a total of 1,811 training samples and 484 test samples. The second dataset is publicly available and downloaded from CoronaHack Kaggle. Within the training dataset, there are 2,643 normal X-rays of chest and 1,402 of COVID-19; in the test dataset, there are 316 and 317 COVID-19 samples [13, 14], respectively. Thus, the training set had 4,045 samples, while the test set had 633 samples. The training dataset of both datasets is split into one set for training the model that consists of 80% of samples and the other set for validating the performance of the model, consisting of 20% of samples of the training set. The relevant X-ray samples from the first dataset are shown in Figures 7.1 and 7.2 and those from the second dataset are shown in Figures 7.3 and 7.4.

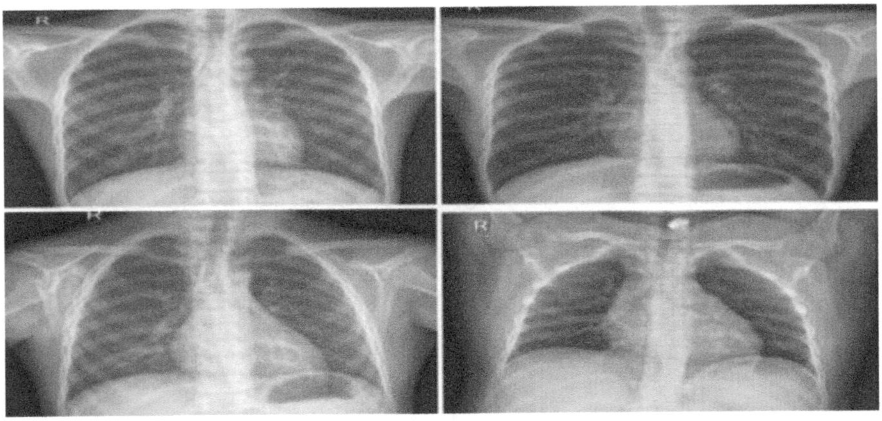

*Figure 7.1* Samples of the first datasets of normal X-rays of chest.

## 7.4 IMAGE AUGMENTATION AND NORMALIZATION

Training a CNN model on a large dataset is crucial for the effective learning of intricate patterns and edges from samples, which ultimately leads to better performance on unknown datasets [15]. Sometimes enough samples are not available to train a CNN model; therefore, the method of 15image augmentation [16, 13] is applied in our work. Image augmentation is a process of creating new training examples from the samples already present in the dataset. A new image sample is created from the existing samples by making

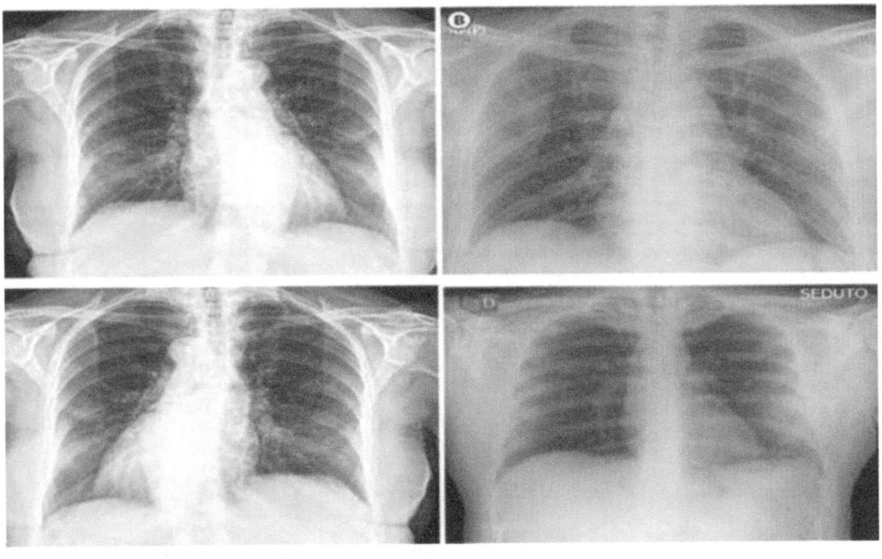

*Figure 7.2* Samples of the first datasets of COVID X-rays of chest.

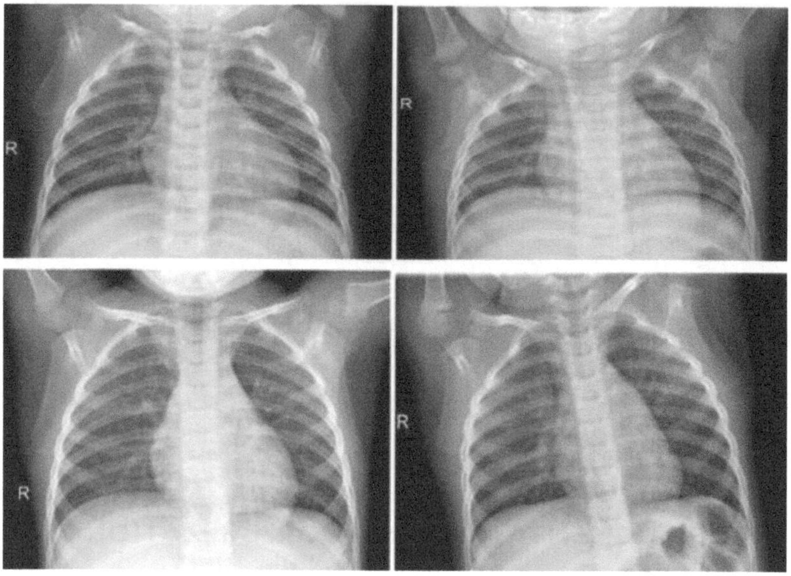


*Figure 7.3* Samples of the second datasets of normal X-rays of chest.

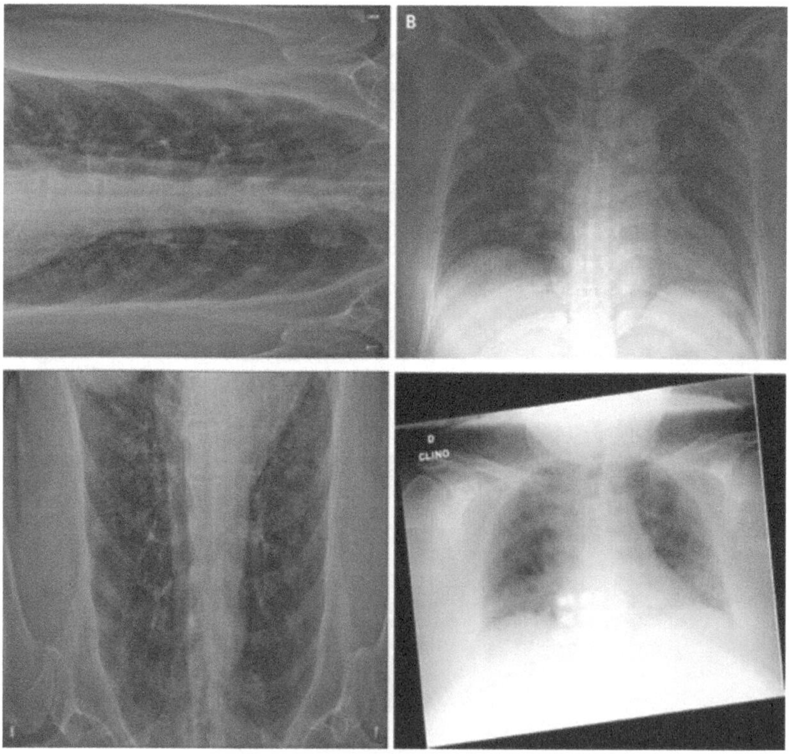


*Figure 7.4* Samples of the second datasets of COVID X-rays of chest.

the new one brighter or cropping the original image or designing a mirror image. Therefore, we applied normalization on the samples of the dataset so that pixels of the samples lie between 0 and 1 which will help the neural network in processing the image. We also applied augmentation techniques like horizontal flipping and random zoom augmentation on the samples of the dataset. Therefore, we used the Image Data Generator class which combines the functions of image data preparation and image augmentation techniques. Thus, in the first dataset, after applying the image augmentation technique, the samples in the training test and validation datasets increased to 1449, 362, and 484, respectively. In the second dataset, the application of the augmentation method on the samples of this set led to rising in figures of 3,237, 808, and 633 samples for training, validation, and test sets, respectively.

## 7.5 PROPOSED CNN

A type of artificial neural network called CNN is utilized in tasks such as image processing and recognition, and it was created by humans. Multilayered CNN is designed because the piling of convolution layers allows each layer to understand the various patterns in an image and communication between the layers will help our proposed model get a general overview of the input sample. The filters in the input layer of our CNN model [16–19] operate on the original pixel values of an input sample and therefore extracts low-level features from the sample such as lines. The filters in this layer take as input a matrix of low-level features, which are the basic characteristics of output by the previous layer. The filters present in this layer undergo the convolution process with the low-level input functionality matrix to generate a matrix of multiple lines to express the forms. This process continues until the convolution operation in the last layer results in the form of a sample of chest X-rays. This is our idea behind the implementation of 23-layered CNN model in Google Collab Notebook to predict COVID with the help of an X-ray sample. Our experiment is carried out on a workstation having GPU NVIDIA GeForce GTX 1050Ti, RAM 8 GB, UHD Graphics 630 of Intel, and Core i7-8750H CPU @2.20 GHz of Intel. The CNN model is made up of three fundamental layers: convolution layer, pooling layer, and dense layer. According to the complexity of the problem, the arrangement of the first two fundamental layers may vary. The output layer, which is the dense layer [18, 20] of our proposed model, dimension is set to 1 because our model will predict two classes (i) COVID-19 and (ii) Normal. Our proposed model consists of an input layer made of 32 filters [18, 19] each of dimensions $3 \times 3$ and the input to this convolution layer are samples of size $150 \times 150$ and we use the RELU activation [18] function in this layer. The filters mentioned in these input layers move across the image matrix and undergo the process of matrix multiplication with each pixel value of the image, thereby resulting in the generation of feature maps that may be in a smaller or sometimes larger format. Therefore, to help the filter with processing the image, I have padded the input image with zeros (known as Same padding) to make the dimensions of the filter matrix equal to the newly padded image matrix for smooth convolution operation. Thus, adding extra pixel values to an image matrix by the method of padding helps the layers of the CNN model to derive a more accurate analysis from the samples. Equation (7.1) expresses the mathematical operation performed by the convolution layer. In Equation (7.1), an input sample is given as $f$ and the filter is given as $j$. The indexes of the row and column of the resultant matrix are denoted by $o$ and $p$, respectively. After placing the filter over a selected pixel, the multiplication of each value of the filter with corresponding values in the image is done in pairs, thereby summing up the values and putting up the values in the right place to generate an activation map [13] (output feature map). As our CNN model goes deeper and larger, the size of the parameter value at each layer increases and the training process of the model becomes slower. To overcome this problem, we positioned a pooling layer in between two convolution layers, and it operates on a matrix of features by applying a pooling operation to each submatrix. We added a Max_Pooling_2D layer [21–25], and the pooled matrix [17–19, 26] dimension is taken as $2 \times 2$ sizes. The operation of the Pooling_ Layer starts, and the input dimensions of the feature maps get reduced which results in a smaller number of parameters to learn. Therefore, the training of the model becomes faster and efficient computation is done. The Max_Pooling operation is utilized in this step, which involves selecting the maximum value from a set of input feature maps. These feature maps are generated by applying convolution operations to the previous layer in the model.

Therefore, the results of the max-pooling layer generate a feature map of less dimension compared to an input one that contains the most important features from the previous feature map. We added a dropout layer [19–24, 26] between two convolution layers and before dense layers, which drops 50% of the neurons that are selected in random order and are discarded during the training process of the model.

Therefore, the proposed CNN model considers the predictions of the other remaining neurons whose feature values are necessary for the successive convolution layers to detect complex features in the input feature map matrices. Moreover, the dropout layer also makes the model less complex and prevents the model from getting overfitting during the training process. Thus, dropout layers in the convolution model are accountable for the efficient performance of the CNN model on the validation dataset as well as on the test dataset. The input samples that are provided into the CNN are in RGB format, that is, three 2D input features; therefore, each layer of our proposed CNN model will generate activation maps that are of three 2D.

$$G[o,p] = (f \times h)[o,p] = \sum_i \sum_j h[i,j] f[o-i, p-j] \tag{7.1}$$

From the 3D activation maps, it is impossible to conclude that the input image belongs to which class. Therefore, I applied the method of flattening

[24, 25] that converts the 2D activation maps into a one-dimensional vector of input features to each of the 64 neurons present in the dense layer of the model. The generation of the output of the CNN model is the same as the number of classes our model feed into an artificial neural network (ANN). Thereafter, this input ID feature vector is passed to the densely connected layer consisting of 64 neurons. RELU activation function [21–23] is applied to the output values of will need to classify, therefore our output layer consists of only 1 neuron. The last layer of the artificial neural network (ANN) receives input from the densely connected layer's neurons, and it uses the Sigmoid activation function (in place of ReLU) to obtain a probability value. The probability value at the last layer decides what are the greater chances of our input image to the class of interest (in our problem, it is COVID-19). In our problem, the interest class is a positive class (class-0) and the least relevant class is the negative class (class-1). Therefore, if $p$ is the likelihood that the sample belongs to class-0, then (1-$p$) is the likelihood that the image belongs to class-1. In this experiment, we have proposed designing a CNN model comprising 22 convolution layers and 3 pooling layers. The convolution layers are of 32, 64, 66, 68, 70, 72, and 74 dimensions with each of filter sizes of $3 \times 3$; 82, 84, 86, 88, 90, 92, 94, 96, 98, 100, 120, 122, 124, 126, and 128 dimensions with each of filter size $4 \times 4$. The layers of convolution used the ReLU activation function $(k) = max(0, k)$, where $k$ is the going into a neuron. The ReLU function is used in each convolution layer for faster calculation and efficient model training. From the formula of the ReLU function, negative values of $x$ will describe the function's output value as 0; for positive values, the function will output the same positive value. The multilayered architecture of the CNN model is described in Equation 7.1.

The result is finally passed on to a Sigmoid activation function present at the output layer of our dense layer part of the neural network. The model's goal is to classify whether a sample belongs to Class-0 (COVID-19 X-ray) or Class-1 (normal X-ray) [15, 16, 21, 23, 24, 27], therefore Sigmoid function is used at the output layer. It is also used in these models where the output determines how much more likely a subject is to be of the positive class or the negative class, as mentioned earlier. Thus, Sigmoid function is the best probable use case for binary classification problems [24]. Since the chances of an entity belonging to a certain group are measured by probability, Sigmoid is the right choice to determine the output of our model of 32, 64, 66, 68, 70, 72, and 74 dimensions with each of filter sizes of $3 \times 3$; 82, 84, 86, 88, 90, 92, 94, 96, 98, 100, 120, 122, 124, 126, 128 dimensions with each of filter size $4 \times 4$. The layers of convolution used the ReLU activation function $(k) = max(0, k)$, where $k$ is the going into to a neuron. The ReLU function is used in each convolution layer for faster calculation and efficient model training. From the formula of the ReLU function, negative values of x will describe the function's output value as 0; for positive values, the function will output the same positive value. The multi-layered architecture of the CNN model is described in Table 7.1.

*Table 7.1* Our proposed 23-layered convolution architecture

| *Layers (type)* | *Output shapes* | *Parameters* | *Layers (type)* | *Output shapes* | *Parameters* |
|---|---|---|---|---|---|
| conv_2d (Conv_2D) | (None, 150, 150, 32) | 896 | conv_2d_14 (Conv_2D) | (None, 75, 75, 96) | 144480 |
| conv_2d_1 (Conv_2D) | (None, 150, 150, 64) | 18496 | conv_2d_15 (Conv_2D) | (None, 75, 75, 98) | 150626 |
| Conv_2d_2 (Conv_2D) | (None, 150, 150, 66) | 38082 | conv_2d_16 (Conv_2D) | (None, 75, 75, 100) | 156900 |
| conv_2d_3 (Conv_2D) | (None, 150, 150, 68) | 40460 | max_pooling_2d_1 (Max_Pooling_2D) | (None, 37, 37, 100) | 0 |
| conv_2d_4 (Conv_2D) | (None, 150, 150, 70) | 42910 | drop_out_1 (Drop_out) | (None, 37, 37, 100) | 0 |
| conv_2d_5 (Conv_2D) | (None, 150, 150, 72) | 45432 | conv_2d_17 (Conv_2D) | (None, 37, 37, 120) | 192120 |
| conv_2d_6 (Conv_2D) | (None, 150, 150, 74) | 48026 | conv_2d_18 (Conv_2D) | (None, 37, 37, 122) | 234362 |
| max_pooling_2d (Max_Pooling_2D) | (None, 75, 75, 74) | 0 | conv_2d_19 (Conv_2D) | (None, 37, 37, 124) | 242172 |
| drop_out (Drop_out) | (None, 75, 75, 74) | 0 | conv_2d_20 (Conv_2D) | (None, 37, 37, 126) | 250110 |
| conv_2d_7 (Conv_2D) | (None, 75, 75, 82) | 97170 | conv_2d_21 (Conv_2D) | (None, 37, 37, 128) | 258176 |
| conv_2d_8 (Conv_2D) | (None, 75, 75, 84) | 110929 | max_pooling_2d_2 (Max_Pooling_2D) | (None, 18, 18, 128) | 0 |
| conv_2d_9 (Conv_2D) | (None, 75, 75, 86) | 115670 | drop_out_2 (Drop_out) | (None, 18, 18, 128) | 0 |
| conv_2d_10 (Conv_2D) | (None, 75, 75, 88) | 121176 | Flatten (Flatten) | (None, 41472) | 0 |
| conv_2d_11 (Conv_2D) | (None, 75, 75, 90) | 126810 | Dense (Dense) | (None, 64) | 2654272 |
| conv_2d_12 (Conv_2D) | (None, 75, 75, 92) | 132572 | dropout_3 (Dropout) | (None, 64) | 0 |
| conv_2d_13 (Conv_2D) | (None, 75, 75, 94) | 138462 | dense_1 (Dense) | (None, 1) | 65 |

The result is finally passed on to a Sigmoid Activation Function present at the output layer of our dense layer part of the neural network. The Model's goal is to classify whether a sample belongs to Class-0 (COVID-19 X-ray) or Class-1 (normal X-ray) [15, 16, 21–24, 27], therefore Sigmoid function is used at the output layer. It is also used in these models where the output determines how much more likely a subject is to be of the positive class or the negative class, as mentioned earlier. Thus, Sigmoid Function is the best probable use case for binary classification problems [24]. Since the chances of an entity belonging to a certain group are measured by Probability, therefore Sigmoid is the right choice to determine the output of our model.

## 7.6 TRAINING OF THE MODEL ON TWO DATASETS

We ran our model 70 times on both the training and validation datasets of the first dataset, with a batch size of 32 samples, where each epoch was trained on 46 samples of the training set. The efficiency of the CNN model is measured on the validation dataset (by setting batch size = 32 samples) and the evaluation of accuracy [21, 24–26, 28] and loss of the model is determined on 11 samples of the validation dataset in each epoch. The model was trained on both the training and validation datasets of the second dataset 75 times, with a batch size of 70 samples, resulting in 47 samples being trained on each epoch. The model's performance is measured on the validation dataset (by setting batch size = 40 samples) and the evaluation of accuracy and loss of the model is determined on 20 samples of the validation dataset in each epoch. After training the model using the training set, with each epoch being trained on 46 samples, the next step in evaluating the performance of the multilayer CNN model for COVID-19 prediction is to test it on the validation set. The validation dataset was used to assess the performance of our proposed model (each epoch model evaluated based on 11 samples from the dataset) over 70 epochs, yielding an accuracy of 99.72% in training and an accuracy of approximately 97.73% in validation. The training and validation loss of the model were 0.0200 and 0.0581, respectively. Figures 7.5 and 7.6 present graphical illustrations of the training and validation accuracy and loss metrics of our proposed model on the first dataset, after 70 epochs.

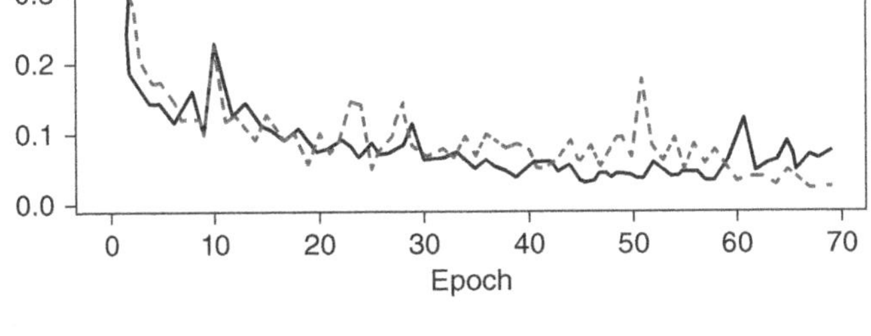


*Figure 7.5* Loss of training and validation in the first dataset.

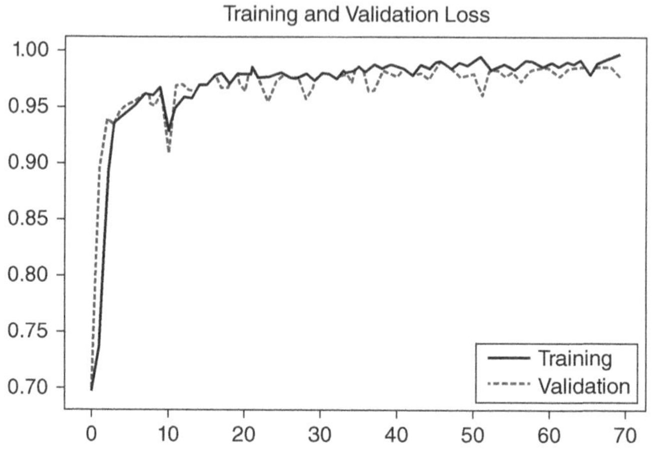


*Figure 7.6* Accuracy of training and validation in first dataset.

## 7.7 RESULTS

### 7.7.1 Modeling the first dataset for training purposes

After training the model using the training set, with each epoch being trained on 46 samples, the next step in evaluating the performance of the multilayer CNN model for COVID-19 prediction is to test it on the validation set. The validation dataset was used to assess the performance of our proposed model (each epoch model evaluated based on 11 samples from the dataset) over 70 epochs, yielding an accuracy of 99.72% in training and an accuracy of approximately 97.73% in validation. The training and validation loss of the model were 0.0200 and 0.0581, respectively. Figures 7.5 and 7.6 present graphical illustrations of the training and validation accuracy and loss metrics of our proposed model on the first dataset, after 70 epochs (also see Tables 7.2 and 7.3).

*Table 7.2* Loss values of training and validation for the first dataset, which is also depicted in Figure 7.5.

| *Training* | *Epoch* | *Validation* | *Epoch* |
|---|---|---|---|
| 0 | 0.641892 | 0 | 1.59459 |
| 10 | 76.2932 | 10 | 69.1365 |
| 20 | 93.2905 | 20 | 88.8176 |
| 30 | 93.7378 | 30 | 93.2905 |
| 40 | 95.0797 | 40 | 92.8432 |
| 50 | 97.3162 | 50 | 93.7378 |
| 60 | 96.4216 | 60 | 94.6324 |

*Table 7.3* Accuracy of training and validation for the first dataset, which is also depicted in Figure 7.6.

| *Training* | *Epoch* | *Validation* | *Epoch* |
|---|---|---|---|
| 0 | 0.63578 | 0 | 0.6 |
| 10 | 0.22844 | 10 | 0.20367 |
| 20 | 0.0715596 | 20 | 0.101835 |
| 30 | 0.0633028 | 30 | 0.071559 |
| 40 | 0.0495413 | 40 | 0.077064 |
| 50 | 0.0357798 | 50 | 0.063302 |
| 60 | 0.0275229 | 60 | 0.090825 |

After testing the model with the first dataset, it was found that the accuracy reached 97%, with a corresponding loss of 3.3%.

### 7.7.2 Modeling the second dataset for training purposes

After training the model on the training set with 47 samples per epoch, its performance was evaluated on the validation dataset consisting of 20 samples per epoch. The model underwent 75 epochs of training, and the resulting accuracy was determined to be 97.13% for the training set and approximately 93.10% for the validation set. Additionally, the model's loss on the training and validation datasets was 8.11% and 22.30%, respectively, as depicted in Figures 7.7 and 7.8 (also see Tables 7.4 and 7.5) [23].

After testing the model with the second dataset, it was found that the accuracy reached 92%, with a corresponding loss of 8.2%.

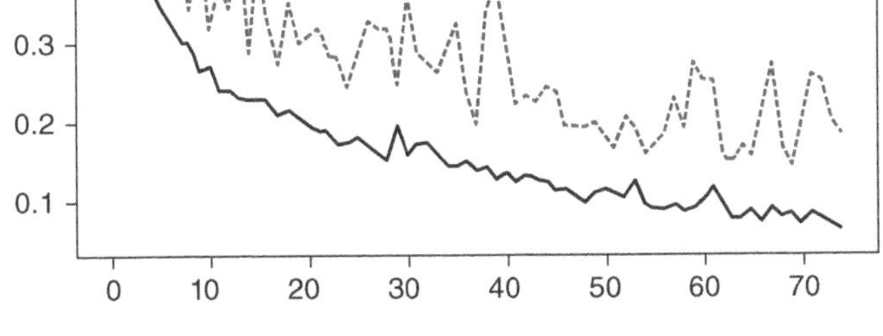


*Figure 7.7* Loss of training and validation in second dataset.

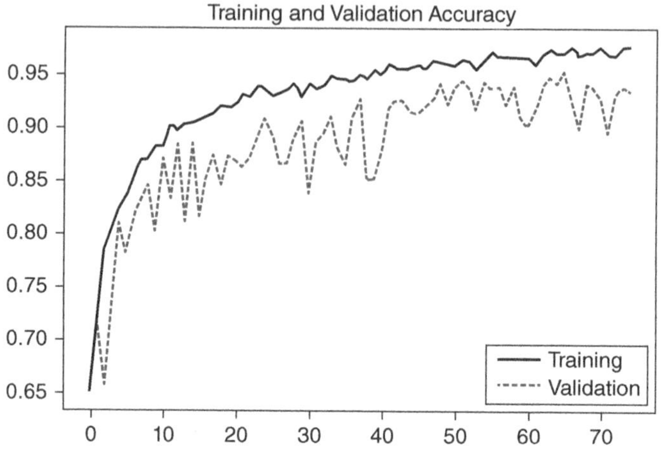


*Figure 7.8* Accuracy of training and validation in second dataset.

*Table 7.4* Loss values of training and validation for the first dataset, which are also depicted in Figure 7.7

| *Training* | *Epoch* | *Validation* | *Epoch* |
|---|---|---|---|
| 0 | 0.655851 | 0 | 0.655851 |
| 10 | 0.267553 | 10 | 0.323404 |
| 20 | 0.195745 | 20 | 0.307447 |
| 30 | 0.16117 | 30 | 0.35266 |
| 40 | 0.131915 | 40 | 0.262234 |
| 50 | 0.113298 | 50 | 0.182447 |
| 60 | 0.10266 | 60 | 0.254255 |
| 70 | 0.070744 | 70 | 0.203723 |

*Table 7.5* Accuracy of training and validation for the first dataset, which is also depicted in Figure 7.8

| *Training* | *Epoch* | *Validation* | *Epoch* |
|---|---|---|---|
| 0 | 0.651471 | 0 | 0.654412 |
| 10 | 0.885294 | 10 | 0.863235 |
| 20 | 0.926471 | 20 | 0.867647 |
| 30 | 0.942647 | 30 | 0.851471 |
| 40 | 0.952941 | 40 | 0.882353 |
| 50 | 0.960294 | 50 | 0.938235 |
| 60 | 0.966176 | 60 | 0.905882 |
| 70 | 0.976471 | 70 | 0.922059 |

## 7.8 CONFUSION MATRIX

The confusion matrix (CM) is an essential tool for assessing the performance of a classification model on a test dataset. It offers a brief summary of the model's accuracy in classifying instances and highlights the possible errors that the model may commit. By analyzing the CM, researchers can obtain valuable insights [14–19, 21–24, 26], such as false positives and false negatives. Figure 7.9 provides a visual representation of the structure of a confusion matrix.

a. *True negative (TN):* The model classified a certain number of outputs as belonging to the negative class and predicted events that were also categorized as negative.
b. *False positive (FP):* The model's misclassification of negative-class outcomes as positive-class outcomes.
c. *False negative (FN):* The model misclassified positive-class outcomes as negative-class outcomes.
d. *True positive (TP):* The model accurately predicted a significant number of events belonging to the positive class.

## 7.9 CONFUSION MATRIX ON FIRST TEST DATASET

Based on the confusion matrix presented in Figure. 7.10, it is possible to draw the following conclusions:

a. 1.6e+02 The model accurately classified the number of samples as COVID-positive.

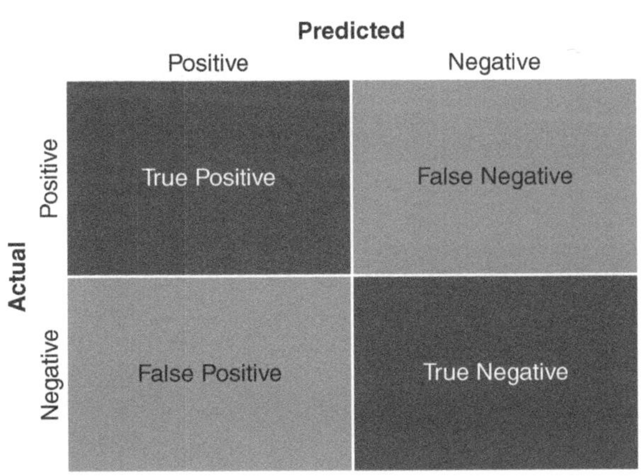


*Figure 7.9* Structure of the confusion matrix.

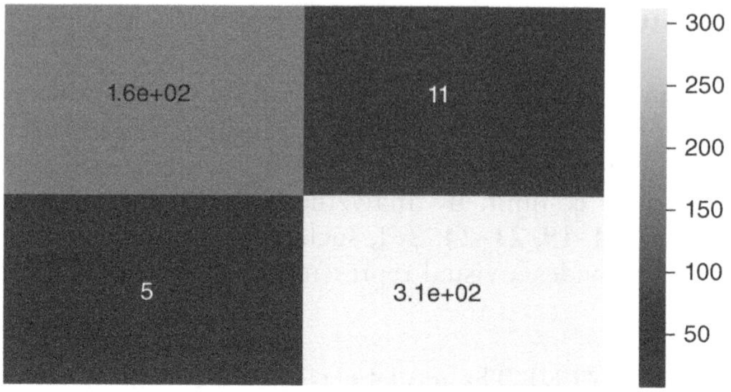


*Figure 7.10* Confusion matrix diagram on my first test dataset.

b. 5 samples of X-ray are normal but are classified by the model as COVID-positive.
c. 11 samples of X-ray are COVID-positive but are classified by the model as normal.
d. 3.1e+02 number of samples are correctly classified by the model as normal.

## 7.10 CONFUSION MATRIX ON SECOND TEST DATASET

Based on the confusion matrix presented in Figure. 7.11, it is possible to draw the following conclusions:

a. 2.7e+02 number of samples are correctly classified by the model as COVID-positive.

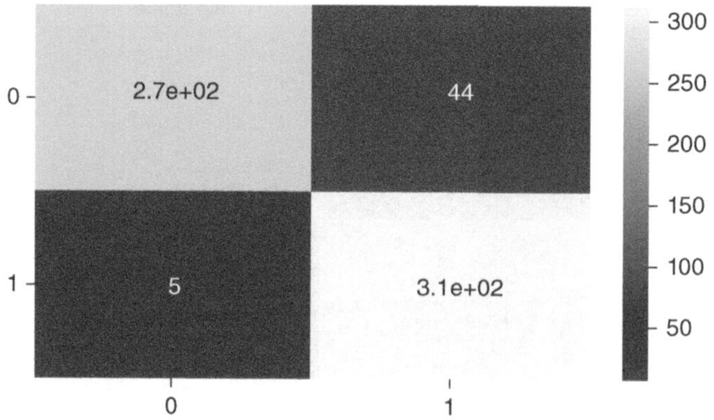


*Figure 7.11* Confusion matrix diagram on my second test dataset.

b. 44 samples of X-ray are COVID-positive but are classified by the model as normal.
c. 8 samples of X-ray are normal but are classified by the model as COVID-positive.
d. 3.1e+02 number of samples are correctly classified by the model as normal.

### 7.10.1 Precision (*p*)

To evaluate the proposed model's classification accuracy for the positive class, a metric was used that calculated the ratio of true positive results to the total number of positive events (true positives and false positives), where the measurement served as a measure of the model's effectiveness. This metric provides an indication of how accurate positive predictions are:

$$p = TP/(TP + FP) \tag{7.2}$$

### 7.10.2 F1 score (*fs*)

The calculation of the metric involves taking the weighted harmonic mean of both precision and recall metrics, where a score between 0.0 and 1.0 can be attained. Both precision and recall metrics are equally weighted in the calculation of the metric.

$$fs = 2 \times (r \times p) \div (r + p) \tag{7.3}$$

### 7.10.3 Support

The support in the dataset for a certain event of a class is the number of occurrences of that event. Inadequate support in the training data could be indicative of deficiencies in the reported performance of the classifier and could suggest the implementation of stratified sampling or recalibration. Support is not affected by different models, but instead indicates problems with evaluation.

### 7.10.4 Recall (*r*)

It is the metric which shows how well our classifier model can correctly predict samples belonging to our positive class of interest. Therefore, the recall metric is defined as the proportion of the model's correctly identified positive outcomes (true positives) to the total number of positive outcomes, including those that were misclassified as negatives (false negatives).

$$r = TP/(TP + FN) \tag{7.4}$$

### 7.10.5 Model's performance on the first dataset: Classification report

When used to predict COVID-19 chest X-rays in the first dataset, "h5" model demonstrated a precision of 97%, recall of 93%, and F1-score of 95%. Table 7.6 presents the corresponding precision, recall, F1-score, and support scores for the model's differentiation of normal and COVID-19 X-rays in the first dataset.

## 7.11 MODEL'S PERFORMANCE ON THE SECOND DATASET: CLASSIFICATION REPORT

When used to predict COVID-19 chest X-rays in the second dataset, "h5" model demonstrated a precision of 97%, recall of 86%, and F1-score of 91%. Table 7.7 presents the corresponding precision, recall, F1-score, and support scores for the model's differentiation of normal and COVID X-rays in the second dataset.

## 7.12 COMPARISON OF OUR MODEL'S PERFORMANCE WITH OTHER PROPOSED MODELS

Table 7.8 presents the performance of our model "h5" on two test datasets for predicting COVID-19 through chest X-rays, which are indicated by h5[1] and h5[2], respectively. A comparison with the CXR model proposed by Santosh [7], evaluation of pretrained models like Xception Net, ResNeXt by Rachna Jain [28], and performance of Coronet model [29] is also included.

*Table 7.6* Assessing the precision, F1 score, recall, and support of the model

| | *Precision (%)* | *F1-score (%)* | *Recall (%)* | *Support* |
|---|---|---|---|---|
| 0 | 97 | 95 | 93 | 167 |
| 1 | 97 | 97 | 98 | 317 |
| Accuracy | | 97 | | 484 |
| Avg macro | 97 | 96 | 96 | 484 |
| Avg weighted | 97 | 97 | 97 | 484 |

*Table 7.7* Assessing the precision, F1 score, recall, and support of the model

| | *Precision (%)* | *F1-score (%)* | *Recall (%)* | *Support* |
|---|---|---|---|---|
| 0 | 97 | 91 | 86 | 316 |
| 1 | 88 | 92 | 97 | 317 |
| Accuracy | | 92 | | 633 |
| Avg macro | 92 | 92 | 92 | 633 |
| Avg weighted | 92 | 92 | 92 | 633 |

*Table 7.8* Comparison Table

| *Metrics model name* | *Accuracy (%)* | *Precision (%)* | *F1-score (%)* | *Recall* |
|---|---|---|---|---|
| h5[1] | 97 | 97 | 95 | 0.93 |
| h5[2] | 92 | 97 | 91 | 0.86 |
| CXR | 96.13 | 93.30 | 96.25 | 0.9940 |
| XceptionNet | 96 | 96 | 96 | 0.95 |
| ResNeXt | 93 | 97 | 86 | 0.78 |
| CoroNet [29] | 89.6 | 84.1 | 83.1 | 0.82 |

## 7.13 CONCLUSION

The time to get the COVID-19 test report ranges from three hours to more than two days and many hospitals in economically developing countries still have no access to the medical kits that give results rapidly. According to a multinational consensus statement published in the Fleischer Society, chest radiography for patients affected by COVID-19 in a resource-limited setting has been deemed an acceptable alternative practice. The expense of medical infrastructure and laboratory supplies utilized for diagnosing patients, particularly in developing and underdeveloped nations, presents a substantial impediment to combating the deleterious virus. Thus, X-ray imaging could be a useful tool for the automated detection of COVID in densely populated countries like India and China, as well as for hospitals in rural areas that do not have access to laboratory kits for COVID-19 testing. Therefore, our proposed CNN model may be incorporated into the website or mobile app so that a patient can upload his/her chest X-ray image and the app will give the result as COVID-19 positive or normal. We have mentioned here the outcomes after detecting COVID-19-positive cases from chest X-rays using a CNN model.

Our h5 model achieved an accuracy of 97% when applied to the test dataset of the first dataset and an accuracy of 92% when applied to the test set of the second dataset. The model significantly outperformed the COVIDPEN model in terms of test-data accuracy (when h5 was applied to first test dataset), precision, recall, and F1 metrics. The results are promising, despite the limited amount of publicly available data. Further validation should be conducted by applying this model to a larger X-ray image dataset and a broader range of COVID-19 clinical trials.

## REFERENCES


1. Novel coronavirus: China, 2020. who-china-joint-mission-on-covid-19-f inal-report.pdf.
2. WHO Report Online, 2022. https://covid19.who.int/
3. COVID-19 symptoms 17. 2025. https://www.cdc.gov/flu/symptoms/fluvs-covid19.html

4. COVID-19 tests recommended by WHO. 2024 https://www.who.int/news/item/07-04-2020-who-lists-two-covid-19-tests-for-emergency-use
5. Advantages and disadvantages of different prescribed COVID-19 tests. https://www.biospace.com/article/prosand-cons-of-the-common-types-of-covid-19-tests/
6. Chest X-rays for diagnosis of COVID-19 disease. https://www.cochrane.org/CD013639/INFECTN_how-accurate-chest-imaging-diagnosing-covid-19
7. Himadri Mukherjee, Subhankar Ghosh, Ankita Dhar, S. K. Md Obaidullah, K. C. Santosh, and Kaushik Roy, 2021. Deep neural network to detect COVID-19: One architecture for both CT scans and chest X-rays. *Applied Intelligence (Dordrecht, Netherlands)*, 51(5), 2777–2789. https://doi.org/10.1007/s10489-020-01943-6
8. Mohamed Elgendi, Muhammad Umer Nasir, Qunfeng Tang, Richard Ribon Fletcher, Newton Howard, Carlo Menon, Rabab Ward, William Parker, and Savvas Nicolaou, 2020. The performance of deep neural networks in differentiating chest X-rays of COVID-19 patients from other bacterial and viral pneumonias. *Frontiers in Medicine (Lausanne)*, 7, 550. https://doi.org/10.3389/fmed.2020.00550
9. Tahmina Zebin and Shahadate Rezvy, 2020. COVID-19 detection and disease progression visualization: Deep learning on chest X-rays for classification and coarse localization. *Applied Intelligence (Dordrecht, Netherlands)*, 51(2), 1010–1021.
10. Andrew C. Li, David T. Lee, Kristoff K. Misquitta, Kajji Uno, and Sasha Wald, 2020. COVID-19 Detection from Chest Radiographs Using Machine Learning and Convolutional Neural Networks. https://doi.org/10.1101/2020.08.31.20175828
11. Amit Kumar Jaiswal, Prayag Tiwari, Vipin Kumar Rathi, Jia Qian, Hari Mohan Pandey, Victor Hugo, and C. Albuquerque, 2020. COVIDPEN: A Novel COVID-19 Detection Model Using Chest X-Rays and CT Scans. https://doi.org/10.1101/2020.07.08.20149161
12. B. Sekeroglu and I. Ozsahin, 2020. Detection of COVID-19 from chest X-ray images using convolution neural networks. *SLAS Technology*, 25(6), 553–565.
13. Pedro R. A. S. Bassi and Romis Attux, 2021. A deep convolutional neural network for COVID-19 detection using chest X-rays. *Research on Biomedical Engineering*, 38, 139–148. https://doi.org/10.1007/s42600-021-00132-9
14. Daniel I. Morís, Joaquim de Moura, Jorge Novo, and Marcos Ortega, "Cycle Generative Adversarial Network Approaches to Produce Novel Portable Chest X-Rays Images for COVID-19 Diagnosis," *2021 IEEE International Conference on Acoustics, Speech and Signal Processing (ICASSP 2021)*, May 13, 2021.
15. Hasib Zunair and A. Ben Hamza, 2021. Synthesis of COVID-19 chest X-rays using unpaired image-to-image translation. *Social Network Analysis and Mining*, 11(1), 23.
16. Alaa S. Al-Waisy, Shumoos Al-Fahdawi, Mazin Abed Mohammed, Karrar Hameed Abdulkareem, Salama A. Mostafa, Mashael S. Maashi, Muhammad Arif, and Begonya Garcia-Zapirain, 2020. COVID-CheXNet: hybrid deep learning framework for identifying COVID-19 virus in chest X-rays samples. *Soft Computing*, 27, 2657–2672.
17. Harsh Panwar, P.K. Gupta, Mohammad Khubeb Siddiqui, Ruben Morales-Menendez, and Vaishnavi Singh, 2020. Application of deep learning for fast detection of COVID-19 in X-rays using nCOVnet. *Chaos, Solitons and Fractals*, 138, 109944.
18. Taban Majeed, Rasber Rashid, Dashti Ali, and Aras Asaad, 2022. Issues associated with deploying CNN transfer learning to detect COVID-19 from chest X- rays. *Physical and Engineering Sciences in Medicine*, 43(4), 1289–1303.

19. Ebenezer Jangam, Aaron Antonio Dias Barreto, and Chandra Sekhara Rao Annavarapu, 2021. Automatic detection of COVID-19 from chest CT scan and chest X-Rays samples using deep learning, transfer learning and stacking. *Applied intelligence (Dordrecht, Netherlands)*, 52(2), 2243–2259.
20. Jianhong Cheng, John Sollee, Celina Hsieh, Hailin Yue, Nicholas Vandal, Justin Shanahan, Ji Whae Choi, Thomas Yi, Terrance Healey, and Jianxin Wang, 2022. COVID-19 mortality prediction in the intensive care unit with deep learning based on longitudinal chest X-rays and clinical data. *European Radiology*, 32(7), 4446–4456.
21. Sohaib Asif, Ming Zhao, Fengxiao Tang, and Yusen Zhu, 2022. A deep learning-based framework for detecting COVID-19 patients using chest X-rays. *Multimedia Systems*, 28(4), 1495–1513.
22. Jalal Rabbah, Mohammed Ridouani, and Larbi Hassouni, "A New Classification Model Based on Stacknet and Deep Learning for Fast Detection of COVID-19 through X Rays Images", *2020 Fourth International Conference on Intelligent Computing in Data Sciences (ICDS)*, November 30, 2020.
23. Mirtha Lucas, Miguel Lerma, and Jacob Furst, 2020. Heatmap template generation for COVID-19 biomarker detection in chest X-rays", in *IEEE 20th International Conference on Bioinformatics and Bioengineering (BIBE 2020)*. IEEE, pp. 438–445.
24. Umair Hafeez, Muhammad Umer, Ahmad Hameed, Hassan Mustafa, Ahmed Sohaib, Michele Nappi, and Hamza Ahmad Madni, 2022. A CNN based coronavirus disease prediction system for chest X-rays. *Journal of Ambient Intelligence and Humanized Computing*, 1–15.
25. Areej A. Wahab Ahmed Musleh and Ashraf Yunis Maghari, "COVID-19 detection in X-ray images using CNN algorithm", *2020 International Conference on Promising Electronic Technologies (ICPET)*. IEEE, January 20, 2021.
26. Alavikunhu Panthakkan, Anzar S. M., Saeed Al Mansoori, and Hussain Al Ahmad, "Accurate Prediction of COVID-19 (+) Using AI Deep VGG16 Model," *2020 3rd International Conference on Signal Processing and Information Security (ICSPIS)*, February 9, 2020. IEEE.
27. Roberto Maroldi, Paolo Rondi, Giorgio Maria Agazzi, Marco Ravanelli, Andrea Borghesi, and Davide Farina, 2020. Which role for chest X-ray score in predicting the outcome in COVID-19 pneumonia? *European Radiology*, 31(6), 4016–4022.
28. Rajeev Kumar Singh, Rohan Pandey, and Rishie Nandhan Babu, 2021. COVIDScreen: Explainable deep learning framework for differential diagnosis of COVID-19 using chest X-rays. *Neural Computing and Applications*, 33(14), 8871–8892.
29. Rachna Jain, Meenu Gupta, Soham Taneja, and D. Jude Hemnath, 2020. Deep learning based detection and analysis of COVID-19 on chest X-ray images. *Applied Intelligence*, 51, 1690–1700.

Chapter 8

# Intelligent diabetes diagnosis using bioinspired algorithms integrated with machine learning and deep learning systems

*S. Sheik Asraf, P. Nagaraj, and V. Muneeswaran*

## 8.1 INTRODUCTION

Diabetes is a long-lasting illness triggered by inadequate insulin synthesis. It eventually affects important physiological systems like blood vessels and neurons and advances blood glucose to dangerous levels. Type 1 and type 2 diabetes are long-lasting circumstances, while pregnancy diabetes and prediabetes can be managed post-birth. Bioinspired optimization algorithms and machine learning (ML), which use deep neural networks (DNNs) to mimic human intellectual methods, have significantly improved the efficiency and precision of diabetes diagnosis.

## 8.2 DIABETES: A CHRONIC HEALTH CONDITION

Diabetes is a severe, long-term condition resulting from the physique's incapability to effectively utilize insulin or generate sufficient glucose (Figure 8.1) [1].

There are three primary forms of diabetes: type 1, type 2, and pregnancy diabetes. Diabetes is a long-lasting disease marked by elevated blood glucose levels and impaired metabolism of fat and protein [2]. Type 2 diabetes prevalence is significantly higher among Asian populations, with higher morbidity,

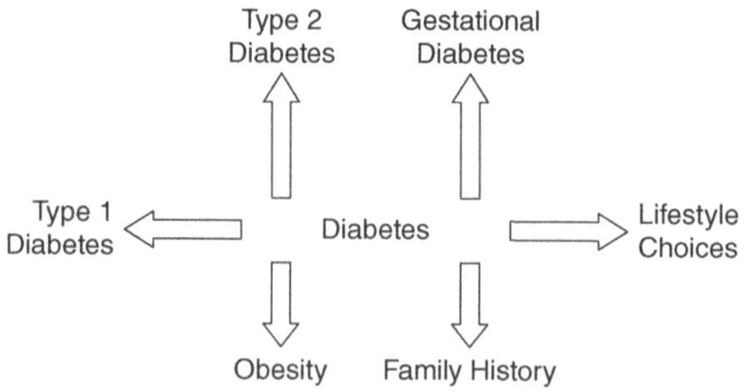


*Figure 8.1* Diabetes: a chronic health condition.


DOI: 10.1201/9781042015931-8

death rates, and complications among young Asians [3]. Diabetes prevalence is rising significantly in the United States, and this will probably continue to rise with the increasing obesity cases [4]. As per the International Diabetes Federation, the global diabetes epidemic is expected to affect almost 640 million people by 2040 [5]. Diabetes is a growing global epidemic, causing significant morbidity and death worldwide [6]. Time of life, culture, personal past, smoking, overweight, and bodily idleness are the main risk factors for diabetes [7]. Diabetes affects 23.3% of Qataris and 90 million Chinese people globally, with 1.3 million deaths in 2011 [8]. Hyperglycemia is a common symptom in a group of autoimmune, metabolic, and genetic diseases, collectively known as diabetes mellitus [9]. Diabetes is expected to kill 592 million people globally by 2035, with causes including high sugar diets, lack of physical activity, genetic predisposition, and lifestyle choices [10].

## 8.3 BIOINSPIRED ALGORITHMS (BIAs): AN OUTLINE

To tackle extremely complicated problems fast, BIAs are required (Figure 8.2).

This is especially true when dealing with problems that include dynamic problem definitions, fluctuating limitations, missing or erroneous information, and limited processing capability. BIAs, such as artificial plant, firefly, cuckoo search, evolutionary, and artificial bee colony algorithms, are used to optimize various aspects of life [11]. BIAs utilize computer architectures inspired by natural phenomena or biosystems, enabling intelligent paradigms capable of handling complex formulations and high-performance computation methods [12]. There are four categories of BIAs [13]. BIAs are effective optimization strategies, with algorithms based on sentient creatures like humans, chimpanzees, or dolphins being the most efficient and optimal solution [14]. BIAs are useful for solving multidimensional nonlinear problems, including those that are typically seen in biosignal analysis [15]. Randomized techniques like BIAs can effectively resolve combinatorial optimization problems with ambiguous parameters, dynamic constraints, and diverse solution representations [16]. Basic components of the neo-Darwinian evolutionary

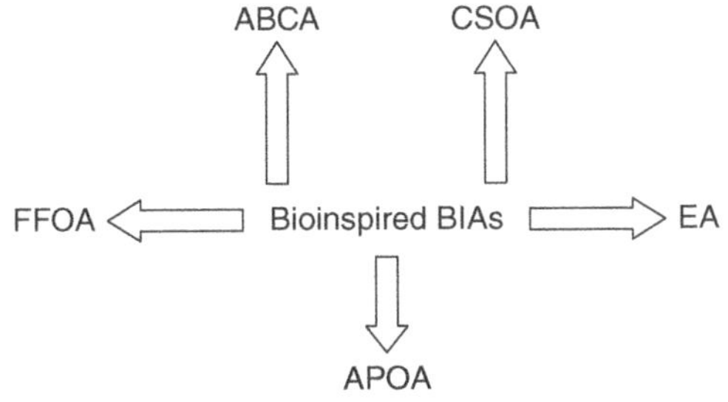


*Figure 8.2* Bioinspired algorithms: an outline.

process can be replicated by BIAs [17]. Researchers are using BIAs to find and optimize solutions to optimization challenges [18]. The swarm behaviors of animal groupings, which continuously exhibit remarkable problem-solving skills, are what cause BIAs [19]. BIAs, derived from biological processes, combine ML techniques with algorithms inspired by biology to effectively tackle complex scientific and technical challenges [20].

## 8.4 MACHINE LEARNING: AN OUTLINE

By gathering information from its surroundings, ML (Figure 8.3), a quickly expanding discipline of computer algorithms, seeks to mimic human intelligence [21].

Recent ML developments have been significantly aided by new learning theories, algorithms, internet data expansion, and low-cost computation [22]. The main benefit of ML is its independence after it has decided how to use the data [23]. There are many examples of common ML algorithm types [24]. ML is the primary focus of artificial intelligence (AI) research [25]. By using ensembles of classifiers, scaling supervised learning algorithms, reinforcement learning, and learning intricate stochastic models, ML improves classification accuracy [26]. Through talks and actions, ML enables computers to simulate human behavior and provide learning tools for later use [27]. Data mining, voice and computer vision learning algorithms, and statistical–computational theories all have greatly benefited from ML during the last 50 years [28]. Without clear guidance, ML extracts general principles from examples it has seen [29]. The computer-based technology, ML enhances method selection and performance in tasks involving categorization and prediction [30].

## 8.5 DEEP LEARNING (DL): AN OUTLINE

DL-based computer models (Figure 8.4) are being utilized in drug development and genomics for tasks like object identification, voice recognition, and visual object recognition [31]. DL employs manifold layers of feature learning

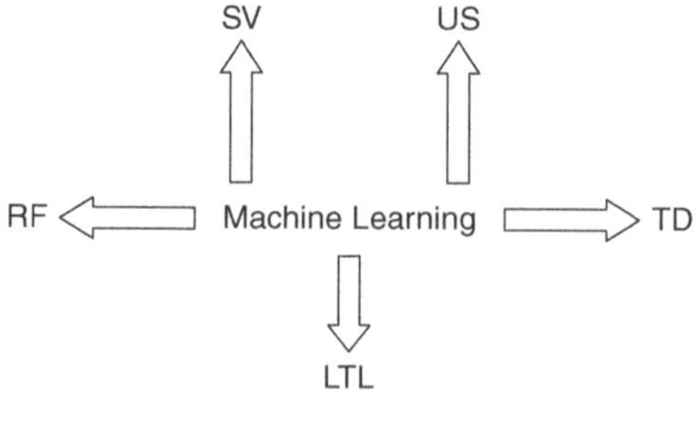


*Figure 8.3* Machine learning: an overview.

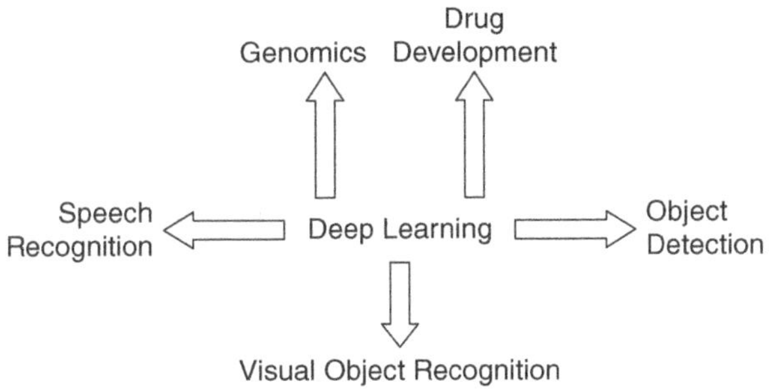


*Figure 8.4* Deep learning: an outline.

from unstructured data to enhance model performance [32]. The authors emphasize the importance of their expertise when selecting suitable DL application sectors [33]. DL, a subfield of ML, is becoming more and more popular. DL models complex data concepts using artificial neural networks (ANNs) [34]. Because it is more useful than classical ML, DL is being used in academic research more and more [35].

Multiple tiers of data distribution representations are automatically generated using AI technology known as DL [36]. To solve the dimensionality problem in shallow systems like support vector machine (SVM), DL employs a two-stage learning technique that makes it possible to extract trustworthy features from data [37]. ANNs are the foundation of the DL area, enabling data interchange and computation through matching input, output, node, and linkages [38]. Researchers are keen on ML and DL due to their potential applications in object recognition, self-driving cars, medication research, and illness diagnosis [39]. Although DL is a popular AI technique for categorizing and assessing biological data, people who are not familiar with it may find it difficult to utilize [40].

## 8.6 UTILIZING BIA-BASED ML MODELS TO DIAGNOSE DIABETES

Diabetes diagnosis (Table 8.1 and Figure 8.5) may benefit from the application of BIAs that are based on ML models.

The optimal diabetes risk prediction algorithm employs various classifiers, including neural network (NN), logistic regression (LR), support vector classifier (SVC), KNN, decision tree (DT), and random forest (RF) [41]. The model for the diagnosis of diabetes was found using DT algorithms, LR, KNN, and SVM [42]. ML algorithms can effectively predict diabetes, potentially reducing annual death rates [43]. Diabetes was diagnosed using ML classifiers [44]. Model designs for predictive programs on diabetes hospital readmissions datasets can create bias when using models like LR, XGBoost, and RF [45]. DT can be used for the diagnosis of type II diabetes [46]. RF has

*Table 8.1* Diabetes diagnosis machine learning models using bioinspired algorithms

| *Sl. No.* | *Machine learning models based on bioinspired algorithms* | *References* |
|---|---|---|
| 1. | Gaussian Naive Bayes (NB), multilayer perceptron (MLP), eXtreme Gradient Boosting (XGB), light gradient-boosting machine (LGBM), *K*-nearest neighbor (KNN), decision tree (DT), random forest (RF), gradient boosting, additional trees, bagging, AdaBoost, neural network (NN), logistic regression (LR), support vector classifier (SVC), and more | [41] |
| 2. | LR, KNN, SVM | [42] |
| 3. | KNN, DT, RF, AdaBoost, NB, XGBoost, multilayer perceptron (MLP) | [44] |
| 4. | LR, XGBoost, RF | [45] |
| 5. | DT | [46] |
| 6. | RF | [47] |
| 7. | NB | [48] |
| 8. | KNN | [49] |
| 9. | SVM | [50] |

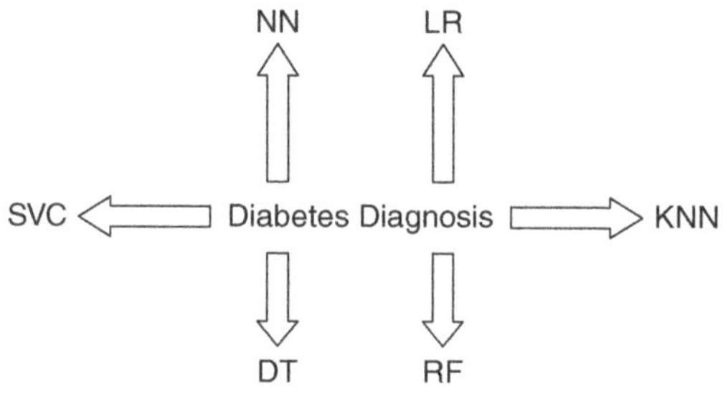


*Figure 8.5* Utilizing bioinspired algorithm-based machine learning models to diagnose diabetes.

proved to be more efficient for the diagnosis of diabetes [47]. The NB classification is employed to forecast diabetes [48]. Diabetes mellitus diagnosis using the *K*-nearest neighbor (KNN) algorithm has been reported recently [49]. SVMs, when properly developed with a clear ruleset, provide a promising method for diabetes prediction, achieving 94% accuracy, 93% sensitivity, and 94% specificity [50].

## 8.7 DIABETES DIAGNOSIS USING DL MODELS BASED ON BIAs

BIAs (Table 8.2 and Figure 8.6), such as DL model-based autoencoders, self-organizing maps, and deep belief networks, have significantly improved diabetes diagnosis through their ability to detect and treat various diseases.

Pattern recognition algorithms have integrated the growing self-organizing map (GSOM) with standard data mining methodology, enabling the diagnosis

*Table 8.2* Deep learning models using bioinspired algorithms to diagnose diabetes

| *Sl. No.* | *Machine learning models based on bioinspired algorithms* | *References* |
|---|---|---|
| 1. | Growing self-organizing map (GSOM) | [51] |
| 2. | DBN | [52] |
| 3. | RBFN | [53] |
| 4. | Deep convolutional neural network | [54] |
| 5. | LSTM | [55] |
| 6. | Dilated recurrent neural network (DRNN) | [56] |
| 7. | GAN | [57] |
| 8. | MLP | [58] |

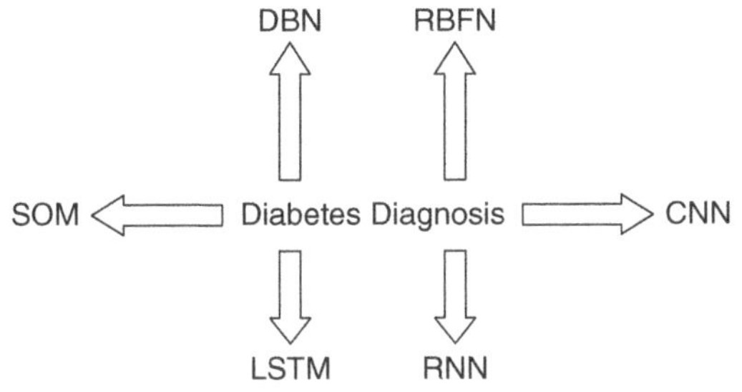


*Figure 8.6* Uses of BIAs-based deep learning models for diabetes diagnosis.

of diabetes [51]. The DBN method outperforms other classifiers in detecting diabetes [52]. The RBF network outperforms other models in diabetes diagnosis [53]. Images of diabetic retinopathy fundus have been graded for severity using DCNNs models [54]. A bidirectional LSTM technique was used in a recent study to identify diabetes mellitus early [55]. Dilated recurrent neural networks (DRNNs)-based DL models have recently generated 30-minute forecasts of future blood glucose levels [56]. The recent report highlights the synthesis of diabetic retinopathy abrasions using a generative adversarial network [57]. Recently, a study on the classification of diabetes using multilayer perceptron was published [58].

## 8.8 OPPORTUNITIES AND DIFFICULTIES FOR DIABETES DIAGNOSIS USING BIAS-BASED AI AND DL MODELS

Diagnosing and treating diabetes (Figure 8.7) in young people poses several special difficulties [59]. Gestational diabetes mellitus leads to maternal and fetal morbidity and mortality, necessitating strict glycemic management and early diagnosis for improved outcomes [60].

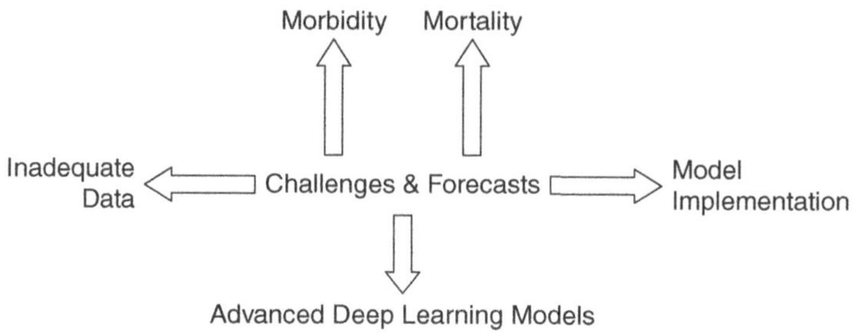


*Figure 8.7* Trials and forthcoming forecasts of BIAs-based ML and DL models for diabetes diagnosis.

Several research problems or difficulties arise with diagnosing diabetes [61]. The main issues for ML-based diabetes diagnosis are those related to inadequate data and model implementation [62]. The recent development of advanced DL models has led to unresolved research on the deployment and performance validation of these systems in diabetes diagnosis [63].

## 8.9 ASSESSING THE PROFITS AND DISADVANTAGES OF DL AND ML MODELS BASED ON BIAs FOR THE DIAGNOSIS OF DIABETES

Diabetes is one of the many medical problems that DL and ML have been shown to be successful in detecting [64]. Data-driven insights, prediction accuracy, personalization, scalability, and real-time decision-making are some of the strong points of ML. Data reliance, black box issues, overfitting, ethical dilemmas, and feature engineering difficulties are some of its drawbacks. In contrast, DL is better at processing complex data, improving accuracy, automating tasks, being flexible, and modeling nonlinear relationships. It is helpful in complicated situations like identifying diabetes from retinal scans or continuous glucose monitoring data since it can manage big datasets and processing capacity. It is computationally demanding, necessitates huge datasets, and is frequently referred to as "black boxes" because of its interpretive challenges.

## 8.10 CRITICAL DISCUSSION: WHEN TO CHOOSE SPECIFIC ML/DL MODELS BASED ON DATASET SIZE, FEATURE COMPLEXITY, OR COMPUTATIONAL EFFICIENCY FOR DIABETES DIAGNOSIS?

A number of parameters, including processing efficiency, feature complexity, and dataset size, must be taken into account when deciding between ML and DL models for diabetes diagnosis [65]. While ML models are appropriate for small- to medium-sized datasets, DL models are best suited for larger datasets [66]. Image data may be handled by RFs, gradient-boosting machines, and DTs.

## 8.11 DIAGNOSING DIABETES EMPIRICALLY USING REAL-WORLD DATASETS

BIA methods like genetics, particle collision optimization (PCO), and ant colony optimization (ACO) are being used in ML and DL models for diabetes diagnosis [67]. These algorithms mimic natural processes, aiding in feature selection, hyperparameter optimization, and ensemble learning [68]. Real-world datasets like the Pima Indians Diabetes Database and Cleveland Heart Disease Dataset provide a 5–10% accuracy boost.

## 8.12 DATA DIFFICULTIES, MORAL DILEMMAS, AND POSSIBLE PREJUDICES IN AI MODELS FOR THE DIAGNOSIS OF DIABETES

The challenges that AI models for diabetes diagnosis must overcome are missing data, data imbalance, privacy and security, heterogeneity, and a lack of longitudinal data. Techniques like imputation and ensemble techniques can be used to close these gaps. By using transfer learning and standardizing data collection and preparation processes, models may be modified for use with different datasets [69]. Explainable AI techniques can assist patients in giving their informed consent, which is crucial [70]. By using techniques like adversarial debiasing and fairness criteria, algorithmic bias may be reduced [71–78]. Open methods and independent audits help reduce confirmation bias in research and model testing.

## 8.13 CONCLUSIONS

Diabetes is a long-lasting disease that impairs physiological systems and raises blood sugar levels due to inadequate insulin synthesis. Gestational diabetes and prediabetes may be treated, but type 1 and type 2 diabetes are long-lasting diseases. Complex optimization problems are addressed by bioinspired optimization algorithms, which are based on biological systems and animal characteristics. By simulating human cognitive processes, ML and DL increase the precision and effectiveness of AI systems in diagnosing diabetes. Accuracy is also improved by other bioinspired algorithms, such as self-organizing maps and deep belief networks. Potential uses for these algorithms exist in a number of industries, including healthcare.

## REFERENCES


1. World Health Organization. (2016). WHO global report on diabetes.
2. Roglic, G. (2016). WHO global report on diabetes: A summary. International Journal of Noncommunicable Diseases, 1(1), 3–8.
3. Ramachandran, A., Ma, R. C. W., & Snehalatha, C. (2010). Diabetes in Asia. The Lancet, 375(9712), 408–418.
4. Mokdad, A. H., Ford, E. S., Bowman, B. A., Nelson, D. E., Engelgau, M. M., Vinicor, F., & Marks, J. S. (2000). Diabetes trends in the US: 1990–1998. Diabetes Care, 23(9), 1278–1283.
5. Papatheodorou, K., Banach, M., Bekiari, E., Rizzo, M., & Edmonds, M. (2018). Complications of diabetes 2017. Journal of Diabetes Research, 2018, 3086167.
6. Papatheodorou, K., Banach, M., Edmonds, M., Papanas, N., & Papazoglou, D. (2015). Complications of diabetes. Journal of Diabetes Research, 2015, 189525.
7. Deshpande, A. D., Harris-Hayes, M., & Schootman, M. (2008). Epidemiology of diabetes and diabetes-related complications. Physical Therapy, 88(11), 1254–1264.
8. Scully, T. (2012). Diabetes in numbers. Nature, 485(7398), S2–S3.
9. Egan, A. M., & Dinneen, S. F. (2019). What is diabetes? Medicine, 47(1), 1–4.
10. Tao, Z., Shi, A., & Zhao, J. (2015). Epidemiological perspectives of diabetes. Cell Biochemistry and Biophysics, 73(1), 181–185.
11. Kar, A. K. (2016). Bio inspired computing: A review of algorithms and scope of applications. Expert Systems with Applications, 59, 20–32.
12. Cuevas, E., Oliva, D., & Osuna, V. (2020). Bio-inspired algorithms and biosystems. Mathematical Biosciences and Engineering, 17(3), 2400–2402.
13. Fan, X., Sayers, W., Zhang, S., Han, Z., Ren, L., & Chizari, H. (2020). Review and classification of bio-inspired algorithms and their applications. Journal of Bionic Engineering, 17, 611–631.
14. Pazhaniraja, N., Paul, P. V., Roja, G., Shanmugapriya, K., & Sonali, B. (2017, March). A Study on Recent Bio-Inspired Optimization Algorithms. In: 2017 Fourth International Conference on Signal Processing, Communication and Networking (ICSCN) (pp. 1–6). IEEE.
15. Wario, F., Avalos, O., & Gálvez, J. (2022). Bio-inspired algorithms. In: Biosignal Processing and Classification Using Computational Learning and Intelligence (pp. 225–248). Academic Press.
16. Assimi, H. (2023). Application of bio-inspired algorithms to selected real-world problems. Doctoral dissertation.
17. Wagner, M. (2013). Theory and applications of bio-inspired algorithms. Doctoral dissertation.
18. Almufti, S., Marqas, R., & Ashqi, V. (2019). Taxonomy of bio-inspired optimization algorithms. Journal of Advanced Computer Science & Technology, 8(2), 23.
19. Duan, H., Li, P., Li, P., & Duan, H. (2016). Bio-inspired computation algorithms. In: Bio-Inspired Computation in Unmanned Aerial Vehicles (pp. 35–69). Springer.
20. Boopalan, K., Shanmuganathan, C., Lokeshwaran, K., & Balaji, T. (2022). A short overview on various bio-inspired algorithms: Machine learning in information and communication technology. Proceedings of ICICT 2021 (pp. 295–301), SMIT.
21. El Naqa, I., & Murphy, M. J. (2015). What is machine learning? In: Machine Learning in Radiation Oncology (pp. 3–11). Springer International Publishing.
22. Jordan, M. I., & Mitchell, T. M. (2015). Machine learning: Trends, perspectives, and prospects. Science, 349(6245), 255–260.
23. Mahesh, B. (2020). Machine learning algorithms: A review. International Journal of Science and Research, 9(1), 381–386.

24. Ayodele, T. O. (2010). Types of machine learning algorithms. New Advances in Machine Learning, 3(19-48), 5–1.
25. Wang, H., Ma, C., & Zhou, L. (2009, December). A Brief Review of Machine Learning and Its Application. In: 2009 International Conference on Information Engineering and Computer Science (pp. 1–4). IEEE.
26. Dietterich, T. G. (1997). Machine-learning research. AI Magazine, 18(4), 97.
27. Alzubi, J., Nayyar, A., & Kumar, A. (2018, November). Machine learning from theory to algorithms: An overview. Journal of Physics: Conference Series, 1142, 012012.
28. Mitchell, T. M. (2006). The Discipline of Machine Learning (vol. 9, p. 3). Pittsburgh: Carnegie Mellon University.
29. Bzdok, D., Krzywinski, M., & Altman, N. (2017). Machine learning: A primer. Nature Methods, 14(12), 1119.
30. Tarca, A. L., Carey, V. J., Chen, X. W., Romero, R., & Drăghici, S. (2007). Machine learning and its applications to biology. PLoS Computational Biology, 3(6), e116.
31. LeCun, Y., Bengio, Y., & Hinton, G. (2015). Deep learning. Nature, 521(7553), 436–444.
32. Mathew, A., Amudha, P., & Sivakumari, S. (2021). Deep Learning Techniques: An Overview. In: Advanced Machine Learning Technologies and Applications: Proceedings of AMLTA 2020 (pp. 599–608).
33. Deng, L., & Yu, D. (2014). Deep learning: Methods and applications. Foundations and Trends® in Signal Processing, 7(3–4), 197–387.
34. Hao, X., Zhang, G., & Ma, S. (2016). Deep learning. International Journal of Semantic Computing, 10(03), 417–439.
35. Du, X., Cai, Y., Wang, S., & Zhang, L. (2016, November). Overview of Deep Learning. In: 2016 31st Youth Academic Annual Conference of Chinese Association of Automation (YAC) (pp. 159–164). IEEE.
36. Lauzon, F. Q. (2012, July). An Introduction to Deep Learning. In: 2012 11th International Conference on Information Science, Signal Processing and Their Applications (ISSPA) (pp. 1438–1439). IEEE.
37. Arnold, L., Rebecchi, S., Chevallier, S., & Paugam-Moisy, H. (2011, April). An Introduction to Deep Learning. In: European Symposium on Artificial Neural Networks (ESANN).
38. Kim, H. (2022). Deep learning. In: Artificial Intelligence for 6G (pp. 247–303). Cham: Springer International Publishing.
39. Khan, M., Jan, B., Farman, H., Ahmad, J., Farman, H., & Jan, Z. (2019). Deep learning methods and applications. In: Deep Learning: Convergence to Big Data Analytics (pp. 31–42). Springer.
40. Webb, S. (2018). Deep learning for biology. Nature, 554(7693), 555–557.
41. Darmawan, I., Gunawan, R. I., & Rahmatulloh, A. (2017, April). Model Accuracy Test for Early Stage of Diabetes Risk Prediction with Data Science Approach. In: International Conference on Engineering, Applied Sciences and System Modeling (pp. 57–76). Singapore: Springer Nature.
42. Shahade, M., Awate, A., Nandwalkar, B., & Kulkarni, M. (2022, November). Diabetes disease prediction using KNN. In: International Conference on Innovations in Data Analytics (pp. 293–306). Singapore: Springer Nature.
43. Mahbub, M. H., Islam, S., & Mahbub, M. A. (2021). Performance assessment of different machine learning algorithms in predicting diabetes mellitus. Doctoral dissertation, Department of Electrical and Electronic Engineering, Islamic University of Technology (IUT).
44. Hasan, M. K., Alam, M. A., Das, D., Hossain, E., & Hasan, M. (2020). Diabetes prediction using ensembling of different machine learning classifiers. IEEE Access, 8, 76516–76531.

45. Ding, T., Madry, A., Raghavan, M., & Wilson, A. (2022). Assessing Machine Learning Models for Predicting Diabetes Hospital Readmission.
46. Al Jarullah, A. A. (2011, April). Decision Tree Discovery for the Diagnosis of Type II Diabetes. In: 2011 International Conference on Innovations in Information Technology (pp. 303–307). IEEE.
47. Benbelkacem, S., & Atmani, B. (2019, April). Random Forests for Diabetes Diagnosis. In: 2019 International Conference on Computer and Information Sciences (ICCIS) (pp. 1–4). IEEE.
48. Priya, K. L., Kypa, M. S. C. R., Reddy, M. M. S., & Reddy, G. R. M. (2020, June). A Novel Approach to Predict Diabetes by Using Naive Bayes Classifier. In: 2020 4th International Conference on Trends in Electronics and Informatics (ICOEI) (pp. 603–607). IEEE.
49. Saxena, K., Khan, Z., & Singh, S. (2014). Diagnosis of diabetes mellitus using k-nearest neighbor algorithm. International Journal of Computer Science Trends and Technology (IJCST), 2(4), 36–43.
50. Barakat, N., Bradley, A. P., & Barakat, M. N. H. (2010). Intelligible support vector machines for diagnosis of diabetes mellitus. IEEE Transactions on Information Technology in Biomedicine, 14(4), 1114–1120.
51. Wickramasinghe, K., Alahakoon, D., Schattner, P., & Georgeff, M. (2011). Self-Organizing Maps for Translating Health Care Knowledge: A Case Study in Diabetes Management. In: AI 2011: Advances in Artificial Intelligence: 24th Australasian Joint Conference, Perth, Australia, December 5–8, 2011. Proceedings 24 (pp. 162–171). Berlin: Springer.
52. Prabhu, P., & Selvabharathi, S. (2019, July). Deep Belief Neural Network Model for Prediction of Diabetes Mellitus. In: 2019 3rd International Conference on Imaging, Signal Processing and Communication (ICISPC) (pp. 138–142). IEEE.
53. Venkatesan, P., & Anitha, S. (2006). Application of a radial basis function neural network for diagnosis of diabetes mellitus. Current Science, 91(9), 1195–1199.
54. Gao, Z., Li, J., Guo, J., Chen, Y., Yi, Z., & Zhong, J. (2018). Diagnosis of diabetic retinopathy using deep neural networks. IEEE Access, 7, 3360–3370.
55. Jaiswal, S., & Gupta, P. (2023). Diabetes prediction using bi-directional long short-term memory. SN Computer Science, 4(4), 373.
56. Yang, H., Chen, Z., Huang, J., & Li, S. (2024). AWD-stacking: An enhanced ensemble learning model for predicting glucose levels. PLoS One, 19(2), e0291594.
57. Abbood, S. H., Abdull Hamed, H. N., Mohd Rahim, M. S., Alaidi, A. H. M., & Abdul Hadi M. Alaidi, H. Th. (2022). DR-LL GAN: Diabetic retinopathy lesions synthesis using generative adversarial network. International Journal of Online & Biomedical Engineering, 18(3).
58. Sivasankari, S. S., Surendiran, J., Yuvaraj, N., Ramkumar, M., Ravi, C. N., & Vidhya, R. G. (2022, April). Classification of Diabetes Using Multilayer Perceptron. In: 2022 IEEE International Conference on Distributed Computing and Electrical Circuits and Electronics (ICDCECE) (pp. 1–5). IEEE.
59. Unnikrishnan, R., Shah, V. N., & Mohan, V. (2016). Challenges in diagnosis and management of diabetes in the young. Clinical Diabetes and Endocrinology, 2, 1–9.
60. Mpondo, B. C., Ernest, A., & Dee, H. E. (2015). Gestational diabetes mellitus: Challenges in diagnosis and management. Journal of Diabetes & Metabolic Disorders, 14, 1–7.
61. Kaul, S., & Kumar, Y. (2020). Artificial intelligence-based learning techniques for diabetes prediction: Challenges and systematic review. SN Computer Science, 1(6), 322.

62. Sharma, T., & Shah, M. (2021). A comprehensive review of machine learning techniques on diabetes detection. Visual Computing for Industry, Biomedicine, and Art, 4(1), 30.
63. Nadeem, M. W., Goh, H. G., Hussain, M., Liew, S. Y., Andonovic, I., & Khan, M. A. (2022). Deep learning for diabetic retinopathy analysis: A review, research challenges, and future directions. Sensors, 22(18), 6780.
64. Nagaraj, P., & Deepalakshmi, P. (2022). An intelligent fuzzy inference rule-based expert recommendation system for predictive diabetes diagnosis. International Journal of Imaging Systems and Technology, 32(4), 1373–1396.
65. Nagaraj, P., Deepalakshmi, P., & Romany, F. M. (2021). Artificial flora algorithm-based feature selection with gradient boosted tree model for diabetes classification. Diabetes, Metabolic Syndrome and Obesity: Targets and Therapy, 14, 2789.
66. Nagaraj, P., & Deepalakshmi, P. (2021). Diabetes prediction using enhanced SVM and deep neural network learning techniques: An algorithmic approach for early screening of diabetes. International Journal of Healthcare Information Systems and Informatics (IJHISI), 16(4), 1–20.
67. Nagaraj, P., Muneeswaran, V., Dharanidharan, A., Balananthanan, K., Arunkumar, M., & Rajkumar, C. (2022, April). A Prediction and Recommendation System for Diabetes Mellitus Using XAI-Based Lime Explainer. In: 2022 International Conference on Sustainable Computing and Data Communication Systems (ICSCDS) (pp. 1472–1478). IEEE.
68. Brintha, N. C., Nagaraj, P., Tejasri, A., Durga, B. V., Teja, M. T., & Kumar, M. N. V. P. (2022, June). A Food Recommendation System for Predictive Diabetic Patients Using ANN and CNN. In: 2022 7th International Conference on Communication and Electronics Systems (ICCES) (pp. 1364–1371). IEEE.
69. Nagaraj, P., Deepalakshmi, P., & Ijaz, M. F. (2022). Optimized adaptive tree seed Kalman filter for a diabetes recommendation system: Bilevel performance improvement strategy for healthcare applications. In: Cognitive and Soft Computing Techniques for the Analysis of Healthcare Data (pp. 191–202). Academic Press.
70. Nagaraj, P., Deepalakshmi, P., Muneeswaran, V., & Muthamil Sudar, K. (2022, July). Sentiment analysis on diabetes diagnosis health care using machine learning technique. In: Congress on Intelligent Systems: Proceedings of CIS 2021 (vol. 1, pp. 491–502). Singapore: Springer Nature.
71. Revathi, B., Elizabeth, S. K., Nagaraj, P., Birunda, S. S., & Nithya, D. (2023, February). Particle Swarm Optimization Based Detection of Diabetic Retinopathy Using a Novel Deep CNN. In: 2023 Third International Conference on Artificial Intelligence and Smart Energy (ICAIS) (pp. 998–1003). IEEE.
72. Deny, J., Rajalakshmi, P., Muneeswaran, V., Sudharsan, R. R., & Nagaraj, P. (2022, August). Automation of Glucose Control for Type-2 Diabetes Mellitus. In: 2022 3rd International Conference on Electronics and Sustainable Communication Systems (ICESC) (pp. 79–83). IEEE.
73. Nagaraj, P., Sudar, K. M., Muneeswaran, V., & Choksi, B. (2024). Case study: Application of ensemble classifier for diabetes healthcare data analytics. In: Deep Learning for Smart Healthcare (pp. 226–243). Auerbach Publications.
74. Revathi, B., Usharani, C., Elizabeth, S. K., & Nithya, D. (2024, April). Deep Learning Classification Techniques on Detecting Diabetic Retinopathy Dataset. In: 2024 International Conference on Inventive Computation Technologies (ICICT) (pp. 663–668). IEEE.
75. Jeyandhan, D., Nagaraj, P., & Jeyanathan, J. S. (2024, September). Investigation of Deep Learning Models for Predicting Diabetic Foot Ulcers in Diabetes Patients. In: 2024 5th International Conference on Smart Electronics and Communication (ICOSEC) (pp. 1356–1363). IEEE.

76. Kumar, D. M., Deny, J., Nagaraj, P., Dhanveer, K., Kumar, V. P., & Arthi, R. (2023, July). Diabetes Prediction Using Machine Learning Algorithms and Hyperparameter Tuning for Expecting Mothers. In: 2023 IEEE World Conference on Applied Intelligence and Computing (AIC) (pp. 325–331). IEEE.
77. Deepalakshmi, M., Deepalakshmi, P., & Nagaraj, P. (2025, June). Efficient Diabetes Detection Using Ensemble Models Optimized by Firefly Algorithm. In: 2025 Second International Conference on Cognitive Robotics and Intelligent Systems (ICC-ROBINS) (pp. 573–578). IEEE.
78. Jeyanathan, J. S., & Nagaraj, P. (2024, November). Implementation of Collaborative Filtering-Based Recommendation System for Grading Diabetic Foot Thermograms. In: 2024 5th International Conference on Data Intelligence and Cognitive Informatics (ICDICI) (pp. 1087–1092). IEEE.

Chapter 9

# Assessing COVID-19 transmission patterns and government responses in India

*Rakesh Kumar, Manik Rakhra, Ramandeep Sandhu, Gagandeep Singh Cheema, and Deepika Ghai*

## 9.1 INTRODUCTION

COVID-19 is a disease that was discovered in Wuhan, China, in December 2019 [1,2]. Severe acute respiratory syndrome coronavirus 2 (SARS-CoV-2) is to blame. It is a very contagious virus. It spreads from person to person by droplets. As a result, big crowds will result in virus transmission from one person to another. The virus will infect a vast number of people. Initially, COVID-19 symptoms include loss of taste, muscle soreness, difficulty breathing, muscle pain, and so on. The virus has a very high mutation rate. Transmissible and mutational viruses can survive for a long time if both conditions are met. If a significant number of people are exposed to the virus, they should all get infected. These afflicted individuals will allow the virus to mutate, making it impossible for scientists to develop a vaccine. Coronavirus mutations delta and omicron are two. The only way to protect yourself from this infection is to be vaccinated.

### 9.1.1 Coronavirus in India

The coronavirus arrived in India from Wuhan with an infected person on January 30 and February 3, 2020, in Kerala. In Jaipur and Hyderabad, there were a few instances of coronavirus. They were also from another country. Following the discovery of the coronavirus's damaging effects, the Ministry of Health and Family Welfare (MoHFW) established foreign travel restrictions and a 14-day quarantine for people returning from abroad. The Ministry also released certain guidelines for individuals to follow, including wearing a mask, keeping a 1-meter distance from others, and washing their hands frequently [3, 4]. To prevent the spread of the coronavirus, India's Prime Minister declared a 21-day statewide lockdown on March 25, 2020. However, this proved insufficient to stop the coronavirus from spreading. Coronavirus cases were on the rise across the country. By March 15, there had been 107 confirmed coronavirus cases. India reported ten confirmed cases in less than 15 days. As a result, on April 14, 2020, India's government enforced the second phase of lockdown. The lockdown was extended until May 17, and then until

DOI: 10.1201/9781042015931-9

May 3 [5–7]. The Epidemic Disease Act of 1897 was effectively employed by the Indian government. This law empowers government officers to inspect citizens of both India and overseas. If they were discovered to be infected with COVID-19, they were placed in quarantine to prevent the coronavirus from spreading. According to the Indian Council of Medical Research, India could have cut COVID-19 transmission by 62%. According to another study, by mid-May, India will have 13 lakh coronavirus cases. India has identified 190,648 cases of coronavirus, including 5,407 deaths, since the lockdown ended on May 31. India had 34,633,255 coronavirus cases as of December 5, 2021. About 34,060,774 individuals recovered, whereas 470,620 died because of the coronavirus. On July 1, 2021, the data used in this paper was last updated. According to it, there are 30,182,402 current coronavirus cases in India.

### 9.1.2 Objectives

The general goal of this chapter is to look at how the coronavirus managed to persist in Indian communities:

- When would the first and second waves of coronavirus appear, and what would the Indian government do to combat them?
- How do individuals in India help the coronavirus survive?
- The positive and negative effects of coronavirus in India.

## 9.2 LITERATURE REVIEW

Numerous research works have been done on assessing COVID-19 transmission patterns and government responses in India. Ghosh et al. [8] conducted their research using state-by-state data from the Indian Population's COVID-19. They looked at the data by state and forecasted the number of cases in each after 30 days. According to them, if the number of instances could be forecast, state governments would be able to allocate healthcare resources better. They analyzed the data using three growth models (susceptible infectious, exponential, and logistic models). In their study, they considered different restrictions and guidelines imposed by the Indian government [9–11]. These limitations and regulations have both beneficial and negative consequences. It had a devastating impact on the Indian economy, human lives, and Indian culture. However, it had a favorable impact on the environment. According to the All-India Association of Industries, India's economic growth slowed by 5–5.6%. Demand in several industries, including soil and power, decreased because of the lockdown. Employees at companies like Swiggy and Zomato were fired. They also revealed how poor people from other states in Maharashtra were forced to return to their homes due to job and food insecurity.

The authors studied the impact of COVID-19 on the Indian educational system. The research focuses on the online classes that began during the COVID-19 pandemic and how they influenced students' learning. Many students in India do not have access to the necessary technology for online learning [12, 13]. Many pupils are unable to participate in online classes. Students and teachers confront a variety of issues during teaching, such as interruptions from family members, a lack of facilities, and so on. The authors discussed the COVID-19 vaccination method in India in their papers [10, 14]. Because India has the world's second-largest population, a strategy to deal with it needed to be devised. Covaxin and Covishield were licensed for emergency usage by the Indian pharma regulator at this critical moment.

On January 16, 2021, the Indian government began a free vaccination campaign for all Indian citizens. Initially, though, the immunizations were given to the country's frontline workers. The Indian government took immediate steps to increase vaccine manufacturing capacity and track and oversee vaccination campaigns across the country. In the Brazilian Northeast, data were obtained from 2,070 persons. They had cold or flu-like symptoms. Only the most serious cases were admitted to hospitals. The goal of the study was to determine COVID-19's survival and mortality. For the survival and mortality of COVID-19, robust Poisson regression, Kaplan–Meier, and Cox regression were utilized. COVID-19 has a survival probability of 87.6%, according to Ref. [15].

They discovered that the elderly were at a high risk of dying during their studies. The authors investigated the effect of poverty on COVID-19 survival in Mexico [16, 17]. COVID-19 had a greater impact on the impoverished in Mexican towns than it had on the rich or middle class according to the study. It was decided to use secondary data. Cox regression was employed to analyze the data. As a result, they discovered that COVID-19 increases the chance of death for poor persons by 9%. The properties were discovered by the authors in Ref. [18]. They perform the correct treatment and provide COVID-19 therapeutic guidance. They gathered information from 239 patients. They split them into two groups: death and improvement. The continuous variables were then tested using a sample *t*-test and a chi-square test. The research article is based on information gathered from COVID-19 patients admitted to tertiary care facilities. The authors investigated COVID-19 patient survival and mortality [19, 20]. The survival of COVID-19 patients will aid professionals in devising a COVID-19 control strategy. Based on the study's findings, individuals who are recognized as high-risk will receive the finest care and treatment. In India, these papers investigated the recovery period of COVID-19 patients [21, 22]. They gathered data from 221 patients for their study. The Kaplan–Meier survival model was then utilized. After the study, they discovered that COVID-19 patients in India recovered in 25 days. There is no difference in the time it takes for a male and a female to recuperate. Experts can design solutions to deal with COVID-19 circumstances using these insights. These papers looked into the difficulties that oncology patients confront throughout therapy [23, 24]. They prepared the questionnaire for the data collection. As a result, they received feedback from 302 patients. They discovered that patients were more concerned about their therapy than the COVID-19 epidemic because of this research. They want to keep doing chemotherapy.

## 15.3 DATASET

The dataset was obtained from Kaggle.com. The dataset includes state-by-state information on COVID-19 patients who are active, discharged, or expired. It includes information on the first, second, and third doses of immunization given to Indian communities, as well as persons who received both. All this information will be useful in determining whether COVID-19 has survived in India. City data includes statistics on the air quality in India's various cities.

## 15.4 PURPOSE OF THE STUDY

Every dangerous virus will persist unless people develop immunity, vaccines, or medicines to combat it. This study was conducted to determine how the coronavirus survives and to assess the virus's negative consequences. They also looked at data on air quality from 2015 to 2020 to see how the lockdown affected air quality.

## 15.5 RESULT ANALYSIS

In April 2020, due to the Kumbha Mela, a massive crowd gathered in Haridwar, Uttarakhand. In March 2020, a large crowd assembled near the mosque in Delhi, in violation of COVID-19 regulations. There was a slew of similar incidents. As a result, we can see that coronavirus cases have been increasing since April 2020. Between May and June of 2020, the first wave of coronavirus arrived. As the number of coronavirus infections decreased after September, the government began to soften the guidelines. COVID-19 vaccinations were made accessible to the public in January 2020. As a result, people believed that because of the vaccine, everything would improve, and they began to break all of the rules. Because of all of these factors, the second wave of coronavirus arrived in March 2021, as shown in Figures 9.1 and 9.2.

It is obvious from Figures 9.1 and 9.2 that confirmed cases and recovered cases are roughly equal. Since the vaccine was introduced in India, the number of recovered cases of COVID-19 has surpassed the number of confirmed cases. We can use the 4, 5, and 6 months of 2021 as a guide. The aggregate of

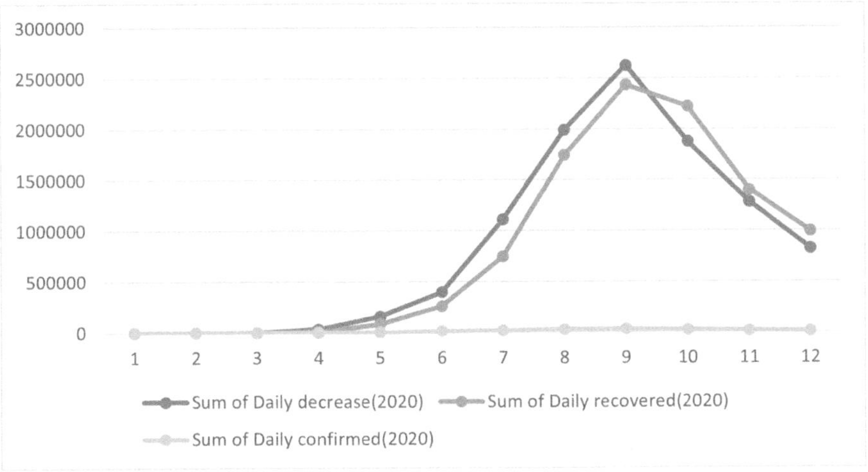


*Figure 9.1* COVID-19 trend in India.

the recovered, confirmed, and deaths of the patient in 12 months and 2 years is shown in Table 9.1. According to Table 9.1, the total number of deaths caused by COVID-19 in India is roughly 393,934. The overall number of confirmed cases in India is estimated to be around 30,182,402. The average number of cases recovered is roughly 29,185,623. Figures 9.1 and 9.2 also show that COVID-19 arrived in India in two waves, the first in October 2020 and the second in May 2021. These are the two months within two years when the coronavirus reaches its maximum value. The Indian people suffered a great deal as a result of this.

Based on the available data, Figure 9.3 displays the number of active cases in each state. The state with the most active cases is Kerala (81,490), followed by Maharashtra. Bihar has the fewest active cases of any of the states. The immunization of people has resulted in a decrease in cases. These are the remaining active cases in India's various states until July 1, 2021, as shown in Figure 9.4.

The size of the circle in each state represents the total number of COVID-19 cases reported to date. The greater the size of the circle, the higher the total count. As a result, Maharashtra currently has the most COVID-19 instances.

The death ratio of the different states is depicted in Figure 9.5. Death is the ratio given by the number of COVID-19 patients admitted to the hospital and the number of people dead due to it. Dadar and Nagar Haveli have the lowest death ratio (0.04). Punjab has the highest death rate of 2.750. The discharge ratios of the various states are depicted in Figure 9.6. This refers to the number of COVID-19 patients who were released versus the number of confirmed COVID-19 cases. Among all Indian states and Union Territories, Lakshadweep has the highest discharge ratio (99.51), and Mizoram has the lowest discharge ratio (91.4).

*Table 9.1* Count of confirmed, recovered, and deaths of patients due to coronavirus

| *Row labels (year)* | *Sum of daily confirmed* | *Sum of daily recovered* | *Sum of Deceased* |
|---|---|---|---|
| **2020** | **10,286,234** | **9,881,565** | **148,427** |
| 1 | 1 | 0 | 0 |
| 2 | 2 | 3 | 0 |
| 3 | 1,632 | 147 | 47 |
| 4 | 33,231 | 8,909 | 1,107 |
| 5 | 155,783 | 82,803 | 4,252 |
| 6 | 395,146 | 255,978 | 12,005 |
| 7 | 1,111,273 | 747,698 | 19,155 |
| 8 | 1,990,885 | 1,741,832 | 28,879 |
| 9 | 2,622,323 | 2,432,623 | 32,677 |
| 10 | 1,873,041 | 2,219,433 | 23,437 |
| 11 | 1,279,861 | 1,399,002 | 15,509 |
| 12 | 823,056 | 993,137 | 11,359 |
| **2021** | **19,896,168** | **19,304,058** | **245,507** |
| 1 | 472,317 | 552,275 | 5,410 |
| 2 | 353,427 | 350,561 | 2,766 |
| 3 | 1,108,656 | 687,933 | 5,766 |
| 4 | 6,936,345 | 4,200,684 | 48,879 |
| 5 | 9,016,687 | 10,266,501 | 120,072 |
| 6 | 2,008,736 | 3,246,104 | 62,614 |
| **Grand Total** | **30,182,402** | **29,185,623** | **393,934** |

Figure 9.7 depicts the number of active cases based on the population of the states. That signifies each state's active ratio is present. Of all the states, Mizoram has the highest active ratio. The active ratio in Mizoram is so high because the population of the state is the lowest in India.

Figure 9.8 illustrates which states were the most successful in vaccinating people. This graph is based on the number of immunizations received. The state of Uttar Pradesh ranked first in the number of persons who received the vaccine. By default, UP has the largest population among all Indian states. We should be pleased that despite its huge population, UP managed to provide vaccine to most of the population. This will aid people in their fight against COVID-19. We can derive from Figure 9.8 that vaccines are delivered according to the need in every region of the country. This signifies that India's leadership made attempts to overcome the coronavirus and return the country to its former position. Figure 9.9 depicts the vaccine dosage trend in India. The Indian government began a vaccination push for frontline employees on January 16, 2021, and it was later made available to all Indians. The government and frontline staff are working hard to get people vaccinated. We can see from the graph above that vaccination in India is doing extremely well and quite quickly.

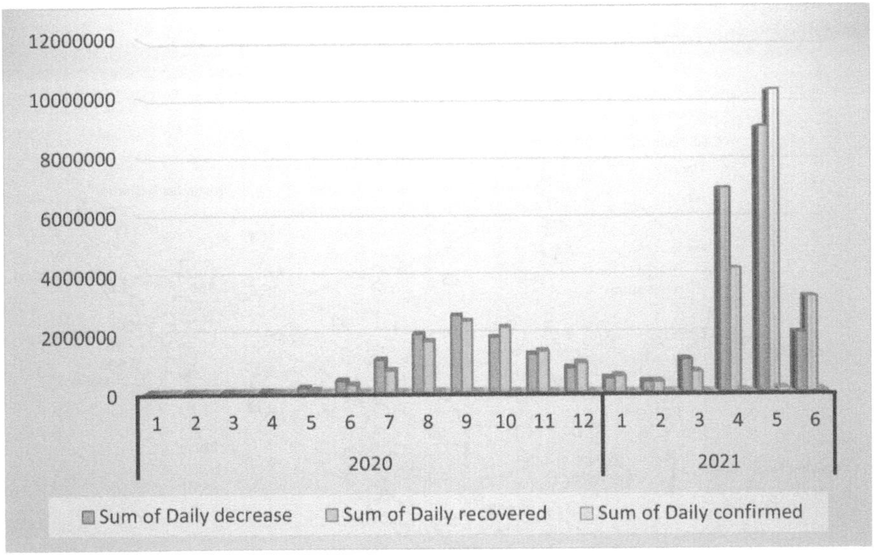


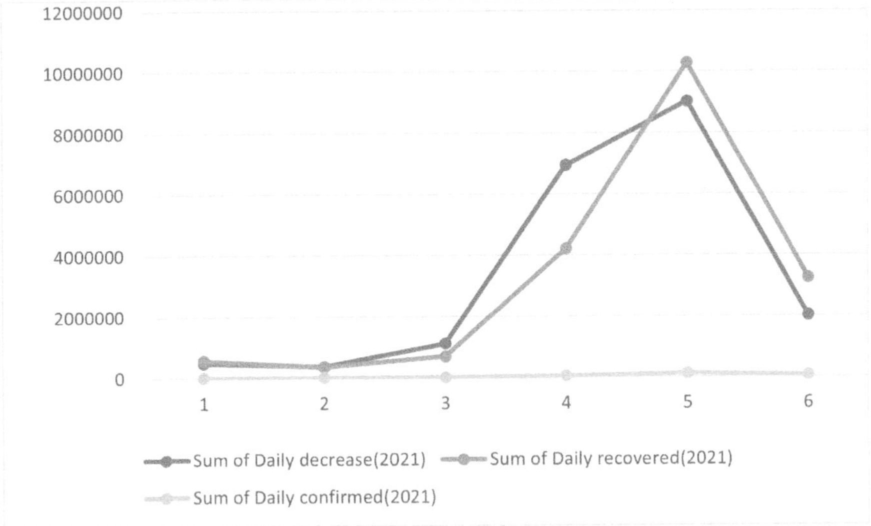


*Figure 9.2* (a) Suitable workplace at home during COVID-19 for the year 2020. (b) Suitable workplace at home during COVID-19 for the year 2021.

The graph in Figure 9.10 depicts the trends in PM10, CO, PM2.5, and other averages. All industries were shut down due to the lockdown. Figure 9.11 shows that air quality is deteriorating year after year. Air quality had improved to a highly moderate level by 2020. As a result, we can argue that India's air quality has improved because of the lockdown.

*Figure 9.3* Active cases of COVID-19 state-wise.

## 9.6 PEOPLE HELPED THE CORONAVIRUS TO SURVIVE OR SPREAD

The survival and spread of coronavirus in India are due to people meeting. To combat the spread of the coronavirus, the government declared a statewide lockdown, but the Kumbh Mela was excluded. In the first 15 days of April, 1,700 people tested positive for COVID-19 due to the large crowds at the Kumbh. A large group of people broke the lockdown regulations. There was a massive gathering of people around the mosque in Delhi in March 2020 [25]. Daily wagers didn't have enough money to survive in the city due to the Indian government's statewide shutdown. They were attempting to return to their village in any way possible. On April 13, 2020, for example, a large number of people congregated at Bandra station. We can see lots of news that migrant workers were gathering around the bus stand and railway

*Figure 9.4* Total cases in different states of India.

station to go to their respective places. Because of all of these things, cases of the coronavirus started rising. These are the reasons that India witnessed its first COVID-19 wave. This continued until September 2020; after that, cases started declining. We can refer to Figure 9.1, which shows a lot of news about migrant workers congregating around bus stops and train stations to get to their destinations. As a result of all of these factors, coronavirus cases began to rise. These are the reasons why India experienced the first COVID-19 wave, which lasted until September 2020 [26]. Cases begin to decline after 2021, as shown in Figure 9.2a.

Until February, it appeared that India had triumphed over the coronavirus. The standards allowed for a lot of latitude, but no one was aware of these waves of coronavirus. People began meeting on a large scale in violation of the coronavirus guidelines when the guidelines were relaxed.

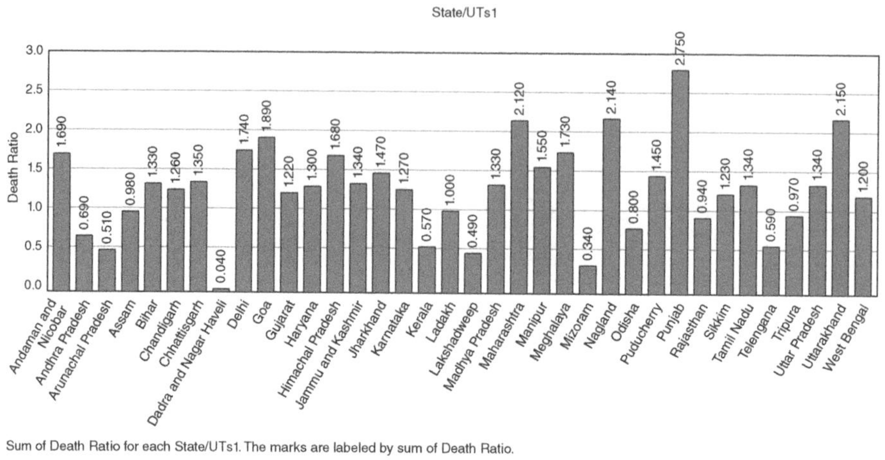


*Figure 9.5* Death ratio of different states in India.

People were also at ease because the immunization campaign began in January: So don't expect the coronavirus to resurface. Cases began to rise again in the month of March. Cases have increased sixfold since the end of March. India was hit very severely by the coronavirus this time. With roughly 400,000 instances every day, B.1.617 is a coronavirus variant discovered by scientists. India has recorded 400,000 deaths so far in the month of July, and the number is continually rising. India had 34,624,360 confirmed cases as of September 27, 2021, with 470,530 deaths owing to the coronavirus [27].

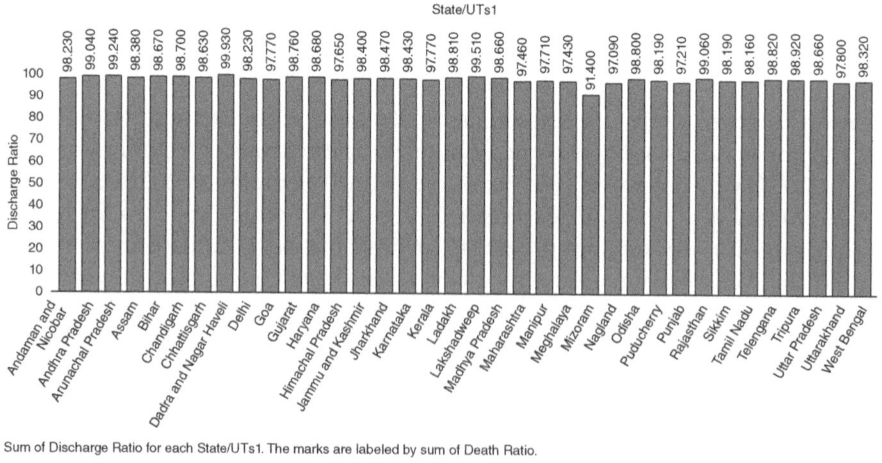


*Figure 9.6* COVID-19 discharge ratio in different states of India.

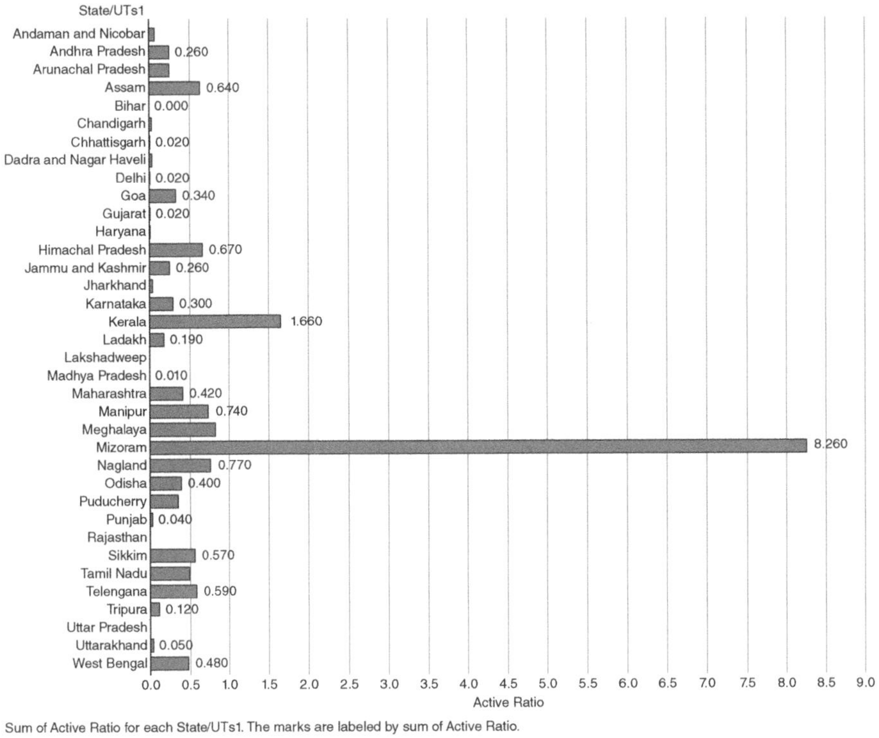


*Figure 9.7* COVID-19 active ratio in different states of India.

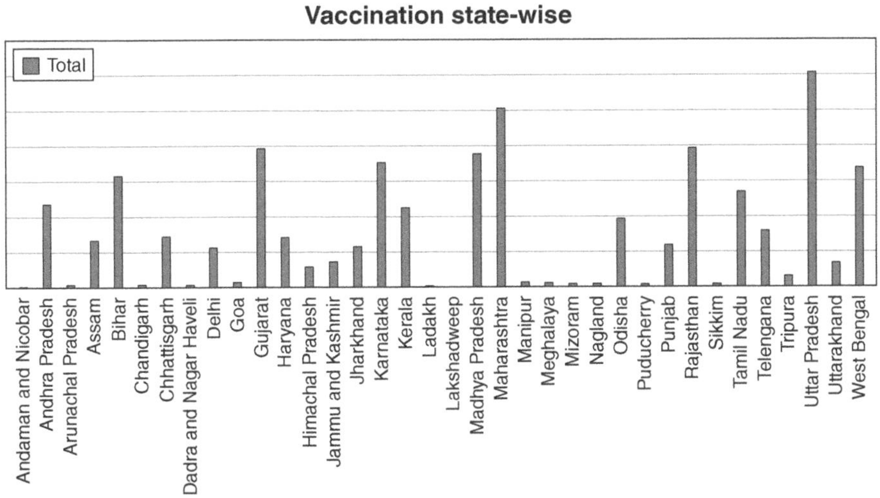


*Figure 9.8* Vaccination state-wise.

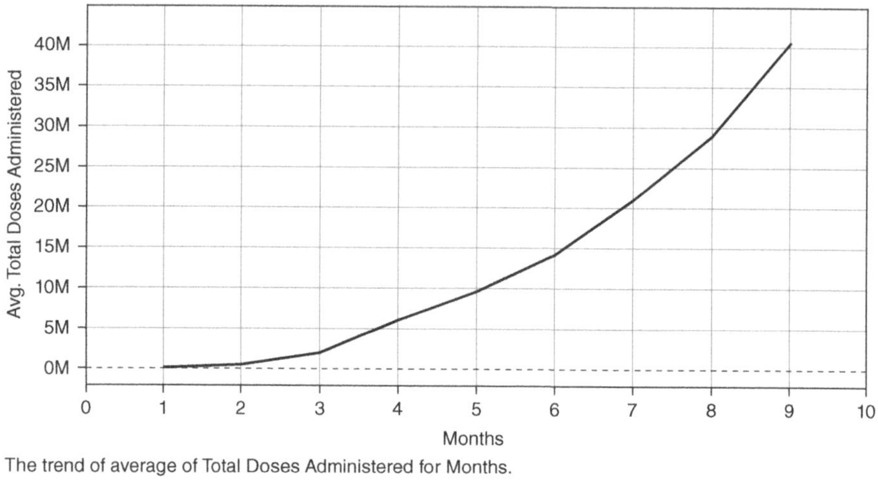


*Figure 9.9* Vaccine dosage average per month.

## 9.7 POSITIVE AND NEGATIVE IMPACT OF CORONAVIRUS

Society has been affected both positively and negatively by the coronavirus pandemic. It has expedited the process of innovation and digital transformation within the healthcare sector. Conversely, it has engendered substantial human casualties, disrupted economic equilibrium, and exacerbated preexisting social inequalities.

### 9.7.1 Positive impacts of coronavirus

- Environment
  1. *Air Quality Better:* The air quality improved in 2020, as shown in Figure 9.1. Because people's movement was restricted, many factories and industries were forced to close due to the lockdown. As a result, all major sources of air pollution have been shut down for an indefinite period of time.
  2. *The Water Is Purified:* People's movement is restricted due to the lockdown. Many factories were forced to close. As a result, rivers around cities are becoming cleaner. The main cause of polluted rivers, according to experts, is sewage [8].

### 9.7.2 Negative impacts of coronavirus

i. Degraded the lifestyle of humans.
ii. The Indian economy falls.
iii. Social distancing norms. Due to this, people are not ready to meet new people.

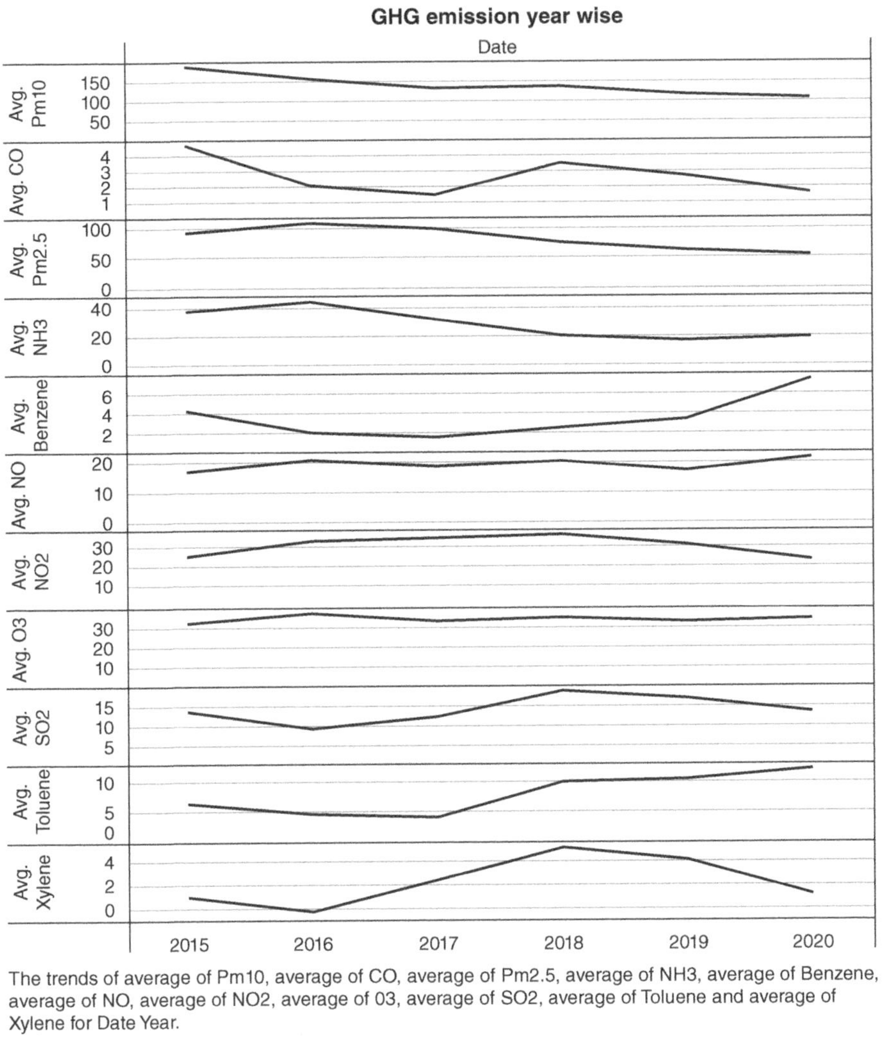


*Figure 9.10* Trends of GHG available in the air.

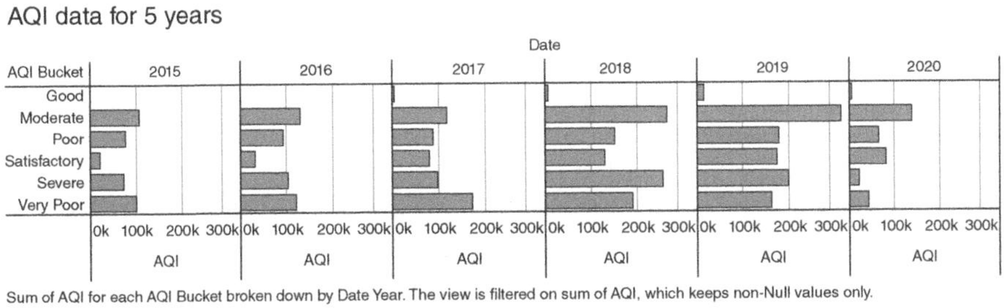


*Figure 9.11* AQI from 2015 to 2020.

## 9.8 GOVERNMENT RULES AND INITIATIVES TO COMBAT THE CORONAVIRUS

Government regulations and efforts to combat the coronavirus encompass requirements for wearing masks, measures for maintaining social distance, and campaigns to promote immunization. These steps are crucial for mitigating the transmission of the virus and safeguarding public health.

### 9.8.1 Lockdown and other restrictions

On January 25, 2020, the government began inspecting passengers at the border and airport. Guidelines on coronavirus were released by the Ministry of Health and Family Welfare. On March 16, 2020, the government imposed a ban on public gatherings. From March 22, 2020, all international flights were halted. On March 25, 2020, the government enforced its first lockdown; on April 14, 2020, the lockdown was extended; on May 1, 2020, the lockdown was extended again; and so on until the first traces of coronavirus began to fade [28].

### 9.8.2 Vaccination

The Indian government began providing free vaccinations on January 16, 2021 [29]. As a result, every Indian will receive the vaccination, assisting India in becoming coronavirus-free as quickly as possible [30–32].

## 9.9 CONCLUSION

This chapter suggests that the primary means by which the coronavirus is transmitted; these are through governmental regulations and negligent individuals. However, India effectively handled the circumstances. It can be inferred that the implementation of closure measures yields favorable environmental outcomes while detrimentally impacting the economy of India. Relatively rare occurrences of the coronavirus are observed in a country of this magnitude. The primary factors contributing to India's triumph were the enforcement of a curfew, stringent regulations, and immunization policies. The vaccination of Native Americans effectively impeded the coronavirus transmission. It is extremely unlikely that the third phase of coronavirus will emerge if all individuals are sufficiently immunized. In addition, several Indian cities experienced an improvement in air quality as a result of the closure. Figures 9.10 and 9.11 provide valuable insights into this matter. Any government can benefit from the findings of this study. Individuals can gain insight from the fact that India encountered two coronavirus outbreaks due to apathy and a lack of knowledge concerning COVID-19. The government can increase public awareness of all diseases via frontline personnel, advertisements, and other means. For the prevention of diseases like COVID-19, vaccination is permissible in all countries, and people are adhering to the instructions precisely. The air quality in some cities improved as a result of the shutdown. Following the COVID-19 pandemic, an expert can gain fresh perspectives and formulate a plan to enhance air quality.

## REFERENCES


1. L.P. Samaranayake, C.J. Seneviratne, and K.S. Fakhruddin, "Coronavirus Disease 2019 (COVID-19) Vaccines: A Concise Review," in *Oral Diseases*, vol. 2, pp. 2326–2336, 2022. doi: 10.1111/odi.13916.
2. S.U. Kumar, D.T. Kumar, B.P. Christopher, and C.G.P. Doss, "The Rise and Impact of COVID-19 in India," in *Frontiers in Medicine (Lausanne)*, vol. 7, p. 250, 2020. doi: 10.3389/fmed.2020.00250.
3. S. Singh, K.S. Parmar, and J. Kumar, "Dynamics of COVID-19 in India Using WANFIS (Wavelet Adaptive Neuro-Fuzzy Inference System) Model," in *Discover Public Health*, vol. 22, no. 1, p. 270, 2025. doi: 10.1186/s12982-025-00670-y.
4. S. Baas, A. Braaksma and R.J. Boucherie, "Constrained Markov Decision Processes for Response-Adaptive Procedures in Clinical Trials with Binary Outcomes," in *Annals of Operations Research*, 2025. doi: 10.1007/s10479-025-06703-8.
5. M. Gupta et al., "Tracing the COVID-19 Spread Pattern in India through a GIS-Based Spatio-Temporal Analysis of Interconnected Clusters," in *Scientific Reports*, vol. 14, no. 1, 2024. doi: 10.1038/s41598-023-50933-4.
6. H. Gao, W. Li, M. Pan, Z. Han, and H.V. Poor, "Modeling COVID-19 with Mean Field Evolutionary Dynamics: Social Distancing and Seasonality," in *Journal of Communications and Networks*, vol. 23, no. 5, pp. 314–325, 2021. doi: 10.23919/JCN.2021.000032.
7. Y. Wang and S.C. Kumbhakar, "COVID-19 Under-Reporting: Spillovers and Stringent Containment Strategies of Global Cases," in *Journal of Productivity Analysis*, vol. 63, no. 1, pp. 87–106, 2025. doi: 10.1007/s11123-024-00741-3.
8. P. Ghosh, R. Ghosh, and B. Chakraborty, "COVID-19 in India: Statewise Analysis and Prediction," in *JMIR Public Health and Surveillance*, vol. 6, no. 3, 2020. doi: 10.2196/20341.
9. Lockdown Leads to Cleaner Water Bodies; Minister Says Time to 'Introspect', Make India Water-Secured, *The Hindu*, May 22, 2022. https://www.thehindu.com/news/national/lockdown-leads-to-cleaner-water-bodies-minister-says-time-to-introspect-make-india-water-secured/article31650233.ece
10. M.K. Arti, "Mathematical Modeling of COVID-19 and Prediction of Upcoming Wave," in *IEEE Journal on Selected Topics in Signal Processing*, vol. 16, no. 2, pp. 300–306, 2022. doi: 10.1109/JSTSP.2022.3152674.
11. S. Patel, M. Kumar, G.C. Sahoo and S.K. Sahu, "Vector-Borne Diseases and Their Role in COVID-19 Dynamics and Death Rates: Focus on India," in *Folia Microbiologica (Praha)*, vol. 70, no. 4, pp. 899–906, 2025. doi: 10.1007/s12223-025-01313-0.
12. M. Rawal, "An Analysis of COVID-19 Impacts on Indian Education System," in *Educational Resurgence Journal*, vol. 2, no. 5, pp. 35–40, 2021. https://www.researchgate.net/publication/370060944
13. S.S. Jatav, "Risk, Perception and COVID-19 Impact on Food Security: Evidence from Bundelkhand Region, India," in *Discover Sustainability*, vol. 5, no. 1, 2024. doi: 10.1007/s43621-024-00267-6.

14. A. Ayoub et al, "Classification and Categorization of COVID-19 Outbreak in Pakistan," in *Computers, Materials and Continua*, vol. 69, no. 1, pp. 1253–1269, 2021. doi: 10.32604/cmc.2021.015655.
15. P. Gulia et al., "A Systematic Review on Fuzzy-Based Multi-Objective Linear Programming Methodologies: Concepts, Challenges and Applications," in *Archives of Computational Methods in Engineering*, vol. 30, pp. 4983–5022, 2023. doi: 10.1007/s11831-023-09966-1.
16. R.O. Millan-Guerrero, R. Caballero-Hoyos and J. Monarrez-Espino, "Poverty and Survival from COVID-19 in Mexico," in *Journal of Public Health*, vol. 43, no. 3, pp. 437–444, 2021. doi: 10.1093/pubmed/fdaa228.
17. P. Sangwan, R. Kumar, A. Kaur and G. Dhiman, A Comprehensive Study of Various Fuzzy C-Means Clustering Algorithms. In *AI-Enabled Multiple-Criteria Decision-Making Approaches for Healthcare Management* (pp. 161–180). Medical Info Science Reference, 2022. doi: 10.4018/978-1-6684-4405-4.ch009.
18. W. Lu et al., "Survival Analysis and Risk Factors in COVID-19 Patients," in *Disaster Medicine and Public Health Preparedness*, vol. 16, no. 5, pp. 1916–1921, 2022. doi: 10.1017/dmp.2021.82.
19. S. Bobdey et al., "An Analysis of Mortality and Survival of COVID 19 Patients Admitted to a Tertiary Care Hospital in Maharashtra, India," in *Medical Journal Armed Forces India*, vol. 77, pp. S353–S358, 2021. doi: 10.1016/j.mjafi.2021.02.004.
20. V. Joshi and R.K. Chandrawat, "Comparative Analysis of Different Structure of the Crisp and Fuzzy Regression Equations Using Realistic Data," in *Think India Journal*, vol. 22, no. 37, pp. 1288–1303, 2019.
21. M.P. Barman, T. Rahman, K. Bora and C. Borgohain, "COVID-19 Pandemic and Its Recovery Time of Patients in India: A Pilot Study," in *Diabetes and Metabolic Syndrome: Clinical Research and Reviews*, vol. 14, no. 5, pp. 1205–1211, 2020. doi: 10.1016/j.dsx.2020.07.004.
22. Pinki et al., "Artificial Intelligence-Enabled Smart City Management Using Multi-Objective Optimization Strategies," in *Expert Systems*, vol. 42, no. 1, p. e13574, 2024. doi: 10.1111/exsy.13574.
23. J. Ghosh, S. Ganguly, D. Mondal, P. Pandey, D. Dabkara and B. Biswas, "Perspective of Oncology Patients during COVID-19 Pandemic: A Prospective Observational Study From India," in *JCO Global Oncology*, vol. 6, pp. 844–851, 2020. https://ascopubs.org/go/authors/open-access
24. S. Slathia, R. Kumar, M. Lone, W. Viriyasitavat, A. Kaur and G. Dhiman, "A Performance Evaluation of Situational-Based Fuzzy Linear Programming Problem for Job Assessment," in *Proceedings of Third International Conference on Advances in Computer Engineering and Communication Systems. Lecture Notes in Networks and Systems* (pp. 655–667). Springer, 2023. doi: 10.1007/978-981-19-9228-5_56.
25. S. Bhattacharya et al., "Deep Learning and Medical Image Processing for Coronavirus (COVID-19) Pandemic: A Survey," in *Sustainable Cities and Society*, vol. 65, p. 102589, 2021. doi: 10.1016/j.scs.2020.102589.
26. G.J.B. Sousa, T.S. Garces, V.R.F. Cestari, R.S. Florêncio, T.M.M. Moreira and M.L.D. Pereira, "Mortality and Survival of COVID-19," in *Epidemiology and Infection*, vol. 148, p. e123, 2020. doi: 10.1017/S0950268820001405.
27. H. Chang and D.Y. Yeung, "Locally smooth metric learning with application to image retrieval," in *Proceedings of the IEEE International Conference on Computer Vision*, 2007. doi: 10.1109/ICCV.2007.4408862.
28. J. Ghosh, S. Ganguly, D. Mondal, P. Pandey, D. Dabkara and B. Biswas, "Perspective of Oncology Patients during COVID-19 Pandemic: A Prospective Observational Study from India," in *JCO Global Oncology*, vol. 6, pp. 844–851, 2020. https://ascopubs.org/go/authors/open-access

29. V.M. Kumar, S.R. Pandi-Perumal, I. Trakht and S.P. Thyagarajan, "Strategy for COVID-19 Vaccination in India: The Country with the Second Highest Population and Number of Cases," in *NPJ Vaccines*, vol. 6, no. 1, p. 60, 2021. doi: 10.1038/s41541-021-00327-2.
30. S.L. Tripathi, N. Mendiratta, D. Ghai, S. Avasthi and K. Dhir, Coronavirus: Diagnosis, Detection, and Analysis. In *Biomedical Engineering Applications for People with Disabilities and the Elderly in the COVID-19 Pandemic and Beyond* (pp. 109–117). Academic Press, 2022.
31. P.K. Shukla, J.K. Sandhu, A. Ahirwar, D. Ghai, P. Maheshwary and P.K. Shukla, "Multiobjective Genetic Algorithm and Convolutional Neural Network Based COVID-19 Identification in Chest X-Ray Images," in *Mathematical Problems in Engineering*, vol. 2021, no. 1, p. 7804540, 2021.
32. S.L. Tripathi, K. Dhir, D. Ghai and S. Patil, (Eds.). *Health Informatics and Technological Solutions for Coronavirus (COVID-19)*. CRC Press, 2021.

Chapter 10

# Wheelchair movements control model implementation with FPGA

*Suman Lata Tripathi and Inung Wijayanto*

## 10.1 INTRODUCTION

Recent advancements in embedded systems and digital design methodologies have substantially broadened the scope for developing assistive technologies aimed at enhancing the quality of life for individuals with physical disabilities. These innovations are increasingly being adopted in healthcare and rehabilitation domains, where the primary objectives include promoting user autonomy, improving mobility, and supporting daily functional activities. Among the various emerging solutions, gesture-controlled systems have gained prominence due to their intuitive, accessible, and user-centric approach to human–machine interaction [1]. Such systems are particularly advantageous in mobility aids, including powered wheelchairs, where traditional control mechanisms such as joysticks or mechanical switches may not be feasible for users with impaired motor function or limited functions. By interpreting deliberate physical gestures, typically involving the hand or arm, these systems convert user intent into control signals capable of actuating motors or other functional components. This paradigm eliminates the reliance on complex physical interfaces, thereby reducing both cognitive and physical strain on the user while enhancing ease of operation and overall user experience [2].

## 10.2 DESIGN METHODOLOGY

To design, simulate, and implement a gesture-based wheelchair control system using pulse-width modulation (PWM) in Vivado and deploy it on an field-programmable gate array (FPGA), hand gestures (simulated by switches) as input and LEDs as output indicators for motor control are used [3–7].

1. To design and implement a PWM module in Verilog/VHDL for motor control simulation based on hand gesture inputs using FPGA switches.
2. To simulate and verify the PWM behavior and gesture recognition logic in Vivado, mapping PWM outputs to LEDs to represent motor actions such as speed and direction.

 DOI: 10.1201/9781042015931-10

3. To deploy and evaluate the real-time performance and accuracy of the complete FPGA-based system in interpreting hand gestures and generating corresponding LED responses.

FPGA-based wheelchair implementation required Xilinx Vivado for HDL-based design simulation and verification. Vivado is a design suite developed by Xilinx (now AMD) used for creating, simulating, and implementing digital circuits on FPGAs, offering tools for synthesis, simulation, timing analysis, and device programming. An FPGA is a reconfigurable integrated circuit that can be programmed to perform specific digital logic functions, allowing for custom hardware designs without manufacturing new chips.

### 10.2.1 Block diagram

The system is designed to control a wheelchair using hand gestures interpreted as 4-bit digital signals. The gesture input is fed into a gesture decoder, which maps specific gesture patterns to movement commands (e.g., forward, backward, stop). Based on the decoded gesture, a 4-bit pwm_enable signal is generated to control the speed and direction of the wheelchair. Figure 10.1 shows the block diagram of wheelchair model with FPGA.

A high-frequency clock (clk) is passed to a clock divider, which reduces the frequency to generate a slower PWM clock suitable for the PWM signals. The four parallel PWM generator modules are driven by clock that are responsible for producing a PWM pulse depending on its enable signal. The generated PWM signal is sent to LED for the motor control of wheelchair system that does a precise control of motion on receiving signals from intuitive hand gestures leading to the actions like forward/reverse movement, turning, and emergency stop.

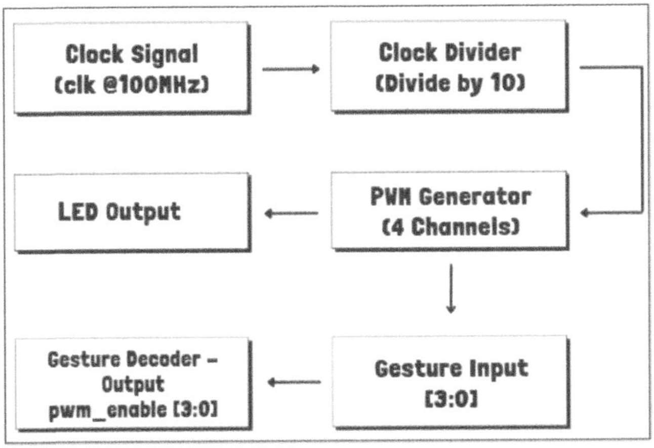


*Figure 10.1* Block diagram of wheelchair model with FPGA.

## 10.3 VERILOG DESIGN MODULES AND VERILOG CODE

Tables 10.1 and 10.2 describe the Verilog design and test modules of wheelchair control through FPGA, respectively.

*Table 10.1* Verilog code

```
module gesture_wheelchair_control(
   input clk,
   input [3:0] gesture,
   output [3:0] led
);

   wire pwm_clk;
   clock_divider #(.DIV_FACTOR(10)) clkdiv (
      .clk_in(clk),
      .clk_out(pwm_clk)
   );

   reg [3:0] pwm_enable;
   wire [3:0] pwm_out;

   genvar i;
   generate
      for (i = 0; i < 4; i = i + 1) begin : pwm_gen
         pwm pwm_inst (
            .clk(pwm_clk),
            .enable(pwm_enable[i]),
            .pwm_out(pwm_out[i])
         );
      end
   endgenerate

   assign led = pwm_out;

   always @(*) begin
      case (gesture)
         4'b0000: pwm_enable = 4'b0000; // Stop
         4'b0001: pwm_enable = 4'b1010; // Forward slow
         4'b0010: pwm_enable = 4'b1110; // Forward fast
         4'b0011: pwm_enable = 4'b0110; // Slight right
         4'b0100: pwm_enable = 4'b1001; // Slight left
         4'b0101: pwm_enable = 4'b0101; // Backward slow
         4'b0110: pwm_enable = 4'b0111; // Backward fast
         4'b0111: pwm_enable = 4'b1001; // Rotate left
         4'b1000: pwm_enable = 4'b0110; // Rotate right
         4'b1001: pwm_enable = 4'b1100; // 180° turn
         4'b1010: pwm_enable = 4'b0011; // Emergency reverse
         default: pwm_enable = 4'b0000; // Stop
      endcase
   end

endmodule
```

*(Continued)*

*Table 10.1 (Continued)*

```
module pwm(
   input clk,
   input enable,
   output reg pwm_out = 0
);

   parameter WIDTH = 8;
   parameter DUTY = 128;

   reg [WIDTH-1:0] counter = 0;

   always @(posedge clk) begin
      counter <= counter + 1;
      if (enable)
         pwm_out <= (counter < DUTY);
      else
         pwm_out <= 0;
   end

endmodule

module clock_divider #(
   parameter DIV_FACTOR = 10
)(
   input clk_in,
   output reg clk_out = 0
);

   reg [$clog2(DIV_FACTOR)-1:0] counter = 0;

   always @(posedge clk_in) begin
      if (counter == DIV_FACTOR/2 - 1) begin
         clk_out <= ~clk_out;
         counter <= 0;
      end else begin
         counter <= counter + 1;
      end
   end

endmodule

module tb_gesture_wheelchair_control;

   reg clk;
   reg [3:0] gesture;
   wire [3:0] led;

   gesture_wheelchair_control uut (
      .clk(clk),
      .gesture(gesture),
      .led(led)
   );
```

*(Continued)*

*Table 10.1 (Continued)*

```
  initial clk = 0;
  always #5 clk = ~clk;
  initial begin
    gesture = 4'b0000;
    #100;

    gesture = 4'b0001; // Forward slow
    #100;

    gesture = 4'b0010; // Forward fast
    #100;

    gesture = 4'b0011; // Slight right
    #100;

    gesture = 4'b0100; // Slight left
    #100;

    gesture = 4'b0101; // Backward slow
    #100;

    gesture = 4'b0110; // Backward fast
    #100;

    gesture = 4'b0111; // Rotate left
    #100;

    gesture = 4'b1000; // Rotate right
    #100;

    gesture = 4'b1001; // 180 turn
    #100;

    gesture = 4'b1010; // Emergency reverse
    #100;

    gesture = 4'b0000; // Stop
    #100;

    $stop;
  end
endmodule
```

Table 10.1 shows the Verilog code which implements a gesture-based control system for a wheelchair using PWM outputs to represent motor control signals. The main module, gesture_wheelchair_control, takes in a fast input clock and a 4-bit gesture signal, and generates four PWM-controlled output signals connected to LEDs that simulate motor behavior. The system includes a clock divider (clock_divider) that generates a slower clock signal (pwm_clk) suitable for driving the PWM modules. The PWM signals are generated using

*Table 10.2* Design constraint (.xdc)

```
## Clock signal (100 MHz)
set_property PACKAGE_PIN W5 [get_ports clk]
set_property IOSTANDARD LVCMOS33 [get_ports clk]

## Gesture inputs (Switches SW0–SW3)
set_property PACKAGE_PIN V17 [get_ports {gesture[0]}]
set_property IOSTANDARD LVCMOS33 [get_ports {gesture[0]}]

set_property PACKAGE_PIN V16 [get_ports {gesture[1]}]
set_property IOSTANDARD LVCMOS33 [get_ports {gesture[1]}]

set_property PACKAGE_PIN W16 [get_ports {gesture[2]}]
set_property IOSTANDARD LVCMOS33 [get_ports {gesture[2]}]

set_property PACKAGE_PIN W17 [get_ports {gesture[3]}]
set_property IOSTANDARD LVCMOS33 [get_ports {gesture[3]}]

## LED outputs (LED0–LED3)
set_property PACKAGE_PIN U16 [get_ports {led[0]}]
set_property IOSTANDARD LVCMOS33 [get_ports {led[0]}]

set_property PACKAGE_PIN E19 [get_ports {led[1]}]
set_property IOSTANDARD LVCMOS33 [get_ports {led[1]}]

set_property PACKAGE_PIN U19 [get_ports {led[2]}]
set_property IOSTANDARD LVCMOS33 [get_ports {led[2]}]

set_property PACKAGE_PIN V19 [get_ports {led[3]}]
set_property IOSTANDARD LVCMOS33 [get_ports {led[3]}]
```

a parameterized PWM module which controls the duty cycle based on an 8-bit counter, with a default 50% duty cycle when enabled. Each bit in the 4-bit pwm_enable signal enables or disables the corresponding PWM module, allowing specific motor actions to be controlled based on FPGA.

The FPGA recognize different gesture codes from 4'b0000 to 4'b1010 that are mapped to PWM enable patterns, which determines active motors and their directions. In this model, a gesture input of 4'b0001 corresponds to a "forward slow" movement where specific motors are enabled via the pwm_enable signal 4'b1010. The forward, backward, turns, and any other emergency actions are dependent on gesture inputs.

Figure 10.3 shows the constraint file for the Verilog code in Figure 10.2. This code specifies the physical pin assignments and electrical standards for the FPGA implementation of the gesture-based wheelchair control system. The clock signal, operating at 100 MHz, is mapped to pin W5 and configured to use the LVCMOS33 I/O standard. The 4-bit gesture input, which is typically provided through switches (SW0–SW3), is assigned to FPGA pins V17, V16, W16, and W17, respectively, with each signal also using the LVCMOS33 standard. Similarly, the 4-bit LED output, used to indicate the

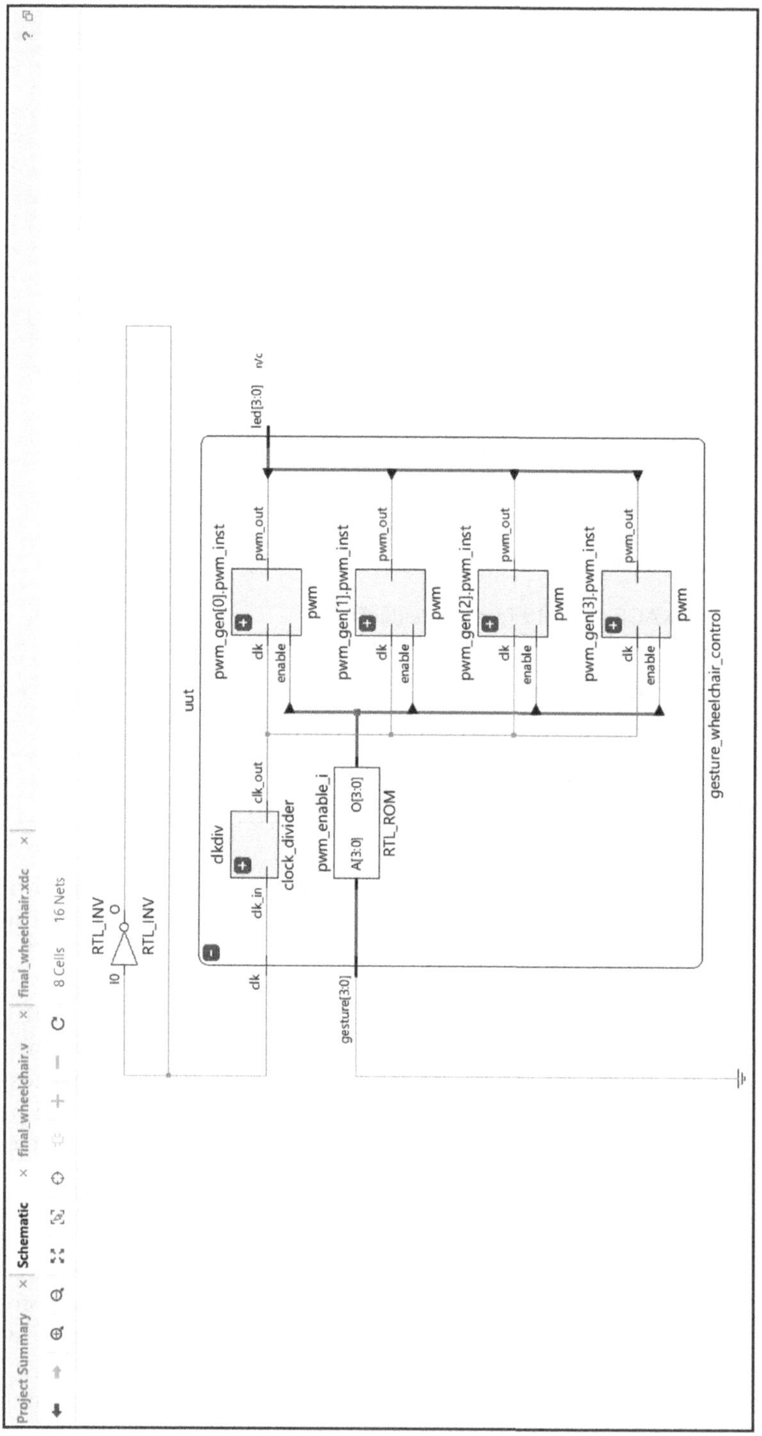


*Figure 10.2* Hierarchical view of the RTL schematic.

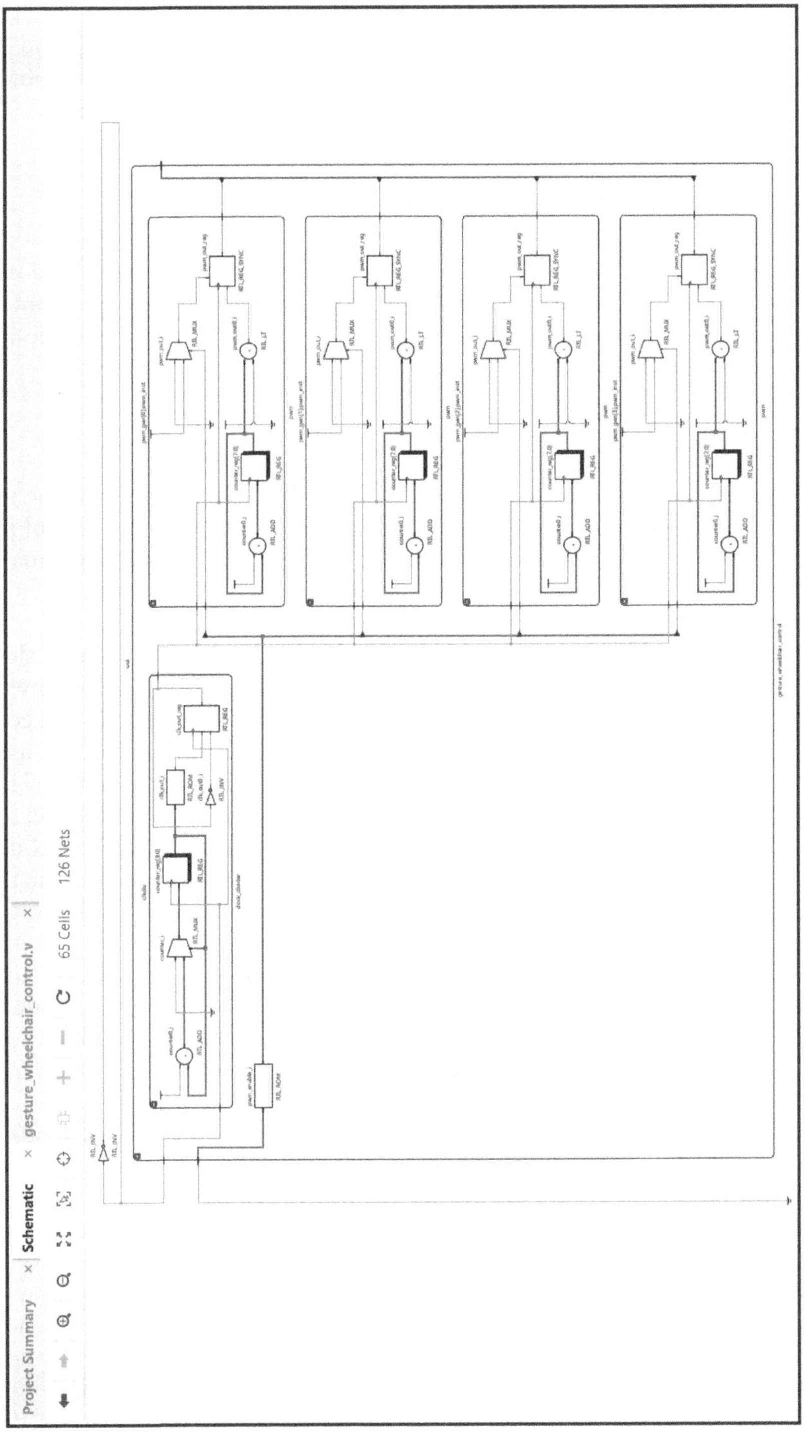


*Figure 10.3* Flattened view of the RTL schematic.

PWM-controlled motor states (LED0–LED3), is mapped to pins U16, E19, U19, and V19. These constraints ensure proper electrical interfacing and signal routing between the FPGA and the external components, enabling reliable hardware implementation and testing of the designed control system.

## 10.4 RESULTS AND DISCUSSION

Register transfer level (RTL) description is obtained through Verilog design and test modules written on Xilinx Vivado platform for wheelchair control. Further, the design is implemented with FPGA by adding top module and design constraint in Xilinx Vivado.

### 10.4.1 RTL schematic

Figure 10.2 is the hierarchical representation of the RTL schematic. The gesture_wheelchair_control system design is structured in well-defined blocks that correspond to separate modules in the Verilog code. Abstraction makes it easier to picture how the top-level module manages the interaction between its subcomponents.

At the top, the clk input is fed into a clock_divider module, which slows down the high-speed system clock into a slower clock (clk_out) named pwm_clk, that is better for driving PWM signals. Then, this slower clock is directed to each of the four identical PWM generator instances which is formed using a generate loop. Each PWM instance is passed its respective enable signal (enable) and gives a PWM output (pwm_out). Another key module seen in this view is an implicitly generated ROM-like logic block for gesture decoding (represented as RTL_ROM). The block captures the case-statement logic that translates each 4-bit gesture to a 4-bit pwm_enable signal. The enable vector dictates which PWM channels are enabled for a specific gesture so that various movement commands may be emulated (e.g., forward, backward, turn). The final output of each PWM module is directed to the led vector, which is the visible indication or control signal for the gesture. Generally, the hierarchical view emphasizes the modular structure, ease of debugging, and clarity of top-level signal flow.

Figure 10.3 shows the flattened view where the RTL schematic expands all hierarchical blocks to show their internal logic elements, which exposes the exact synthesized gates and registers that will be mapped to hardware.

The clock_divider module is shown with counters and combinational logic used to toggle the output clock. The logic includes an adder (RTL_ADD), a comparator (RTL_EQ), and a multiplexer for toggling the clk_out when the counter attains half the division factor. Each instance of PWM generator is separated into its core logic: an 8-bit counter (counter_reg), a comparator (RTL_LT) that compares the counter with a specified duty cycle (128), and logic to output a high or low PWM signal conditionally based on the enable input. This conditional output logic is made up of a multiplexer (RTL_MUX) and a register (RTL_REG_SYNC) to provide synchronous output with respect to the pwm_clk. Here, the gesture decoder logic, which was abstracted as a ROM in the hierarchical view, appears as a detailed combinational logic network. It takes the 4-bit gesture input and produces the resulting 4-bit pwm_enable signal based on the hardcoded case structure. This mapping logic controls the enable lines for each of the PWM instances.

## 10.4.2 Simulation waveform

Figure 10.4 illustrates the dynamic response of the Verilog-based wheelchair control system to various gesture inputs over time. It provides a visual confirmation that the system correctly interprets each 4-bit gesture input and generates the corresponding PWM signals to control the simulated motor outputs, represented by the LED signals.

In the waveform, each gesture input change is clearly timestamped and sequentially applied such as 0001 for forward slow, 0010 for forward fast, 0101 for backward slow, and so on. For each change, the output signals (led[3:0]) respond by toggling according to the internal PWM logic, which is based on the enabled bits specified in the design's case statement. These toggling patterns confirm that the PWM modules are functioning correctly and modulating the signal at a consistent duty cycle, providing visual proof that motor directions and speeds are being properly simulated. The simulation shows a predefined behavior of proposed FPGA-based wheelchair model meeting the timing requirement and on-chip power consumptions.

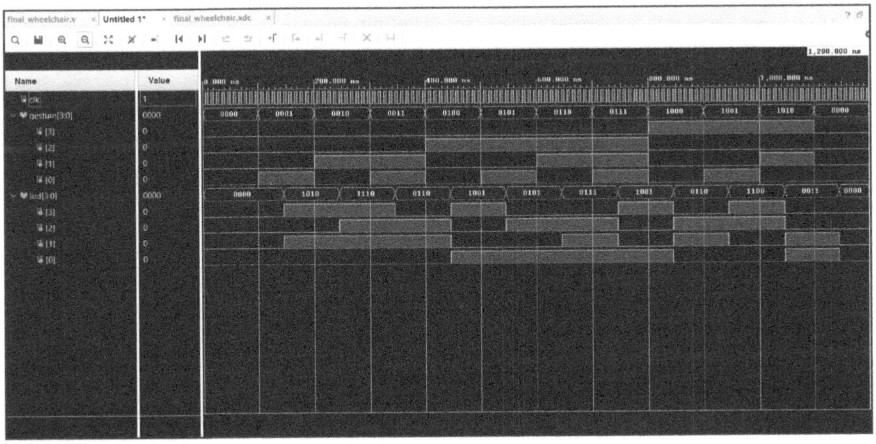


*Figure 10.4* The simulation waveform of Verilog-based wheelchair control system.

### 10.4.3 Power and timing analysis

Figures 10.5 and 10.6 show the power and timing analysis of the design, respectively, revealing that the system is both thermally efficient and meets all timing constraints. The total on-chip power consumption is reported to be 0.1 W, with the majority, approximately 97% or 0.097 W, attributed to static power. This means leakage currents control the power consumption, as it is typical in contemporary FPGA-based systems, particularly when there is little logic activity. The dynamic power is low, merely 0.003 W or 3% of the total, and is spread across clocking (16%), signal routing (3%), logic computation (4%), and I/O operations (77%). The significant portion of dynamic power consumed by I/O suggests high activity or continuous toggling on output interfaces, such as LEDs or communication pins.

The design is no violation to timing constraints. The setup timing shows a worst negative slack (WNS) of 7.456 ns and total negative slack (TNS) of 0.000 ns, indicating no setup timing violations. The hold timing shows a worst hold slack (WHS) of 0.174 ns with no endpoints failing, ensuring the correct data latching. Also, the pulse width check passes with a worst pulse width slack (WPWS) of 4.500 ns across all evaluated endpoints. These results confirm that the design is working properly within the clock domain, with no violations.

### 10.4.4 FPGA implementation of wheelchair model

Figure 10.7a shows the FPGA board responding to the gesture input "0111", which represents the "Left" command. The input is given via slide switches, and the illuminated LEDs "1001" confirm that the design correctly interprets

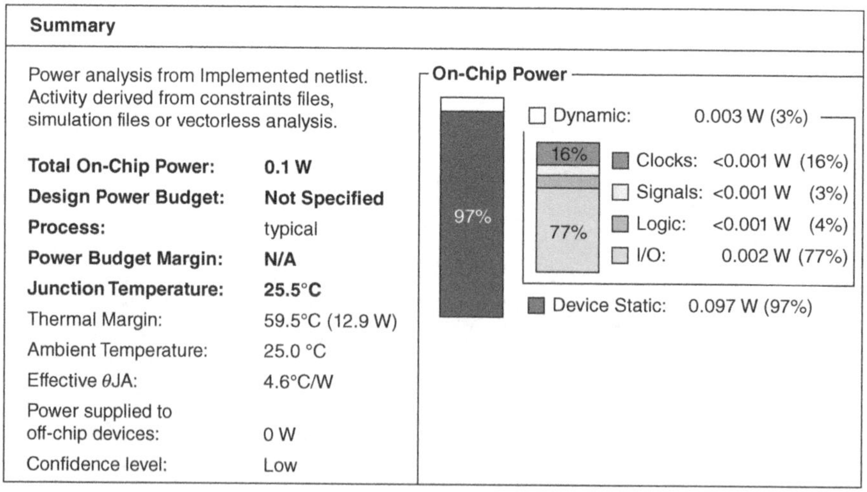


*Figure 10.5* Power performance.

**Design Timing Summary**

| Setup | | Hold | | Pulse Width | |
|---|---|---|---|---|---|
| Worst Negative Slack (WNS): | 7.456 ns | Worst Hold Slack (WHS): | 0.174 ns | Worst Pulse Width Slack (WPWS): | 4.500 ns |
| Total Negative Slack (TNS): | 0.000 ns | Total Hold Slack (THS): | 0.000 ns | Total Pulse Width Negative Slack (TPWS): | 0.000 ns |
| Number of Failing Endpoints: | 0 | Number of Failing Endpoints: | 0 | Number of Failing Endpoints: | 0 |
| Total Number of Endpoints: | 9 | Total Number of Endpoints: | 9 | Total Number of Endpoints: | 6 |

**All user specified timing constraints are met.**

*Figure 10.6* Timing report.

the gesture and activates the appropriate control logic. This demonstrates successful real-time hardware execution of the gesture-based control system.

Figure 10.7b shows the FPGA board responding to the gesture input "0010", which corresponds to the "Forward Fast" motion command. The binary input is set using the slide switches, and the illumination of specific LEDs "1110" on the board confirms that the system has correctly recognized the gesture. This verifies that the high-speed forward PWM control logic is functioning as intended, demonstrating the system's ability to differentiate between speed levels in directional movement.

*Figure 10.7* FPGA implementation of wheelchair model.

*Table 10.3* Wheelchair responses.

| *Simulation waveform* | | *Experimental results* | |
|---|---|---|---|
| *Gesture* | *LED* | *Gesture* | *LED* |
| 0111 (Left) | 1001 | 0111 (Left) | 1001 |
| 0010 (Forward Fast) | 1110 | 0010 (Forward Fast) | 1110 |
| 0001 (Forward Slow) | 1010 | 0001 (Forward Slow) | 1010 |
| 1001 (180° Turn) | 1100 | 1001 (180° Turn) | 1100 |

Figure 10.7c illustrates the gesture input "0001", mapped to the "Forward Slow" command. When this input is applied via the switches, the FPGA activates the slower PWM signal, which is indicated by the change in "1010" LED patterns. This confirms that the system can successfully distinguish and respond to varied motion speeds, ensuring controlled forward movement at a reduced pace.

Figure 10.7d presents the FPGA board executing the "180° Turn" operation in response to the binary input "1001". This command simulates a full directional reversal. The corresponding "1100" LED indicators light up, demonstrating that the FPGA logic correctly interprets and processes the input to produce the desired turning behavior, validating the functionality of turn-related motion control.

Table 10.3 shows simulation and experimental results of a gesture recognition system, where each gesture is represented by a 4-bit binary code and linked to a specific LED pattern. Gestures such as "Left," "Forward Fast," "Forward Slow," and "180° Turn" were tested. For every case, the LED outputs from the simulation match exactly with those obtained in experimental results. For instance, the gesture "0010" (Forward Fast) consistently produced the LED pattern "1110" in both scenarios. This consistency confirms that the system accurately detects gestures and reliably translates them into the correct LED outputs, demonstrating its effectiveness and real-world applicability.

## 10.5 FUTURE SCOPE

Advent of artificial intelligence and machine learning (AI ML) further helps in hardware customization with low latency, energy efficiency, and real-time data processing with FPGA [7–10]. AI ML also helps in design parameter optimization, placement and routing. Several AI ML–based AMD simulation based open access software are available for such design, verification, and implementation. Audio and image data processing through FPGAs, medical devices, and surgical robots are the upcoming areas that can be targeted with FPGA using AI ML techniques.

## 10.6 CONCLUSION

In this work, a gesture-controlled wheelchair control system was successfully developed and implemented using Verilog, deployed on an FPGA. The system was designed and simulated in Vivado, where hand gestures, represented by 4-bit switch inputs, were effectively translated into motor control commands using PWM. A modular PWM module was developed in Verilog and instantiated multiple times to simulate independent motor actions. The simulation results confirmed the accuracy and consistency of the PWM signals in response to different gesture inputs, validating the functionality of the control logic.

## REFERENCES


1. García-Gil, G., Del Carmen Del Carmen López-Armas, G., & De Jesús Navarro, J. (2025). Human–Machine Interaction: A Vision-Based Approach for Controlling a Robotic Hand through Human Hand Movements. *Technologies*, 13(5), 169. https://doi.org/10.3390/technologies13050169
2. Dritsas, E., Trigka, M., Troussas, C., & Mylonas, P. (2025). Multimodal Interaction, Interfaces, and Communication: A Survey. *Multimodal Technologies and Interaction*, 9(1), 6. https://doi.org/10.3390/mti9010006
3. Uppal, K., & Uppal, K. (2025, February 28). FPGA: Accelerating Innovation in Technology. Fidus Systems. https://fidus.com/blog/fpga-accelerating-innovation-in-technology.
4. Tripathi, S.L., Kumar, A., & Mahmud, M. (2022). FPGA for Secured Hardware & IP Ownership. In: *2022 IEEE International Conference of Electron Devices Society Kolkata Chapter (EDKCON)*, Kolkata, India. IEEE, pp. 184–189. doi: 10.1109/EDKCON56221.2022.10032935.
5. Badiganti, P.K., Peddirsi, S., Rupesh, A.T.J., & Tripathi, S.L. (2022). Design and Implementation of Smart Healthcare Monitoring System Using FPGA. In: Rawat, S., Kumar, A., Kumar, P., & Anguera, J. (Eds.) *Proceedings of First International Conference on Computational Electronics for Wireless Communications*, Lecture Notes in Networks and Systems, vol. 329. Springer, Singapore. https://doi.org/10.1007/978-981-16-6246-1_18
6. Tripathi, S. L., Saxena, S., Sinha, S. K., & Patel, G. S. (2022). *Digital VLSI Design and Simulation with Verilog*. John Wiley & Sons, Inc., New York, NY.
7. Prasanna, D.L., Tripathi, S.L. (2024). A Review on Tongue Based Assistive Technology, Devices and FPGA Processors Using Machine Learning Module. *Wireless Personal Communications*, 134, 151–170. https://doi.org/10.1007/s11277-024-10897-8
8. Tripathi, S. L., Prasanna, D. L., & Mahmud, M. (2023). A Low-Power Audio Processing Using Machine Learning Module on FPGA and Applications. In: *Explainable Machine Learning Models and Architectures*, John Wiley & Sons, Inc., New York, NY, pp. 49–63. https://doi.org/10.1002/9781394186570.ch4
9. Tripathi, S. L., & Mahmud, M. (Eds.). (2023). *Explainable Machine Learning Models and Architectures*. John Wiley & Sons, Inc., New York, NY.
10. Tripathi, S. L., Prakash, K. B., Balas, V. E., Mohapatra, S. K., & Nayak, J. (Eds.). (2021). *Electronic Devices, Circuits, and Systems for Biomedical Applications: Challenges and Intelligent Approach*. Academic Press, London.

Chapter 11

# Overview of emerging AI technologies and their potential impact on healthcare

*Shiva Khemka, Kapil Kumar Gupta, and Namrata*

## 11.1 INTRODUCTION

The architecture of modern healthcare is gradually being transformed by artificial intelligence (AI), which now transcends the boundaries of experimental systems. Entering the last decade, the machine learning (ML) models evolved from isolated systems, functioning as diagnostic aides, to platforms that may interpret intricate medical data, providing real-time support for clinicians and expediting scientific discoveries. The latest advancements in deep learning (DL), generative models, and large language systems have opened a new wave of excitement for AI, the application of AI in drug design, hospital workflows, genomics, and decision support.

These shifts are not just stand-alone innovations, they reflect the larger transformation of the healthcare system, which is increasingly beginning to operate less as a static service provider and more an adaptive learning environment underpinned by data and predictive intelligence. In such a scenario, AI allows early disease identification, tailored treatment, and widespread operational effectiveness. At the same time, explainable AI (XAI) and federated learning are solving age-old problems related to transparency, privacy, and bias-focused concerns as AI systems become more prevalent in patient-facing apps.

The advent of quantum-enhanced modelling, digital twins, and real-time multimodal platforms points to a future where synthetic worlds, cross-domain inference, and reasoning tools will increasingly inform clinical decisions. Though still in its infancy, the technology's path is obvious, AI is ceasing to play a supplementary role and becoming an integral part of medicine foundation. Understanding what these technologies are, how they are used, and their limitations is central to traversing the next phase of innovation in healthcare [1, 2].

### 11.1.1 Understanding AI and its role in healthcare

AI is a popular subfield of computer science that aims to develop intelligent systems that accomplish tasks that normally require human intelligence


DOI: 10.1201/9781042015931-11

functions like learning from experience, pattern recognition, problem-solving, and adaptive learning. In just a few past decades, AI has arguably progressed from a thought experiment to a fledgling powerhouse that is reshaping industries. Hardly anywhere is this transformation more evident than in the healthcare field.

At its most basic level, AI consists of algorithms that use data to direct their learning large datasets. Examples of such datasets are patients' records, diagnostic images, gene sequences, and study outcomes. The system, via AI, uses the data to look for patterns and make predictions or recommend actions. Significantly distinct from other software, the AI system can not only perform tasks based on preset instructions but also enhance them as new information is fed to them over time. The thing to consider is that AI is not working in isolation in the healthcare industry, but in an ecosystem where data is collected, analysed, and interpreted to guide clinical and operational decisions. Broadly speaking, this architecture consists of three interconnected layers:

- *Obtain and Process Data*: The information is gathered from numerous sources like electronic health records (EHRs), imaging equipment, biosensors, and laboratory tests, and then it has to get cleaned, standardized, and formatted for analysis. It is the most significant process in healthcare because data quality has a direct link to the safety of patient safety and the accuracy of AI outputs.
- *Modelling and Learning*: By applying ML techniques, possible patterns and correlation will be identified. DL is used for the interpretation of medical scans, and natural language processing (NLP) helps in analysing clinical notes. The models can be trained through supervised, unsupervised, or reinforcement methods through learning according to the application's requirement.
- *Decision Support and Output*: This layer provides evidence-based recommendations from AI models to assist clinicians in making informed decisions without replacing their judgement.

AI-driven applications in healthcare can be categorized into several broad types based on their functions:

- ML is used for making predictions such as patients' imminent risk for complications and forecasting disease outbreaks via data analysis.
- DL is a ML technique which is used to analyse complex inputs, such as genomic data.
- NLP is the part of AI which interprets clinical documents, and extracts data from unstructured records.
- Computer vision automates the interpretation of radiology images, pathology slides, or surgery video.
- Robotic process automation (RPA) is utilized to eliminate or simplify repetitive administrative tasks.

- GANs is a generative technique which enables the creation of synthetic medical images for training, thereby addressing the issue of insufficient data and privacy concerns.
- Transformers: Advanced DL architectures like M3H (multimodal multitask multitarget hierarchical) are capable of analysing image and genomic data in EHR to derive insights across tasks.

AI systems have revolutionized modern medical practice beyond the technical level that they possessed. The huge data resources that are fed into AI create practicable insights that support early diagnosis, specific therapy, prevention of medical complications, and much more. This frees up more time for patient care by alleviating the administrative burden on healthcare workers.

Bureaucracy, which is a major part of the healthcare system, is getting smoothened by a seamless infrastructure that operates in an interoperable manner, with data flowing into algorithms and producing clinically relevant outputs. Through AI, we can analyse the real-world data and obtain insights which are imperceptible to the human eye. For example, in radiology, AI can diagnose early disease manifestations with high sensitivity related to the imaging studies. In pathology, it reveals that some tiny structures within the cells exhibit minute changes in cases of specific conditions. Additionally, the genomics field positions AI to expedite the identification of mutations and therapeutic target candidates.

What makes AI unique is the fact that it never stops learning. The model's performance improves with every data point introduced, to transform and evolve as medical knowledge shifts and progresses across distinct patient groups.

### 11.1.2 Defining "Emerging" in the AI space

While some AI tools are being used in hospitals nowadays, "emerging" AI is a cutting-edge new technology which is still in its development phase, typically at the junction of several fields. Consider AI tools that can decode immune responses, create drug molecules from scratch, or foresee mental health shifts from the way someone talks. These are not mainstream (yet). Emerging AI is not so much about what exists today as about what is becoming really fast.

### 11.1.3 Healthcare data types

The success of AI in the healthcare sector relies on the diversity and abundance of the training data. The categorical types of data for healthcare are as follows:

- *Electronic Medical Records (EMR):* Structured information such as demographics, medications, lab results. and histories of treatment for prediction purposes.
- *Medical Imaging:* Imaging data (e.g. RAD) processed through CNNs on X-rays, CT, and MRIs for disease and anomaly detection.

- *Genomics/Proteomics:* High-dimensional sequencing data for precision medicine such as therapies tailored to genetic characteristics.
- *Audio Data:* Patient's voice and cough sounds for model development with NLP and audiovisual DL methods to evaluate and screen for diseases such as Parkinson's disease and different respiratory disorders.
- *Time-Series Data:* Monitoring data from wearable devices, for example ECG, heart rate, and glucose levels in real time to facilitate timely interventions.

This multi-modal data is then used by AI systems for constructing models employable for actionable insights across clinical and operational settings.

### 11.1.4 Why healthcare needs emerging AI solutions

The modern healthcare landscape is encountering significant challenges such as enormous complicated patient data and limited human resources. Traditional systems are falling short when it comes to efficiently processing the large volumes of medical records, imaging, and genomic data in real time. That's why the adoption of AI technologies is no longer just a choice; it's becoming a necessity for tackling this complexity, facilitating faster decisions, and enabling scalable, data-driven care.

### 11.1.5 Evolution of AI in healthcare

The journey of AI in healthcare has come a long way, evolving from simple rule-based systems to sophisticated multimodal agents, showcasing a remarkable increase in transformative power (Figure 11.1).

1. *Early Foundations: Rule-Based Systems and ML (1970s–2000s)*
   - *MYCIN (1976):* This was one of the first expert systems that suggested antibiotics based on fixed, hand-coded rules. While it was quite accurate, it had its limitations when it came to scalability.

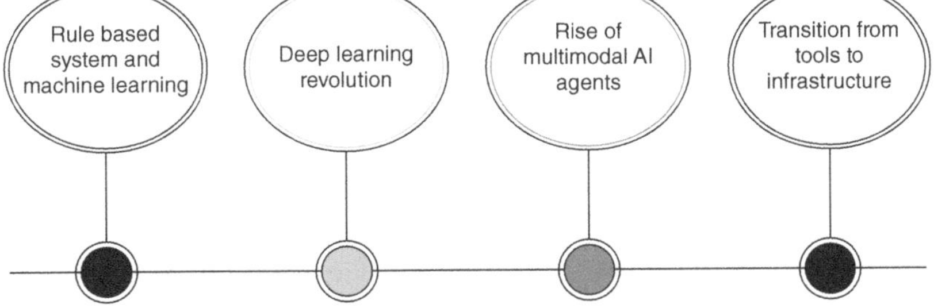


*Figure 11.1* The timeline of AI evolution in healthcare.

- *Emergence of ML:* By the 2000s, statistical models could analyse structured EHR data for tasks such as risk prediction. However, they were not able to deal with unstructured data like imaging.

2. *DL Revolution (2010s)*
   *CNNs:* Revolutionizing medical imaging by achieving human-level accuracy (e.g. 99.3% in breast cancer specificity).
   - NLP applications for clinical texts to gain insights into the links between symptoms and diagnoses.
   - Nevertheless, successful as they were, these early models were largely task-oriented and failed primarily on cross-domain data integration.
3. *Rise of the Multimodal AI Agents (2020s to Present)*
   The current systems integrate multiple data streams:
   - AlphaFold is expected to accelerate drug discovery by predicting protein structures, with atomic precision [3].
   - M3H (imaging, genomics, and EHRs) achieves 11.6% better performance than single-task models across 40 disease diagnoses [4].
   - Generative AI tools use synthetic medical data to generate outputs ensuring patient privacy.
4. *Transitioning from Tools to Infrastructure*
   AI is now firmly grounded inside healthcare systems:
   - *Diagnostics:* Anomaly detection in imaging systems in real time is reducing missed findings by 31% [4].
   - *Operations:* AI is supporting bed management in ICUs and resource allocation, resulting in a 37.5% reduction in waiting time [6].
   - *AIoT Ecosystems:* Wearable devices streaming real-time data for proactive care, for instance detecting arrhythmias with an accuracy of 98.7% [7].

The increasing complexity of healthcare, combined with technological advancements, has led us to a pivotal moment:

- *Global Challenges:* With aging populations, a rise in chronic diseases, and pandemics, there's a pressing need for scalable solutions.
- *Digital Maturity:* AI is harnessing the massive growth of healthcare data (projected to reach 165 zettabytes by 2025) and the power of computation for large-scale applications.

### 11.1.6 Milestones and achievements of AI in healthcare

Here are just a few amazing milestones achieved in the health sector through AI:

1. *MYCIN Expert System (1972):* Created at Stanford, MYCIN was one of the first AI systems aimed at helping with the diagnosis of bacterial infections, reaching accuracy levels rivalling that of human specialists,

and constituting the initial step into debating rule-based clinical decision support systems.

2. *IBM Watson for Oncology (2011):* In 2011, IBM Watson combined NLP with clinical data to offer cancer treatments. This resulted in the development of massive datasets that reduced diagnostic errors in the clinical trial. AI's role in precision medicine is increasingly backed by such compelling evidences.
3. *AlphaFold by DeepMind (2020):* AlphaFold accomplished a scientific coup in the prediction of protein structures with breathtaking accuracy. This step carries drug discovery into new territory and might very well represent a new dawn in computational biology [8].
4. *IDx-DR (2018):* In 2018, IDx-DR was the very first of its kind, a completely autonomous AI system to get FDA approval to provide a diagnosis to diabetic retinopathy patients. It worked independently without any input from specialists and improved access to diabetes screening in primary care settings.
5. *In Silico Drug Discovery (2021):* In silico, AI-developed drug ISM001-055 became the fastest to enter clinical trials, completing the distance in only 18 months while consuming the least possible budget. This demonstrated AI's capability to speed up and cut costs of drug innovation.

## 11.2 RESEARCH MOTIVATION

Healthcare systems are experiencing an increase in the number of patients, further driving complexity and constraints on resources. While traditional medical practices succeeded in offering primary healthcare services, they would not do well with this unprecedented scale, diversity, and speed of data. In such a situation, the use of AI promises to bring a new revolution by drawing on complex datasets and finding patterns and converting them into actionable insights for clinical decision-making.

However, the majority of current AI applications in healthcare are restricted to very specific tasks or functions, such as those related to diagnosis and administration. The true promise of this emerging technology lies in its potential to link diagnostic treatment planning, patient management and monitoring, as well as operational and resource management in smart data-driven systems.

This research is precisely motivated by the need to examine how AI's emerging technologies can fill the critical care delivery voids and positively impact clinical outcomes for the purpose of achieving better efficiency at scale. This study will also look into the ethical, infrastructure, and regulatory challenges of implementing and scaling similar technologies. While as global health evolves towards more predictive, personal, and preventive care models, it is urgent for companies to understand and leverage the role of emerging AI technologies in the business world.

## 11.3 LITERATURE REVIEW

Recent advances in AI for healthcare have witnessed key interdisciplinary research efforts in several domains, including diagnostics, therapeutics, operations, and policy. Over the past decade, AI has evolved from theoretical concepts to clinical validation and implementation in healthcare. Now, as AI systems become more sophisticated and autonomous in their reasoning, several interconnected issues concerning interpretability, bias, infrastructure, and governance have emerged as pressing topics in both academic and policymaking circles.

A recurrent theme in the literature is the struggle to balance AI's predictive power and transparency. The latter issue is known as explainable AI and seeks to address the looming lack of trust and accountability in clinical settings. Establishing frameworks that employ post hoc tools like SHAP and LIME to enhance model interpretability in clinical diagnostics was suggested by Saraswat et al. [9]. The researchers report that the sensitivity of the diabetic retinopathy detection system is comparable to that of less explicit DL systems. The same principle is advocated in the mental health field by Mamalakis et al. [10] for AI systems design which can potentially give psychiatric patients specific reasoning.

Generative AI has also moved from the experimental phase to real-world applications. Chen and Esmaeilzadeh [11] illustrated how GANs generated synthetic chest radiographs with 92% diagnostic equivalence to the original datasets. This advancement resolves several persistent challenges, including data scarcity, patient privacy, and the modelling of rare diseases. Ethics and governance issues in companies are a pervasive theme in literature. Nasir et al. [12] considered ethical frameworks under professional oversight, auditability, and fairness principles. These align with the FUTURE-AI principles developed by Lekadir et al. [13].

The operational advantages of AI are becoming more evident from studies done in practice. For instance, reinforcement learning has been used for bed allocation in ICUs, leading to a reduction in the average wait time of over 20% (Hassan Ali, 2022). [14] The swarm learning frameworks reviewed by Tajabadi et al. (2024) [15] promote decentralized clinical machine learning across independent institutional silos, satisfying strict data privacy and sovereignty regulations while allowing collaborative peer-to-peer model training without a central server.The systems of triage based on AI reduced overcrowding in the emergency room by 44% during the COVID-19 pandemic. NLP systems played a critical role in analysing evidence from over half a million scientific papers to enhance the speed of therapeutic discovery. The catch, though, is that most of these incredible advancements are currently benefiting high-income countries the most.

Beyond these immediate applications, the research also hints at some truly exciting possibilities that are slowly but surely entering clinical conversations. Think about "digital twins," as explored by Corral-Acero et al. [16] – these are like personalized simulations of individual patients that could revolutionize treatment planning. Wierzbicki et al. (2021) have even shown early success with 3D mixed-reality holographic imaging in surgery, helping surgeons pinpoint gastrointestinal and liver tumors more accurately in real time. [17] While these innovations are still in their early stages, they clearly point towards a future of medicine that's more personalized, predictive, and heavily supported by computational power.

In a nutshell, the body of literature on AI in health is constantly growing, incredibly diverse, and increasingly focused on delivering tangible results. While foundational research in diagnostics and therapeutics has undeniably proven AI's enormous potential, we're still grappling with significant challenges around ethics, infrastructure, and global inequalities. Overcoming these hurdles will demand strong collaboration across different fields, robust regulatory frameworks, and inclusive data practices. This isn't just about building new technology; it's about how we design, implement, and maintain these systems to ensure they prioritize fairness, transparency, and genuine clinical effectiveness for everyone.

## 11.4 CONCEPTUAL FRAMEWORK TO EVALUATE EMERGING AI IN HEALTHCARE

The increasing adaptation of AI in healthcare, from diagnostics and drug development to hospital management and digital therapeutics, requires close coherent framework for evaluating their performance, safety, and accessibility within healthcare systems in real world. The existing evaluation methods tend to focus narrowly on accuracy or other performance metrics, leaving aside larger, overarching issues about ethical sustainability, clinical relevance, and patient impact.

Here we propose a multidimensional conceptual framework for the evaluation of emerging AI systems in healthcare. This framework, based on principles from internationally accepted guidelines (FUTURE-AI, WHO/ITU AI4Health, DECIDE-AI), encompasses five interrelated domains: technical validity, clinical usability, systemic integration, ethical soundness, and patient-centred value. These domains are not independent; they are mutually interrelated and reinforcing. These are the layers that demonstrate the intricacies of an ecosystem where AI operates.

### 11.4.1 Technical validity

An AI must demonstrate excellent technical performance under clinically relevant conditions. This encompasses the following:

- The model's accuracy on datasets from different demographics and geographic locations and diversities.

- Interoperability mechanisms to interrogate and explain model behaviour.
- Adaptability over time to dynamic patterns of input (e.g. seasonality, mutations of various diseases).
- Ability to generalize across modals and validate on external datasets.

Performance must be measured and compared against the metrics applicable in the clinical field: AUC–ROC, recall, sensitivity and specificity, false positive burden, and calibration curves. Technical perfection is a must, but many models that perform perfectly in the lab cannot replicate those results in real world.

### 11.4.2 Clinical usability

No matter how sophisticated AI capabilities appear, if they are isolated from work flow and product usage frontlines, they will not be implemented. Clinical usability refers to the extent AI systems actually assists human decision-making rather than interfering with it.

The following factors should be taken into account:

- Seamless integration into EHRs, PACS, or other clinical systems
- Interface designs that allow interpretation in time-constrained settings
- The relevance of automated task (e.g. diagnostic triage, drug dosage prediction)
- Output actionability, that is, insight can be trusted and acted on by clinicians

Measuring clinical usability includes assessing engagement with users, uses overridden, training time needed, and feedback from practicing clinicians. This domain also stands for human–AI collaboration, not just automation.

### 11.4.3 Systemic integration

AI applications will also extend beyond individual clinicians to encompass broader institutional and societal structures. Interoperability focuses on how AI integrates with other parts of healthcare infrastructure, policies, and economic realities.

Such considerations include the following:

- Interoperability between hospitals, devices, and architectures
- The scalability of this technology from tertiary hospitals to community clinics
- Cost-effectiveness over time (implementation and maintenance costs)
- Compatibility with the country's health data governance policies and international best practices

The emerging technologies, such as federated learning, edge computing, or AIoT (AI of Things), will require assessment under this domain for the intended implementation in underserved areas.

### 11.4.4 Ethical soundness

The ethical implications of use of AI in healthcare are profound. Issues related to fairness, accountability, privacy, and transparency need to be considered at all stages of an AI system's life cycle, from development through deployment.

The ethical landscape needs to keep the following points under consideration:

- Representation of various demographics in the training datasets themselves (race, gender, age comorbidity profiles)
- The possibility of algorithmic biases and the need for continuous audits
- Privacy preservation laws and policies (e.g. GDPR, HIPAA, or data localization)
- Evidence of governance protocols to know who is responsible when an AI system does something wrong
- Standards for explainability to make patients/clinicians understand the reasoning behind a recommendation

Examples include the WHO guidance on ethics and governance of AI for health (2021) and the EU AI Act. They, therefore, form the foundation of assessment and analysis.

### 11.4.5 Patient-centred value

More than anything, the central question in any innovative healthcare project is whether it improves patient care. Ultimately, AI needs to improve health, accessibility, and experience of the patient.

These effects may include the following:

- Quantifiable improvements in morbidity, mortality, quality of life, or progression of disease
- The interventions in health equity in terms of technology reach and ability to penetrate populations with limited health distribution historically
- Aid in patient self-determination and an informed consent process in AI-enabled care
- Trust and transparency, does this make sense to patients in the pathway of their care with AI?

This attribute therefore needs metrics and outcomes (readmission rates, survival curves) as well as qualitative feedback (satisfaction surveys, digital

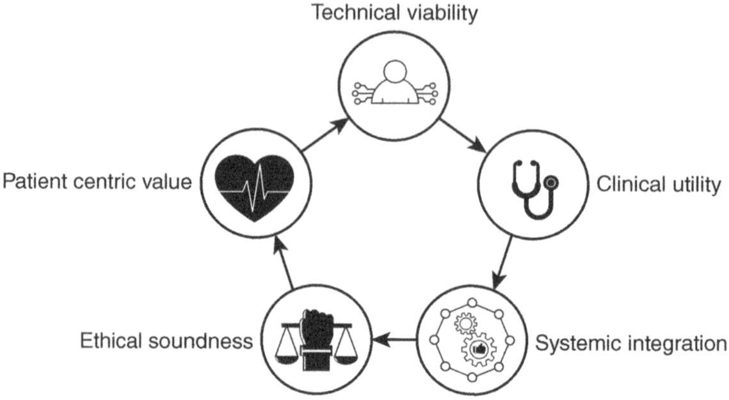


*Figure 11.2* The conceptual framework required to evaluate emerging AI in healthcare.

literacy assessments). Such concerns also raise the questions of co-design: Should patients be involved in designing the tools that influence their health?

Therefore, these five domains are not linear or isolated. High technical performance could increase clinical utility which can further boost systemic integration. Likewise, ethical increases patient trust, which drives adoption and ultimately an outcome. Accurate evaluation is continuous, not a one-time process (Figure 11.2).

### 11.4.6 Stakeholder considerations

Different stakeholders weigh these domains differently:

- Developers measure the tool's effectiveness and environmental alignment and applications.
- Clinicians look after safety and usability while demanding interpretability for aiding their decision-making.
- Hospital administrators interested in returns and integration costs.
- Policymakers interested in having regulatory compliance with readiness of health systems.

A successful evaluation framework will translate into a common language across these perspectives.

## 11.5 LANDSCAPE OF EMERGING AI TECHNOLOGIES IN HEALTHCARE

Healthcare is evolving rapidly, thanks to the advancement of AI. AI can improve diagnosis accuracy, help physicians personalize treatment, and make hospital operations more efficient.

These feats are feasible due to the following new technologies.

## 11.5.1 Multimodal learning architectures

Multimodal learning is a major frontier in healthcare AI that aims to combine various data formats, such as images, clinical notes, genomics, and time-series signals, into a unified predictive model. Traditional methods faced scalability issues since the model interactions needed to be established for every pair of modalities.

Recent activities, such as M3H, support significant advances to overcome this challenge. The model in M3H multitasks learning across all 40 diseases. An attention mechanism is proposed to balance task-specific and common information optimally. This novel method performs 11.6% better. In addition, task interdependence is measured by TIM score, which affects the model's interpretability [18].

The following key components drive this evolution:

- *3D convolutional networks* for high-resolution imaging diagnostics [19].
- *Graph neural networks* for phenotyping across clinical and genomic data [4].
- *Transformer-based models* for dynamic balancing of multimodal inputs.

The other important factor that dictates *learnable synergies* is liberty of models to select only clinically relevant cross-modal interactions while retaining memory for key patterns. This method has shown great promise on radiology and ophthalmology datasets, often outperforming conventional architectures [18]. As multimodal systems progress towards real-time clinical workflow-enabled implementation, they are anticipated to reach levels of clinical practice yet unseen in the era of intelligent integrated care (Figure 11.3).

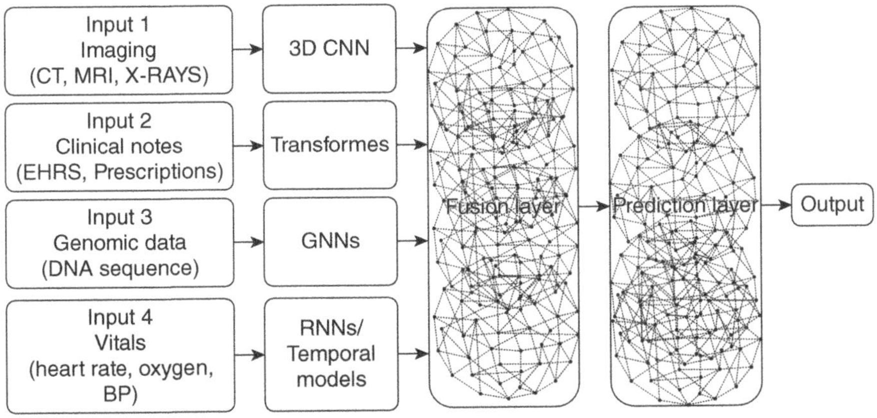


*Figure 11.3* Showing how multimodal takes in various input and returns an output.

## 11.5.2 Scalable AI foundations

As we implement AI in real-world healthcare environment, it is crucial that it can operate within current infrastructure setups. Four paradigms, namely Edge AI, federated learning, neuro-symbolic integration, and quantum machine learning, are revolutionizing how AI adapts to different constraints and possibilities within medical settings.

### *11.5.2.1 Edge AI and on-device intelligence*

Edge computing enables AI models to function independently on local hardware without the need for high-bandwidth links to a centralized server (Figure 11.4). The real-time diagnostics in clinics can be performed by quantized neural networks on low-power devices. For example, architectures designed for wearables achieved 98.7% accuracy for ECG signal interpretation on ARM Cortex-M7 targets [7]. With these advancements, continuous monitoring, early intervention, and decentralized diagnostics are now made possible even in rural or resource-limited settings [4].

### *11.5.2.2 Federated learning and decentralized collaboration*

Federated learning allows a cross-institutional collaborative model-building process without transferring sensitive data from its owner, thus providing data privacy against centralized training programmes. This approach has found successful applications in global health programmes like the 22-nation TB detection programme, ensuring security and autonomy for data and the institutions having it. Thus, as mounting regulatory concerns over data sharing grow, federated systems are set to play an increasingly crucial role in scaling ethically AI applications for healthcare (Figure 11.5).

### *11.5.2.3 Neuro-symbolic integration for clinical reasoning*

Traditional DL models do not perform well on an explainable and systematic logic. A neuro-symbolic model that enhances a neural network with

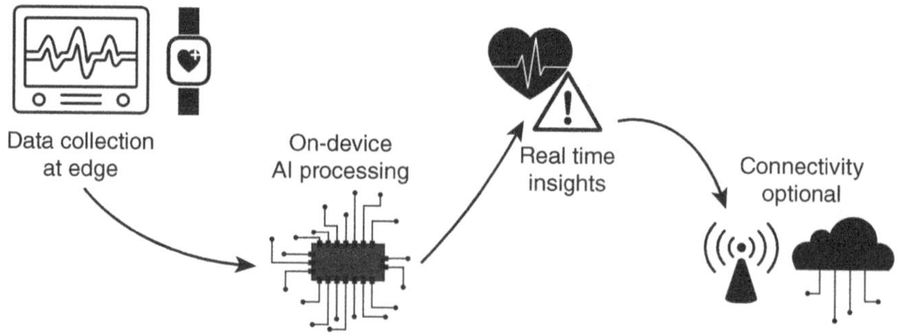


*Figure 11.4* Showing how edge AI works on edge devices without requiring a connection.

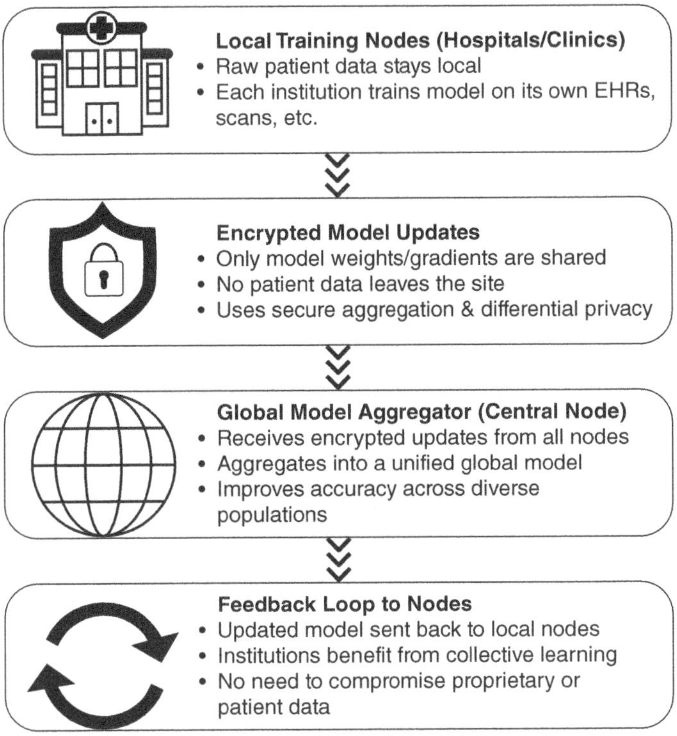


*Figure 11.5* Showing how fragmented learning AI overcomes regulatory concerns.

structures such as a medical knowledge graph now promises to eliminate hallucinations (false results generated by large language models) (Figure 11.6) [20]. They offer high prediction accuracy and transparency that is especially required in multistep clinical reasoning.

#### *11.5.2.4 Quantum machine learning for high-complexity tasks*

Quantum machine learning leverages quantum processors to tackle biomedical challenges which are beyond the reach of classical computers (Figure 11.7). It is speculated that future error-corrected quantum computers might speed up certain precise simulations of molecules [21]. Such capabilities may radically change drug discovery procedures, especially where complex molecular structures are involved and accelerate development of cure for rare diseases, where the traditional methods are way too slow or simply inefficient.

### 11.5.3 Generative models and large language systems in healthcare

Generative models and large language systems are transforming the way healthcare information is created, interpreted, and applied to generate novel

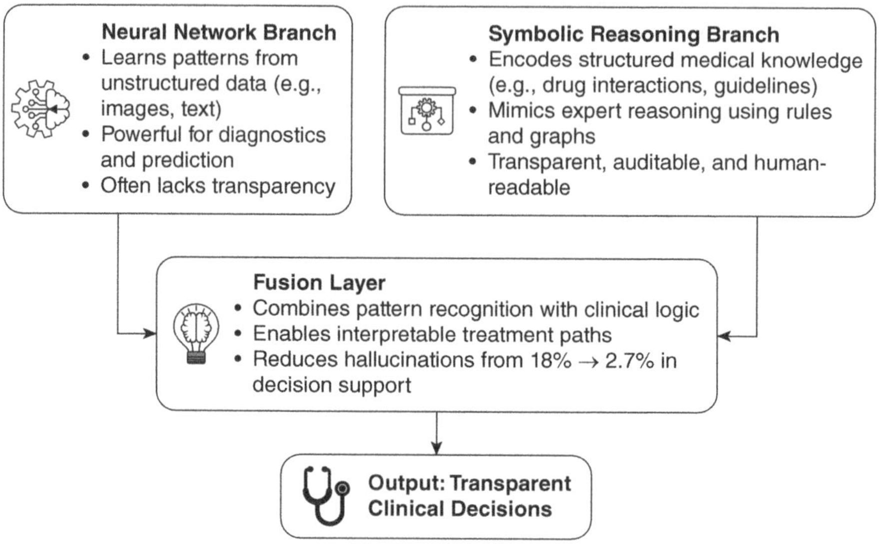


*Figure 11.6* Showing how neural network combined with symbolic reasoning eliminates hallucinations.

content like molecules, synthetic patient records, or clinical notes based on biomedical and institutional training datasets on a large scale. As their use widens, they become increasingly embedded into research workflows, clinical infrastructure, and operational support systems.

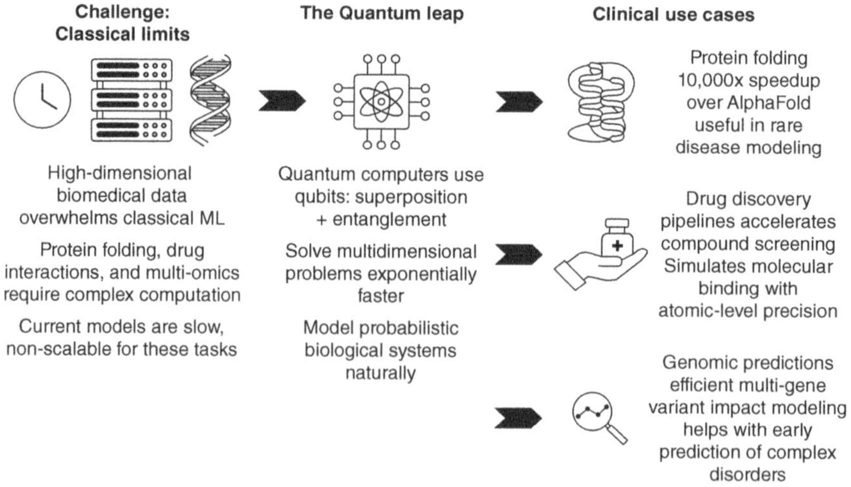


*Figure 11.7* Showing which challenges can be overcome due to quantum machine learning, how, and its use cases.

Techniques such as generative adversarial networks (GANs) and diffusion models that denote generation of realistic outputs are examples of generative AI. There is substantial evidence of value of these models in biomedical discovery pipelines. For example, the ability of AI to generate molecular structures has reduced the time required for exploratory screenings since the structures of valid compounds are generated computationally. Furthermore, these models are being adapted to simulate rare disease presentations and synthetic perturbations, forming the basis for personalized compound generation and biologically designed applications.

Synthetic data generation is one of the most rapidly advancing applications. MedGAN and related models create statistically valid synthetic EHRs that minimize privacy loss when compared to real clinical datasets and have been used to enable cross-institutional research and privacy-compliant model training, particularly in areas where law around data protection is very stringent.

Language models of great magnitude (which include transformer-based architectures such as GPT-4) are now known to exhibit functionality over a plethora of language-centric operations in healthcare, including the generation of clinical summaries, transcription of physicians' notes, standardization of medical coding, and formatting of unstructured documentation. Initial implementations have shown that systems of this kind can assist in automating routine communication and alleviating the administrative load. However, to ensure the safety and factual accuracy of content generated by LLMs, most implementations are now resorting to retrieval-augmented generation (RAG) pipelines that allow the models to interact with datasets before they produce their answers. This alleviates the likelihood of hallucinations, where the AI gives inaccurate or fabricated responses that seem reasonable but are in fact wrong.

### 11.5.4 Causal representation learning and counterfactual modelling

There is a new class of AI model that not only focuses on understanding what is plausible but also what would happen if circumstances were different. This shift from mere pattern recognition to causal reasoning is called causal representation learning, and models try to identify how variables interact and, therefore, try to investigate alternative scenarios particularly in cases where real-world experiments can hardly be conducted.

Recent advances have integrated DL and causal inference frameworks to establish counterfactual scenarios, in which one aspect of the same case has been altered. Neural networks are today trained to disentangle cause–effect relationships from complex observational data, while often using a graphical or latent variable model to convey relationships. By doing this, AI systems can not only reason on the level of statistical correlations but also consider interventions and mechanisms responsible for the effects on behaviours.

Causal modelling is still in development but offers considerable promise for interpretable, safe, and adaptable AI in high-stake areas such as healthcare.

### 11.5.5 Self-supervised learning (SSL) and foundation model pretraining

Much of the healthcare data available is unlabelled mines of text, images, and signals that remain without structured annotations. SSL is becoming important for the training of AI in such an environment. It lets models learn useful patterns by predicting parts of the data from one part of the data to another, such as filling in masked words and aligning multiple views of an image.

It is now being implemented to train large foundation models for subsequent reuse. In clinical settings, models like BioBERT, GatorTron, and MedCLIP are trained on large collections of text, radiographies, or timelines of patients, and fine-tuned with very little additional data for a specific task. The basic medical knowledge obtained by the model (the AI) through training assists them to be more efficient and versatile.

As the field evolves, self-supervised pretraining becomes the default route to building powerful, flexible AI systems that can learn from healthcare environments where labels are scarce but information is abundant.

### 11.5.6 Explainable AI

XAI is a related field that uses techniques to make the decisions (process and internal logic) of AI systems understandable to humans. In clinical settings, where trust, accountability, and safety are fundamental, explainability ensures that the healthcare professionals can comprehend not only what an AI model predicts but also why it makes such predictions.

Contemporary techniques such as SHAP and Grad-CAM identify which parts of the input data are most influential for a model's output, which helps align AI outputs and clinicians' expectations. For instance, XAI-assisted models have achieved 94.8% sensitivity in diabetic retinopathy detection while also generating the interpretable heatmaps closely matching expert annotations [5].

Newer models like M3H have intrinsic explainability due to TIM (task interaction modulation) scores, which show how inputs and tasks affect one another during the learning process. This independence stems from the fact that the records, which are open for everyone to see, do not need to be confirmed by an outside authority [22].

Some countries within the European Union have already mandated explainability audits for high-risk AI tools, which means XAI is becoming essential not only for trust but also for regulatory compliance. AI will continue to need XAI for it to be viable in real-world healthcare applications as it becomes increasingly autonomous.

## 11.6 IMPACT ZONES OF AI IN HEALTHCARE

We are beginning to see the emerging AI technologies that we discussed earlier actually have real measurable effects in the real-world healthcare setting (Figure 11.8). In other words, this identifies the key areas where AI is fundamentally altering how systems detect illnesses, manage operations, and assist patients. The changes are early, but they're quite significant and point to a much broader shift in how care is provided through intelligent systems.

### 11.6.1 Diagnosis and early detection

AI enables faster, earlier, and more accurate diagnoses. DL models can analyse medical images with the precision of leading experts. MobileNet detected COVID-19 based on chest image with 99.7% accuracy and BiGRU obtained over 99% accuracy while classifying the cough and text data [23].

In a practical sense, AI identifies diabetic retinopathy with 94.8% sensitivity and forecasts the Alzheimer's disease six years prior to the symptom's onset. In pathology, AI assists in identifying subtle discrepancies that might be overlooked by human analysts [18, 19].

The latest versions integrate data from imaging and genomic platforms as well as clinical parameters to minimize diagnostic errors on rare diseases by more than 40%. This transition positions AI as an independent tool, rather than just a complementary one which forms the backbone of early diagnostics in healthcare [4].

### 11.6.2 Drug discovery and development

AI has been hailed as the next big thing to substantially enhance the drug development process, a mechanism that normally requires over a decade and

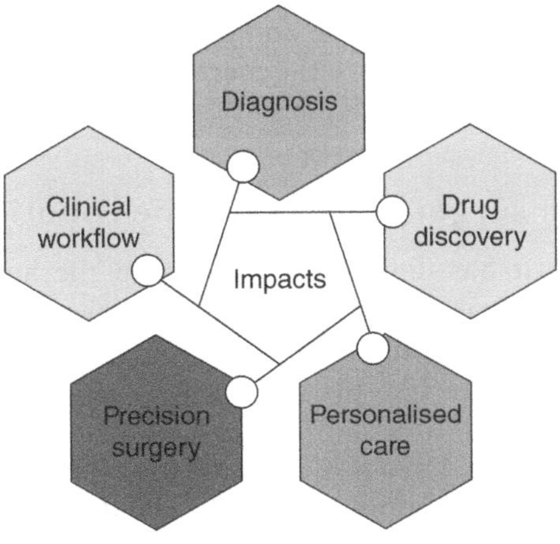


*Figure 11.8* Showing the impacts of AI in healthcare.

costs more than $2 billion per approved drug. Due to the numerous failures of potential new treatments for rare and complex diseases, timelines for implementing the screening and clinical trials are challenging and expensive.

Generative AI models have the capability to design drug-like molecules within weeks. For instance, a potential fibrosis drug was designed in 21 days by Insilico Medicine (over 50% faster than traditional techniques) and later it moved into phase I trials. AI-based tools such as AlphaFold are reshaping the way researchers interpret protein structures, reducing prediction times from months to hours [3].

AI screening and toxicity prediction limit the failed compounds that never reach trials. These are still developing, but preliminary findings suggest that AI can trim a few years from drug development, leading to a new era of quicker, more accurate, and less costly therapeutic discovery [24].

### 11.6.3 Personalized and predictive care

Healthcare has long relied on general treatment guidelines – the same drugs or therapies are given to most patients with the same diagnosis. But responses often vary, and side effects are common. AI is now enabling a shift towards more personalized care: treatments tailored to an individual's genetics, history, behaviour, and environment [2].

Today, AI models can already identify rare genetic mutations, predict how a patient may respond to a drug, or flag early signs of mental health decline through speech patterns. In rehabilitation, AI-powered exoskeletons adjust movement in real time, improving stroke recovery by up to 37% [25].

Emerging systems are going further. "Digital twins," virtual versions of individual patients, are being developed to simulate treatments before they're given. Closed-loop AI models are being tested to automatically adjust medication doses or therapy settings based on real-time data. As more health systems integrate wearable data and harmonize records across institutions, personalized care is becoming more scalable and predictive.

This marks a growing shift, from generalized medicine to data-informed, precision care that adapts to each patient.

### 11.6.4 Precision surgery and real-time robotics

While traditionally it has depended solely upon the human precision with which the operation was performed, even experienced hands have speed, accuracy, and consistency limits. AI is beginning to change all that. Hints of modern robotic systems like the da Vinci Xi support surgeons in an even better way, with under-millimetre precision. These tools, in the microvascular procedures guided by AI, have achieved an operational accuracy of 0.2 mm, reducing operative time by more than 42% [4]. Robots are fed real-time AI feedback on blood flow and tissue elasticity that helps them alter their actions on the fly, minimizing complications.

Emerging systems go beyond such possibilities: in early trials, automated suturing displayed an accuracy of 98.7%. Using tools like HoloLens, holographic overlays are now guiding brain surgeons in removing tumours more accurately. AI also learns from synthetic data how to predict how soft tissue responds during surgery, thus reducing unintended damage. AI innovations expand from robotic assistance to self-adaptive instruments and offer surgery not only with greater precision but also with smarter intelligence – switching once manual techniques into adaptive data-based systems.

### 11.6.5 Clinical workflow and operational intelligence

For ages, hospitals had depended on manual or late decision-making scheduling that crowded their ICUs, fatigued their staff, and resulted in a shortage of medications. Now, AI is coming out to be the most important tool in this management for a more intelligent way of conducting these systems.

Currently, reinforcement learning has been successful in using a model of allocating ICU beds and surgical scheduling, cutting down the number of patients waiting by 22% and increasing the general output. For example, a tool such as Sepsis Watch has improved compliance to sepsis treatment by 32% and reduced mortality associated with the condition by 18%. Additionally, tools for processing natural language have been saving physicians over 6 hours of work each week in documentation, while predictive staffing algorithms have enhanced hospitals' forecasting for peak demand during the flu seasons and other outbreaks.

What lies ahead goes beyond all that. Ambient AI systems are already being piloted to record clinical environments in real time, monitoring speech, vital signs, and actions of staff in the efforts to flag risks early. Logistics driven by AI are reducing stock-out occurrences in medications by over 80% and facilitating vaccine integrity in remote regions.

These systems are still maturing, but evidence already demonstrates that AI can transform healthcare operations from reactive to predictive, offering better care at a faster pace and with fewer obstacles.

### 11.6.6 Public health surveillance and predictive response

Public health data reporting was always behind, forcing the system to rely on static and tightly coordinated data. Not surprisingly, AI is helping public health to become more real time and predictive. The most recent systems, which rely on mobility data and social signals, cut detection time of outbreak incidences by more than 50%. India witnessed lowered death rate during COVID-19 in high-risk areas because AI-based ventilators were allocated there.

AI is monitoring wastewaters for heightened pathogen levels, scanning social media for flu-related trends, and scrutinizing wearable data for early warning signals. This will truly boost public health by enabling proactive responses to possible emerging threats rather than just responding to them but the impact will depend on how the data is obtained and how the deliverables are distributed in rural areas.

## 11.7 ETHICAL, SOCIETAL, AND REGULATORY CONSIDERATIONS

As AI becomes integrated with healthcare services, it poses several ethical, societal, and legal challenges. While the potential benefits are immense, we need to ensure that its implementation is fair, transparent, and accountable.

### 11.7.1 Ethical concerns in clinical AI

AI tools must adhere to the fundamental values of medical practice such as beneficence, non-maleficence autonomy, and justice. Most of the time, the black-box models tend to offer cases of explanation gaps. Researching AI-generated recommendations shows merely 23% patient comprehension unless using visual or human interpretive assistance. New dynamic consent models allow patients to specify how their data is used during the full life cycle of AI development. Synthetic datasets created by GANs allow for increasing adoption, achieving more than 90% diagnostic accuracy while preserving identity.

Equity failure algorithmic bias is a known threat for health equity.

- Models for melanoma detection have 34% less accuracy with darker skin tones [26].
- Detectors of pneumonia developed based on urban datasets fail to perform as well in rural areas (AUC drops from 0.91 to 0.67) [27].
- Only 12% of the genomic data has been collected for populations outside Europe [26].

Some suggestions by FAIR4Health for auditing models include SHAP value and parity metrics. Such tools are meant to surface and fix the demographic skewness during development and deployment.

### 11.7.2 Societal impacts

Where does AI now lead the load-sharing between humans in the clinical context? Shifts in professional responsibility occur by automated documentation systems, triage systems, and AI-assisted decision-making. Healthcare professionals will require training to help them understand the constraints of these tools, without which even the best tools might be resisted by patients and clinicians.

### 11.7.3 Regulatory fragmentation

Global regulations vary widely:

- The EU's AI Act mandates transparency and risk assessments for health AI-deemed high risk [2].
- US frameworks, such as HIPAA, necessitate compliance measures that can introduce up to a 23% computational overhead for AI systems [27].
- In India, Health AI systems must store health data locally under the Digital Health Act.
- China requires total interpretability of diagnostic AIs.

All these barriers limit the ability of multinationals to deploy together due to complex compliance laws.

### 11.7.4 Accountability and governance

As mechanical AI systems become more autonomous, the traditional models are struggling to keep up. Several studies highlight the need for of an integrated system of governance, which would include monitoring of AI decision-making by humans, maintenance of decision logs to audit systems, and identifying the legal ownership of AI-based actions.

## 11.8 INFRASTRUCTURE AND DATA INTEGRATION CHALLENGES

The lack of infrastructure will affect the clinical success of AI models independently from their performance. Regrettably, most healthcare institutions still utilize outdated information technology systems that are not AI workflow-friendly. Additionally, because clinical data types (e.g. EHRs, imaging, and genomics) are siloed in different platforms, they cannot be analysed in real time. Global surveys indicate that less than 20% of hospitals have fully interoperable digital systems in 2023. Even the best models will perform poorly when data pipelines are disorganized and too slow. In critical care, even a millisecond's delay in data access would negate the empirical advantage of predictive algorithms. Limited edge computing and slow secure API deployment also impede AI implementation for point-of-care diagnosis. But without properly set up infrastructure and data flow, AI is just another silo: precise but not ready for actual clinical implementation.

## 11.9 LIMITATIONS OF CURRENT AI SYSTEMS IN HEALTHCARE

Despite increasingly likely predictions, AI in healthcare is still hampered by several limitations:

- *Generalizability*: Most models perform poorly when applied outside the datasets against which they had been validated. A model developed in a tertiary hospital may misfire in rural clinics.
- *Missing Data*: In a clinical setting, data completeness styles are usually violated; most models are, however, usually operated on the assumption of complete data input.
- *Hallucinations*: Generative AI sometimes hallucinates incorrect output, which, while being medically plausible, presents considerable risk in a high-stakes situation such as intensive care.
- *Explainability*: Many such devices do not clearly justify their suggestions, further impairing trust from the clinicians and invoking concerns of ethical and legal standing.

In 2024, fewer than 25% of the AI tools cleared by the FDA underwent post-market validation within a real-world setting, an emblem of precaution and oversight.

## 11.10 STRATEGIC ROADMAP FOR RESPONSIBLE AI INTEGRATION

In order for AI to meaningfully improve healthcare, the following priorities need to be addressed:

1. **Strengthen Infrastructure**
   It should adopt interoperable health data standards and scalable cloud-edge architectures that will enable AI deployments.
2. **Real-World Validation**
   Prospective multisite clinical trials should be required for high-risk AI tools. The performance of the devices should be tracked post-implementation through global registries.
3. **Trust through Explainability**
   Models should be built in a way that ensures output transparency, and clinician feedback should be incorporated in the design and decision loops.
4. **Precision Regulations**
   We need to cross the line into binding international, sectoral AI regulations on safety, privacy, and liability issues.

5. **Extend Access**
   Encourage public–private partnerships to bring AI to low-resource settings for affordable diagnosis and federated learning. AI should be able to do so with the advancement of infrastructure and corresponding trust and governance.

## 11.11 CONCLUSION

Emerging AI technologies not only facilitate efficiency but also alter the very structure of healthcare. From generative models that aid drug development to multimodal systems enhancing diagnostics, AI significantly impacts clinical decision-making, operational efficiency, and patient care pathways. The likes of XAI, federated learning, and real-time robotics have exited the labs and are forming the bedrock for next-generation healthcare systems.

As these tools become autonomous and enter the safety-critical market, ensuring interoperability, scalability, and continuous validation is vital. The future success of AI technologies in healthcare will depend on their ability to work together as a system rather than independently. The immediate struggle is to bridge the gap between pilot programmes and real-world implementation. AI needs to become a foundational tool for delivering efficient healthcare broadly rather than just an accessory.

## REFERENCES

1. Michael van Hartskamp, Sergio Consoli, Wim Verhaegh, Milan Petković, Anja van deStolpe (2019) Artificial Intelligence in Clinical Health Care Applications: Viewpoint. *Interactive Journal of Medical Research*, 8(2), e12100. https://arxiv.org/pdf/1906.02090
2. Balaji Shesharao Ingole, Vishnu Ramineni, Nikhil Kumar Pulipeta et al. (2024) The Dual Impact of Artificial Intelligence in Healthcare: Balancing Advancements with Ethical and Operational Challenges. *European Journal of Computer Science and Information Technology*,12(6), 35–45. https://doi.org/10.37745/ejcsit.2013/vol12n63545
3. J. Jumper, R. Evans, A. Pritzel et al. (2021) Highly Accurate Protein Structure Prediction with AlphaFold. *Nature*, 596, 583–589. https://www.nature.com/articles/s41586-021-03819-2
4. Yihao Liu, Xu Cao,Tingting Chen, Yankai Jiang, Junjie You, Minghua Wu, Xiaosong Wang, Mengling Feng, Yaochu Jin, Jintai Chen (2025) A Survey of Embodied AI in Healthcare: Techniques, Applications, and Opportunities. *Information Fusion*, 119, 103033. https://arxiv.org/html/2501.07468v1
5. Ahmad Momani (2025) Implications of Artificial Intelligence on Health Data Privacy and Confidentiality. https://doi.org/10.48550/arXiv.2501.01639
6. Yao Fehlis, Paul Mandel, Charles Crain, Betty Liu, David Fuller (2024) Accelerating Drug Discovery with Artificial: A Whole-Lab Orchestration and Scheduling System for Self-Driving Labs.
7. Sarfraz Iqbal (2023) Artificial Intelligence Tools and Applications for Elderly Healthcare: Review. https://doi.org/10.1145/3594315.3594347

8. Firas Khader, Gustav Mueller-Franzes, Tianci Wang, Tianyu Han, Soroosh Tayebi Arasteh, Christoph Haarburger, Johannes Stegmaier, Keno Bressem, Christiane Kuhl, Sven Nebelung, Jakob Nikolas Kather, Daniel Truhn (2022) Medical Diagnosis with Large Scale
9. Deepti Saraswat, Pronaya Bhattacharya, Ashwin Verma, Vivek Kumar Prasad, Sudeep Tanwar, Gulshan Sharma (2024) Explainable AI for Healthcare 5.0: Opportunities and Challenges. *IEEE Access*, 10, 84486–5984517. doi: 10.1109/ACCESS.2022.3197671.
10. Michail Mamalakis, Héloïse de Vareilles, Graham Murray, Pietro Lio, John Suckling (2024) The Explanation Necessity for Healthcare AI. https://doi.org/10.48550/arXiv.2406.00216
11. Yan Chen, Pouyan Esmaeilzadeh (2024) Generative AI in Medical Practice: In-Depth Exploration of Privacy and Security Challenges. *Journal of Medical Internet Research*, 26, e53008. doi:10.2196/53008
12. Sidra Nasir, Rizwan Ahmed Khan, Samita Bai (2024) Ethical Framework for Harnessing the Power of AI in Healthcare and Beyond. *IEEE Access*, 12, 31014–31035. https://ieeexplore.ieee.org/document/10445375
13. K. Lekadir et al. (2025) FUTURE-AI: International Consensus Guideline for Trustworthy and Deployable Artificial Intelligence in Healthcare. *BMJ*, 388, e081554. https://doi.org/10.1136/bmj-2024-081554
14. Hassan Ali (2022) Reinforcement Learning in Healthcare: Optimizing Treatment Strategies, Dynamic Resource Allocation, and Adaptive Clinical Decision-Making doi: 10.7753/IJCATR1103.1007
15. Mohammad Tajabadi, Roman Martin, Dominik Heider (2024) Privacy-Preserving Decentralized Learning Methods for Biomedical Applications. *Computational and Structural Biotechnology Journal*, 23, 3281–3287. https://doi.org/10.1016/j.csbj.2024.08.024
16. Jorge Corral-Acero, Francesca Margara, Maciej Marciniak et al. (2020) The 'Digital Twin' to Enable the Vision of Precision Cardiology. European Heart Journal, 41(48), 4556–4564. https://doi.org/10.1093/eurheartj/ehaa159
17. Ryszard Wierzbicki, Maria Pawłowicz, Józefa Job, Robert Balawender, Wojciech Kostarczyk, Maciej Stanuch, Krzysztof Janc, and Andrzej Skalski (2021) 3D mixed-reality visualization of medical imaging data as a supporting tool for innovative, minimally invasive surgery for gastrointestinal tumors and systemic treatment as a new path in personalized treatment of advanced cancer diseases doi: 10.1007/s00432-021-03680-w
18. Shuroug A. Alowais, Sahar S. Alghamdi, Nada Alsuhebany, Tariq Alqahtani, Abdulrahman I. Alshaya, Sumaya N. Almohareb, Atheer Aldairem, Mohammed Alrashed, Khalid Bin Saleh, Hisham A. Badreldin, Majed S. Al Yami, Shmeylan Al Harbi, Abdulkareem M. Albekairy (2023) Revolutionizing Healthcare: The Role of Artificial Intelligence in Clinical Practice. *BMC Medical Education*, 23, 689. doi: 10.1186/s12909-023-04698-z
19. Ahmed Al Kuwaiti, Khalid Nazer, Abdullah Al-Reedy, Shaher Al-Shehri, Afnan Al-Muhanna, Arun Vijay Subbarayalu, Dhoha Al Muhanna, Fahad A Al-Muhanna (2023) A Review of the Role of Artificial Intelligence in Healthcare. *Journal of Personalized Medicine*, 13(6), 951. doi: 10.3390/jpm13060951
20. Maximilian Frenzel, Marie Cherasse, Joanna M. Urban, Feifan Wang, Bo Xiang, Leona Nest, Lucas Huber, Luca Perfetti, Martin Wolf, Tobias Kampfrath, Xiaoyang Zhu, Sebatian F. Maehrlein (2023) How Neuro-Symbolic AI Bridges the Trust Gap for Physicians. https://www.medicaleconomics.com/view/how-neuro-symbolic-ai-bridges-the-trust-gap-for-physicians
21. Maximilian Frenzel, Marie Cherasse, Joanna M. Urban, Feifan Wang, Bo Xiang, Leona Nest, Lucas Huber, Luca Perfetti, Martin Wolf, Tobias Kampfrath, Xiaoyang Zhu, Sebatian F. Maehrlein (2023) Nonlinear Terahertz

Control of the Lead Halide Perovskite Lattice. *Nature*, 9(21), eadg3856. https://doi.org/10.48550/arXiv.2301.03508
22. Dimitris Bertsimas, Yu Ma (2024) M3H: Multimodal Multitask Machine Learning for Healthcare by Dimitris Bertsimas. https://doi.org/10.48550/arXiv.2404.18975
23. Md Shofiqul Islam, Khondokar Fida Hasan, Hasibul Hossain Shajeeb, Humayan Kabir Rana, Md Saifur Rahmand, Md Munirul Hasan, AKM Azad, Ibrahim Abdullah, Mohammad Ali Moni (2025) Multimodal Marvels of Deep Learning in Medical Diagnosis: A Comprehensive Review of COVID-19 Detection. *AI Open*, 6, 12–44. https://doi.org/10.48550/arXiv.2501.09506
24. Dong Chen, Jian Jiang, Nicole Hayes, Zhe Su, Guo-Wei Wei (2025) Artificial Intelligence Approaches for Anti-Addiction Drug Discovery. *Digital Discovery*, 4(6), 1404–1416. https://arxiv.org/abs/2502.03606
25. Mousumi Khanam, Sume Akther, Iffath Mizan, Fakhrul Islam, Samsul Chowdhury, Nayla Mehereen Ahsan, Deepa Barua, Sk K. Hasan (2024) The Potential of Artificial Intelligence in Unveiling Healthcare's Future. *Cureus*, 16(10), e71625. doi: 10.7759/cureus.71625.
26. Sribala Vidyadhari Chinta, Zichong Wang, Xingyu Zhang, Thang Doan Viet, Ayesha Kashif, Monique Antoinette Smith, Wenbin Zhang (2024) AI-Driven Healthcare: A Survey on Ensuring Fairness and Mitigating Bias. *PLOS Digital Health*, 4(5), e0000864. https://doi.org/10.48550/arXiv.2407.19655
27. África Periáñez, Ana Fernández del Río, Ivan Nazarov, Enric Jané, Moiz Hassan, Aditya Rastogi, Dexian Tang (2024) The Digital Transformation in Health: How AI Can Improve the Performance of Health Systems. *Health Systems and Reform*, 10(2). https://doi.org/10.48550/arXiv.2409.16098

Chapter 12

# Pharmacological frontier for drug discovery system

## A new path towards personalized medicine

*S. Sharmila, V. Praveena, S. Velmurugan, M. Menagadevi, and S. Anbumani*

## 12.1 PERSONALIZED MEDICINE

Personalized care in drug delivery aligns treatments with a person's unique characteristics, improving outcomes and reducing risks. By analysing a patient's genetic profile, clinicians can optimize drug dosages and reduce trial-and-error, leading to faster and more effective outcomes. This approach also lowers the risk of adverse drug reactions by considering unique pharmacokinetics and pharmacodynamics. Tailored treatments enhance patient compliance, especially in chronic diseases requiring long-term management, by aligning with individual preferences and needs. Emerging technologies like nanotechnology and 3D-printed drug formulations further improve targeted delivery, reducing systemic exposure and enhancing therapeutic outcomes [1]. Personalized drug delivery marks a transformative shift, promoting precision medicine and better health outcomes.

### 12.1.1 Introduction

Precision medicine, also known as personalized medicine, customizes medical decisions and treatments to individual patients. This approach minimizes adverse effects and enhances recovery by utilizing a patient's medical and genetic profile to predict responses to medications. It aims to improve diagnostic accuracy, prevent disease progression, and provide personalized care. Core principles include genomic integration, where pharmacogenomics identifies gene variants that influence drug efficacy and safety, supported by advanced techniques like multigene analysis. Pharmacokinetics and pharmacodynamics evaluate factors such as absorption, distribution, metabolism, excretion (ADME), and biological effects on drug targets like receptors or enzymes, exemplified by mutations in beta-2 adrenergic receptors that influence responses to treatments for respiratory issues and hypertension. Biomarkers further guide tailored treatment plans by predicting, diagnosing, and monitoring diseases [2]. Personalized medicine offers numerous benefits, including enhanced treatment efficacy, informed patient choices about genetic predispositions and therapeutic options, and proactive strategies for early intervention


DOI: 10.1201/9781042015931-12

and disease prevention [3]. However, significant challenges remain, such as obtaining regulatory approval, safeguarding data privacy, addressing the high costs and scalability of developing tailored therapies, and educating medical professionals to adopt these approaches. Innovations driving the field forward include personalized digital therapeutics, advanced diagnostic methods for early disease detection, patient-derived cell and organoid models for optimizing treatments, and customized infertility treatments. Though challenges exist, precision medicine can transform healthcare with its personalized, efficient, and forward-looking solutions, setting the stage for a healthier future.

## 12.2 FRONTIERS IN PHARMACOLOGY

The interactions between medications and biological systems, aimed at preventing and addressing human diseases, are a key focus in the field of pharmacology [4]. This discipline encompasses pharmaceutical chemistry, toxicology, clinical applications, and foundational studies to understand the mechanisms behind drug effects and disease treatment. The broad scope of this field includes areas such as genetics, cardiovascular and smooth muscle pharmacology, metabolic processes, transport mechanisms, drug outcomes, research advancements, and policy development.

### 12.2.1 Molecular based

Metal ions play a vital role in biomolecules, enabling essential biological processes. Small-molecule inhibitors that target the metal components of these biomolecules can effectively modify their systems. Cisplatin's discovery set the foundation for creating metal-based drugs extensively used in treating various conditions through their metal–ligand interaction mechanisms. Consequently, the exploration and application of metal-coordinating pharmaceuticals remain a crucial yet complex area in medicinal chemistry. Increasingly, atomic-level simulations are being employed to address the limitations of traditional drug discovery methods and to complement wet-lab research by elucidating the mechanisms of drug action [5]. Through multiscale simulations, the behaviour of metal-containing drugs and metal-binding inhibitors can be better understood, offering valuable insights into the roles of metal complexes within their natural environments.

### 12.2.2 Artificial intelligence (AI) and machine learning (ML)

The process of discovering or developing new drug candidates involves many factors, making it a costly and time-consuming endeavour that may sometimes result in failure. In recent decades, significant breakthroughs in computer technology, including advancements in software, computing infrastructure,

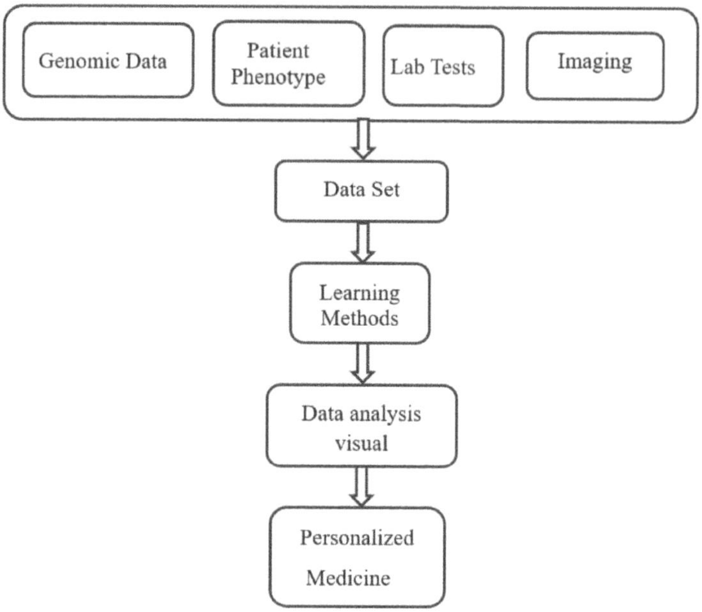


*Figure 12.1* AI in drug discovery and development.

and the generation of vast amounts of data, have occurred. As a result, more sophisticated algorithms, as depicted in Figure 12.1, are now needed to handle this increasing volume of data. This highlights the potential of computational intelligence (CI) and AI as effective tools for analysing medicinal chemistry data. AI's role in drug development has revolutionized data science, driven by the need to process large datasets and the availability of high-performance processors, particularly GPU computing [6]. This challenge is not limited to drug development; the abundance of scientific data and the automation of processes have created an ideal environment for AI applications.

In medicinal chemistry, computational techniques are used to link biological attributes, such as molecular descriptors, physical and chemical properties, to drug pharmacokinetics (absorption, distribution, metabolism, and excretion), as well as bioactivity and molecular target affinity. ML, an area within AI, empowers computers to derive insights from data, a capability that is transforming drug discovery. CI also includes the development and application of computing paradigms based on language and biology, encompassing three main approaches: evolutionary computation, fuzzy systems, and neural networks (NN). Artificial neural networks, in particular, have shown considerable success in medicinal chemistry.

The therapeutic potential of pharmacological treatments targeting epigenetic dysfunction, which has been linked to numerous diseases, especially cancers, highlights the promise of these approaches. Computational techniques are crucial for the design and discovery phases of epi-drugs and probes.

These techniques include virtual screening, scaffold hopping, molecular dynamics simulations, quantum chemistry calculations, pharmacophore modelling, homology modelling, stability prediction, and 3D quantitative structure–activity relationships.

### 12.2.3 Protein

Proteins involved in modifying epigenetic marks on nucleosomes are classified into three main categories based on their roles: writers, which add covalent modifications; erasers, which remove these modifications; and readers, which detect specific modifications and recruit chaperones. Over recent years, significant advancements have been made in creating epi-probes for epigenomic research. Epigenetic writers, including specific enzymes, transfer methyl or acetyl groups from cofactors to DNA, histones, or non-histone proteins. These targeted modifications influence chromatin structure and control gene expression. For example, DNA methyltransferases (DNMTs) transfer methyl groups to the 5-carbon position of cytosine, a key step in DNA methylation. In mammals, this function is carried out by five distinct DNMT family members.

Since the early 2000s, the field of computer-aided pharmaceutical research has greatly benefited from the use of AI, and ML is increasingly being applied in drug design, discovery, and repurposing throughout different stages of the drug development process [7]. This interdisciplinary approach has attracted the attention of experts from fields such as computer science, medicine, biochemistry, and pharmaceuticals. Drug design is often considered a distinct phase within the broader drug discovery process, involving activities like simulation, screening, prediction, and exploration of drug–target interactions. The main focus is on lead optimization, which entails the creation and refinement of potential medicinal compounds.

Computer-aided drug discovery (CADD) has become a vital element in various scientific fields, playing a key role in identifying and optimizing compounds that advance through later stages of drug development or make it to the market. Despite its continued significant contributions, some CADD methods have been caught up in the hype cycle, leading to unrealistic expectations, disappointments, and, in some cases, successful outcomes. Disillusionment often stems from insufficient training in data analysis, trends, and excessive use. The process of evaluating advancements in drug development and making informed decisions has increasingly depended on the concept of chemical space and its representation. The searchable chemical space provides abundant opportunities to identify potential hit molecules.

CADD encounters various persistent challenges, such as increasing the success rates of virtual screening, refining methods for covalent and protein–protein docking, improving molecular docking for flexible compounds and macrocycles, modelling large and intricate systems, and accurately identifying druggable pockets and binding sites, including allosteric sites. Data curation remains a critical and ongoing global challenge, focusing on ensuring the quality of data, particularly that which is published in public databases [8]. This type of data is essential for creating precise models and generating valuable insights. Other challenges include the disruption of drug development progress caused by inadequate communication between research teams and stakeholders, which often contributes to project failures. One persistent issue that must be addressed is the lack of effective communication among principal investigators, graduate and undergraduate students, and other involved parties.

In science, precise documentation and peer review are crucial for the effective dissemination of knowledge. One key issue being addressed today is the sharing of data and methodologies to ensure transparency and reproducibility. Adequate education and formal training are crucial for students, as well as for early-career and experienced researchers from diverse fields who are moving into CADD [9]. This is important for fostering interdisciplinary collaboration and maximizing the advantages of CADD. While CADD has proven effective, there is still room for improvement in addressing both emerging and ongoing challenges. Technological advancements have the capacity to revisit and broaden the chemical space critical for drug discovery, facilitate the repurposing of approved drugs and existing chemical libraries, and reveal valuable hits concealed in SAR data from corporate sources. CADD-based approaches hold promise for exploring new chemical and biological domains linked to drug development. However, creating and sharing high-quality data remains a significant challenge for data-driven technologies like ML and deep learning. Researchers using CADD, whether working independently or with computational teams, should keep realistic expectations to ensure that fundamental principles are upheld despite prevailing trends and hype.

Fragment-based drug design (FBDD) is an approach for creating potent molecules by assembling smaller molecular fragments. FBDD generally results in molecules with low molecular weight, simple chemical structures, and moderate binding affinity to the target. While FBDD may not completely replace high-throughput screening (HTS) in drug discovery, it provides several benefits, including reduced experimental costs, a wider range of potential hits, and various strategies for developing new molecules [10]. These benefits have driven scientists to adopt this approach in designing inhibitors for various targets. As structural biology evolves and new screening techniques are developed, FBDD has become a vital component of target-based drug discovery.

Isothermal titration calorimetry (ITC) is an effective method for measuring the binding stoichiometry, enthalpy changes, and binding affinity of molecular interactions between a protein and a protein/ligand in solution. The process usually starts by adding one molecule to a solution containing another. Next, temperature changes in the mixture are monitored. The final step involves analysing the data to determine the necessary parameters. ITC is frequently used to assess binding affinities, making it a valuable tool for ranking a molecule's binding strength to a protein, which is crucial when selecting potential hits for further study.

## 12.3 EMERGING TECHNOLOGIES IN DRUG DISCOVERY

Pharmaceutical chemistry involves the study and optimization of drug candidates to enhance their affinity, selectivity (minimizing adverse effects), efficacy, metabolic stability (to prolong half-life), and bioavailability. Key emerging technologies in drug discovery include nanotechnology, proteomics, 3D bioprinting for improved drug models, data-sharing tools in biomedical research, lab automation, quantum theory and quantum computing, AI and big data, as well as cryo-EM and cell-free DNA techniques for drug discovery and development [11]. Let's briefly explore each of these cutting-edge technologies used in the drug discovery industry.

### 12.3.1 Nanotechnology in drug discovery

Nanomedicines, as indicated by their name, are pharmaceutical products based on nanotechnology, used for diagnosing, treating, and preventing various diseases. Materials in the nanoscale range of approximately 1–100 nm display unique chemical, physical, or biological properties compared to their larger-scale counterparts. A typical nanomedicine consists of a carrier and a drug, with the drug serving as the active pharmaceutical ingredient (API). In certain cases, the API itself can be produced at the nanoscale, transforming it into a nanomedicine.

Nanomedicines are recognized for their effectiveness in treating chronic conditions such as diabetes, cancer, and dementia, as they improve the body's ability to process drugs. The demand for nanomedical products, including DNA probes, nanosensors, and nanoparticle-based imaging agents, is increasing. However, the rapid growth of the nanomedicine industry encounters challenges, including the high costs associated with production and research and development. Focusing on nanotechnology for precision drug delivery can help overcome challenges associated with traditional delivery methods, such as intracellular trafficking and biodistribution. The aim of nanoparticles is to pass through biological membranes, stabilize the active ingredients they carry, and ultimately improve the overall safety and efficacy of treatments.

#### *12.3.1.1 3D bioprinting to create better drug models*

A drug must be tested on a model that closely simulates its intended use to gain approval for clinical trials. Technological innovations, such as 3D tissue printing, are advancing our ability to create these accurate models. In addition to existing three-dimensional models like organoids and organ-on-chip systems, 3D printed tissues provide several advantages over traditional two-dimensional cell cultures. Organoids, commonly used in drug discovery, are three-dimensional tissue structures derived from stem cells [12]. Although they resemble organs, organoids are not perfect models, as they often lack essential components such as immune system elements and blood vessel formation. Organoids are more affordable and easier to use compared to more complex 3D tissue models, although they often exhibit heterogeneity, which should be considered as a limitation. To replicate the ion and metabolite exchange seen in real organs, cells of various phenotypes are organized into coordinated 3D cultures, forming tissue-like organoids with interconnections. An early example of this approach is the creation of a liver-on-a-chip by assembling cells that represent the liver acinus. The next step in developing a "body-on-a-chip" would involve integrating simulations of various organs, such as the liver, intestines, and pancreas, on the same chip, ultimately leading to a modern version of the homunculus.

#### *12.3.1.2 Data-sharing technologies in biomedical research*

Generating biological data is crucial, but having technologies that enable access to this data in a way that is practical for developing ML pipelines, while also adhering to commercial confidentiality and regulatory standards required for handling sensitive information, is equally important. This challenge becomes even more complex when dealing with real-time data access, as models must adapt and adjust "on the fly" to real-world events and deliver timely, relevant predictions [13]. Organizations with secure data-sharing capabilities may have a greater opportunity to capitalize on their data.

Insufficient data used to train large-scale deep learning models contributes to the poor generalizability of AI algorithms, despite the vast amounts generated daily in pharmaceutical research. However, strict protectionism, secrecy between competing pharmaceutical companies, and government regulations on personal and medical information limit access to much of this valuable information. While data creation remains essential, equally important are the technologies that enable secure and efficient access for ML, while ensuring compliance with regulatory requirements. Furthermore, real-time data access presents additional challenges, as models must swiftly adapt to real-world events and deliver timely predictions.

Drug discovery is being revolutionized by emerging technology that make more accurate and efficient methods possible, as shown Figure 12.2. The figure with multiple emerging technologies is explained in detail.

#### *12.3.1.3 Lab robotization and automation*

The development of more cooperative robots equipped with safety sensors and basic AI is enabling scientists to use "human-friendly" equipment without the need for cumbersome safety measures. This advancement allows for closer interaction with the equipment, facilitating better monitoring during

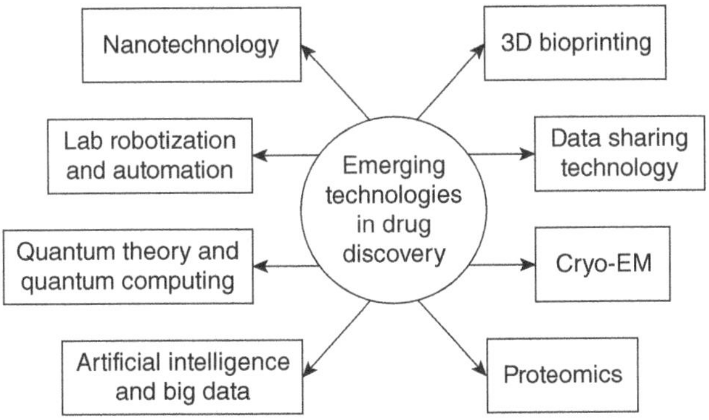


*Figure 12.2* Various emerging technologies in drug discovery.

screening processes and faster error correction. The integration of user-friendly and efficient interfaces makes it easier to interact with complex automation systems. In addition to hardware infrastructure, software scheduling plays a key role in optimizing assay performance and maintaining system stability [14]. To enhance efficiency and minimize downtime, careful protocol development and limiting robot movements are essential. Incorporating critical decision points, such as placing "failed" microtiter plates in quarantine to ensure proper processing, further streamlines the process. When handling large numbers of microtiter plates, system scheduling and barcoded labware help track protocol stages, ensuring consistency throughout. Automated quality control can further enhance assay performance.

#### *12.3.1.4 Quantum theory and quantum computing*

Molecular mechanics (MM) is a well-established computational technique used in medicinal chemistry, synthetic organic chemistry, and various areas of drug design. However, MM methods have limitations, particularly when it comes to analysing electron-related properties in the drug-receptor environment. Although quantum mechanics (QM) approaches are often resource-intensive and costly, they offer the potential for significantly improved prediction accuracy and provide more accurate models of chemical and biological compounds and their interactions.

#### *12.3.1.5 AI and big data technologies*

AI has become a key term in drug discovery, with concepts like "big data," "data-driven algorithms," and "intelligent systems" gaining attention in both academia and the media. These technologies are expected to improve patient stratification, biomarkers, and make drug development more efficient, faster, and cheaper. "Big data" refers to large, diverse datasets that need to be organized for easy access, search, and analysis to extract meaningful insights [15].

Driven by the global surge in data creation, "big data" has become a strategic concept in AI drug development. ML algorithms allow computers to learn tasks without explicit programming, using known examples to adapt and perform complex tasks. This encompasses supervised learning, where algorithms utilize labelled data, as well as unsupervised learning and "trial and error" methods that do not rely on a training dataset.

#### *12.3.1.6 Cryo-EM*

Cryogenic electron microscopy (cryo-EM) refers to a group of three similar methods aimed at "freezing and fixing" molecules by imaging them at very low temperatures. Recently, structural biologists have shown increasing interest in single-particle cryo-EM, with some biotech start-ups focusing on this technique [16]. Cryo-EM starts by flash-freezing protein solutions and other biomolecules to create vitreous ice, followed by electron bombardment and computer data analysis. Unlike traditional electron microscopy, cryo-EM bypasses chemical fixation, dehydration, and staining, maintaining the molecules in their native hydrated form. A key advantage of cryo-EM over other methods is its ability to avoid the need for crystallization, making it especially valuable for studying molecules that are challenging to crystallize, like membrane proteins.

### 12.3.2 Challenges that occur due to emerging technologies in drug discovery

Developing technology in the industry offers several benefits but also presents certain challenges. Firstly, because technology is heavily regulated, any application must comply with constantly evolving rules, making adherence to regulations a complex task. Additionally, ensuring data integrity adds another layer of difficulty. Secondly, managing sensitive data and relying on digital solutions makes data privacy and security critical. Protecting confidential and proprietary information from cyberattacks remains an ongoing concern. While technology can increase efficiency, the initial investment and ongoing operational costs can be significant, requiring a balance between cost control and innovation [17]. By staying competitive and avoiding obsolescence due to rapid technological advancements, companies must consistently update their infrastructure. Alternatively, they may choose to partner with third-party tech firms to streamline operations for specific tasks.

## 12.4 CLINICAL IMPLEMENTATION OF PERSONALIZED MEDICINE

Personalized medicine involves using diagnostic tests, including genetic markers, to determine the most appropriate treatments for individual patients. By integrating genetic data with medical history, physicians can create customized treatment and prevention plans. Despite the increasing availability of targeted therapies and molecular diagnostics, the healthcare system has been slow to implement personalized medicine in clinical settings. A recent survey found that only 4 in 10 consumers are familiar with personalized medicine, and just 11% of patients have had discussions about it with their doctors. This delay in adoption is due to challenges in adapting healthcare systems to new rules and standards. The impact of genetic factors on disease progression and treatment effectiveness is driving the growth of personalized medicine. Genetic differences can cause adverse reactions to certain medications, requiring adjusted dosages or alternative therapies. Additionally, tumour genetics can guide personalized treatment in cancer care. Implementing personalized medicine requires careful planning, preparation, and a systematic approach to ensure sustainability.

Pharmaceutical companies now have opportunities to develop molecularly targeted drugs and optimize existing medications, thanks to personalized medicine. It will change disease diagnosis and treatment, encouraging patient participation. For example, prostate cancer patients may choose between immediate treatment with side effects or waiting for disease progression [18]. Personalized medicine takes into account individual circumstances, offering a more tailored approach to healthcare. Despite its potential, challenges remain, such as limited understanding of disease mechanisms, economic factors, and privacy concerns in research. Policy issues also complicate the relationship between government research and regulatory bodies.

### 12.4.1 Strategies for implementation

Pharmaceutical companies must invest in emerging technologies and collaborate with university research teams to facilitate the development and swift adoption of personalized medicine. A proactive approach requires identifying more precise biomarkers [19]. A recent advancement is the detection of circulating DNA in the blood through liquid biopsies, offering a non-invasive, lower-risk alternative to traditional biopsies for early disease detection. Techniques like liquid biopsies enable medical professionals to predict tumour progression and switch patients to alternative therapies as soon as medication resistance is identified. Personalized medicine optimizes treatment plans, enhancing the effectiveness of existing therapies.

Pharmaceutical companies are adopting various strategies for implementing personalized medicine. Firstly, they recognize that transitioning from traditional methods to personalized treatments is critical, with therapies tailored for specific patient groups rather than the general population. Secondly, personalized medicine is seen as a way to improve healthcare outcomes and reduce overall healthcare costs by integrating digital technologies and utilizing advanced tools like single-cell omics for high-throughput cell analysis. Thirdly, integrating personalized medicine into existing healthcare systems

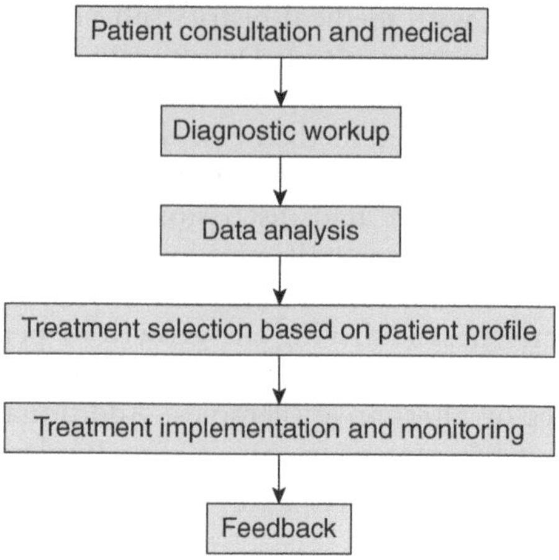


*Figure 12.3* Flowchart of implementation.

requires collaboration among stakeholders from diverse fields, aiming to leverage scientific advancements to enhance patient care. Finally, using biomarkers in research and development helps reduce trial sizes, speed up time-to-market, and support the development of promising therapeutic drugs that may otherwise be overlooked (Figure 12.3).

Collaborations provide opportunities to access new capabilities. For example, companies partnering with assay developers and other sectors can gain access to advanced diagnostic technologies. Another important aspect is the need for sales teams to be well-informed, equipped with the latest knowledge. In addition to understanding diagnostic and treatment procedures, sales teams must be familiar with patient histories, disease pathways, and molecular analysis [20]. Post-market surveillance also plays a crucial role in personalized medicine, enabling more targeted clinical trials for pharmaceuticals. As awareness of personalized medicine grows, there will be a greater demand for it in the coming years, helping to shift from traditional practices to modern diagnostic and treatment approaches. While clinical trials are currently labour-intensive and time-consuming, regulatory approval will streamline future trials, making them more efficient and better suited for personalized therapy.

In conclusion, personalized medicine offers the potential to reduce healthcare costs, lower drug development expenses, and shorten development timelines, all while improving patient outcomes. Achieving this will require active participation from patients and consumers in clinical trials, innovation in tools and genetic data analysis, education for both providers and consumers, and policy and regulatory changes. Physicians will need to understand diseases at the molecular level, academic researchers must collaborate with innovators to uncover new disease insights, and the IT sector should create tools for managing and protecting patient data [21]. Furthermore, stakeholders, payers, and policymakers must investigate new business models, diagnostic tools, and targeted therapies. The potential advantages of personalized medicine for the healthcare system are significant.

### 12.4.2 Challenges

Personalized treatment and prevention initiatives must be approached as a comprehensive, cross-sector challenge, considering their legal and social implications. The expansion of health information raises concerns about privacy and the potential for discrimination. Additionally, issues such as underrepresentation, cost, and accessibility in genetic research could exacerbate existing health inequities. To ensure that scientific advancements benefit all groups, research should include diverse populations. By leveraging personalized medicine, health disparities can be reduced and ultimately eliminated. Further research is needed on initiatives that can help achieve these goals, particularly for younger populations, as many preventive treatments rely on behavioural changes.

Equally important is the active involvement and commitment of all stakeholders – healthcare professionals, citizens, and patients – in the successful implementation of personalized medicine. Investing in healthcare workers' education, literacy, and capacity-building is essential, which includes integrating personalized medicine into core healthcare degree programmes and offering continuous learning opportunities [22]. Citizens and patients must also be prepared to adopt new habits and practices concerning their healthcare. This will only be feasible if educational, socio-economic, and cultural barriers are appropriately addressed and managed.

## 12.5 DRUG DESIGN AND DEVELOPMENT FOR PERSONALIZED THERAPY

Personalized medicine leverages an individual's genetic profile to inform decisions related to prevention, diagnosis, and treatment. This approach has the potential to revolutionize drug development and healthcare delivery by utilizing diagnostic markers and customized treatments based on molecular profiles. By understanding the molecular causes of disease, new drug targets can be identified, toxicogenomic indicators used for screening, and better patient selection for clinical trials achieved. However, challenges remain in transitioning from the traditional trial-and-error approach to personalized treatment.

Personalized medicine can help address discrepancies in medication safety between clinical studies and real-world experiences, focusing on patient needs rather than just disease specifics. Genomics plays a key role, especially in pharmacogenomics and genetic mutations in targeted molecular therapeutics. However, the evolution of personalized medicine also depends on other factors, like chromosomal variations in target proteins, requiring changes in clinical trial design and drug development financial models [23]. While randomized controlled trials will remain the gold standard, alternative methods are being explored to speed up drug development.

Personalized medicine challenges the traditional drug development model and can improve medication safety by prescribing the right drug to the right patient. Drawing parallels with the aviation industry, safety standards in drug prescribing could be improved with redundant warning systems for potential adverse reactions, along with proper training for healthcare professionals. This would allow for timely identification of issues and adjustments to reduce risks, maximize benefits, and save costs. Monitoring treatment and creating targeted drug delivery systems can further enhance safety and effectiveness, benefiting both existing and future pharmaceutical products.

## 12.5.1 Prerequisites for customized treatment

Our personalized medicine approach is built on the ability to select patients for treatments based on their unique characteristics, adjust dosing schedules, monitor drug movement within the body non-invasively, store individualized drug response data, and create simulation models to guide the planning and evaluation of drug safety clinical trials. To achieve this, personalized medication must fulfil the following four key requirements.

### *12.5.1.1 Biomarkers*

Developing disease-specific biomarkers is essential for determining the disease stage at early diagnosis and tracking its progression over time. Equally important is the design of physiologically responsive drug carriers to optimize therapeutic benefits and tailor drug release profiles to individual conditions. Validating disease biomarkers would help manage the complexity of the disease and enable early detection, allowing for the timely initiation of treatment. One factor contributing to the high attrition rate of phase III treatment candidates is the use of late-stage disease models, which reflect the point at which symptoms appear but often show limited therapeutic response. The ongoing challenges in developing effective treatments for Alzheimer's disease illustrate this issue [23]. However, there is hope for improved treatment, as recent reports have outlined an algorithm based on serum and protein biomarkers for early Alzheimer's disease detection.

### *12.5.1.2 Drug monitoring*

Regular updates on medication therapy effectiveness are rare for both patients and doctors, but drug monitoring is essential in personalized medicine. Developing point-of-care diagnostics to assess treatment response will offer valuable insights to patients and healthcare teams, especially in the absence of clear signs of the condition. This approach can identify patterns in toxicity and efficacy, enabling therapy adjustments. Non-invasive molecular imaging will also aid in mapping drug distribution, especially for targeted sites [24]. The integration of novel biomaterials and nanotechnology will guide the creation of focused drug delivery systems. Incorporating these technologies early in the drug development process enhances therapeutic benefits, minimizes side effects, and increases success rates. Storing clinical and pharmacokinetic data will also improve illness simulation models and serve as a key tool for monitoring adverse medication responses.

#### *12.5.1.3 Smart drug delivery system*

The development of new drug delivery systems, moving beyond traditional fixed-dose tablets and capsules, is essential. Innovations such as microneedles and modified inkjet printers are being designed to make it easier for patients, particularly the elderly, to manage multiple doses of medication. A patient's genetics, lifestyle, existing health conditions, and other medications all play a role in how their treatment is prescribed. For example, elderly patients may take ten or more medications simultaneously. To conclude, personalized medicine represents more than just a novel treatment strategy; it calls for proactive leadership and innovation to redefine drug development and meet the unique needs of various patient groups [25]. Similar to how aircraft revolutionized business travel and scientific progress, personalized medicine has the potential to greatly enhance the efficiency of drug development. However, the true impact of personalized medicine remains unclear due to delays in analysing and interpreting the vast amounts of genetic data collected.

### 12.5.2 Challenges

The protection of patients is arguably the most significant concern regarding the commercialization of personalized medicine. One of the major issues is the fear and potential consequences for patients who, through genetic testing, are found to have a predisposition to certain conditions or who do not respond to specific therapies. This also includes the psychological impact that genetic test results can have on individuals [26]. Another concern is the rights of family members who cannot provide consent directly, considering that genetic risks and predispositions are inherited. Additionally, the effects on specific ethnic groups and the presence of common genetic markers must be taken into account. Personalized medicine also heavily relies on data biases. Ensuring that the gene sample under examination originates from diverse populations is crucial. In order to prevent the samples from reflecting the same human biases that influence our decision-making, this is done because certain people do not have access to genetic testing, biased algorithms meant for customized treatment will inevitably lead to biased algorithmic results.

## 12.6 CONCLUSION AND FUTURE WORK

The evolution of personalized medicine represents a significant transformation in healthcare, focusing on individualized treatments and the integration of innovative technologies. With breakthroughs in AI, nanotechnology, and 3D bioprinting, therapy development is becoming more precise, reducing side effects and enhancing patient outcomes. However, challenges like data privacy, regulatory barriers, and ethical considerations must be carefully addressed to ensure successful implementation.

Looking ahead, integrating multi-omics data will be key to refining predictive models and improving therapeutic outcomes. The future of personalized medicine drug discovery systems presents a scenario where innovation and ethical responsibility go hand in hand. To ensure fair healthcare progress, addressing data bias and ensuring the inclusion of diverse populations in clinical research is vital. Further development of advanced drug delivery systems and non-invasive monitoring tools will improve patient care. Collaborative initiatives, backed by strong ethical guidelines and thorough patient education, will be crucial in realizing the full potential of personalized medicine, ensuring it is accessible and sustainable for everyone.

## REFERENCES

1. Farrokhi, M., Taheri, F., Khouzani, P. J., Rahmani, E., Tavakoli, R., Fard, A. M., & Sadoughi, K. (2023). Role of precision medicine and personalized medicine in the treatment of diseases. Kindle, 3(1), 1–164.
2. Sadee, W., Wang, D., Hartmann, K., & Toland, A. E. (2023). Pharmacogenomics: Driving personalized medicine. Pharmacological Reviews, 75(4), 789–814.
3. Mancinelli, L., Cronin, M., & Sadée, W. (2000). Pharmacogenomics: The promise of personalized medicine. AAPS pharmSci, 2, 29–41.
4. Showbharnikhaa, S., Vijayalakshmi, M. K., & Akshaya, T. (2024). Unlocking the pharmacogenomics and the potential of personalized medicine. Journal of Pharma I Research, 2(1), 50–55.
5. Sharmila, S., Nirmala, M., Somasundaram, D., & Menagadevi, M. (2024). Advancements in modelling, imaging, and simulation of cardiovascular diseases: A technological revolution in modern healthcare. In: Computational Intelligence and Blockchain in Biomedical and Health Informatics (pp. 258–274). CRC Press.
6. Cichońska, A., Ravikumar, B., & Rahman, R. (2024). AI for targeted polypharmacology: The next frontier in drug discovery. Current Opinion in Structural Biology, 84, 102771.
7. Sunildutt, N., Chethikkattuveli Salih, A. R., & Choi, K. H. (2023). Revolutionizing drug development: Harnessing the potential of organ-on-chip for disease modeling discovery. Frontiers in Pharmacology, 14, 1139229.
8. Sharmila, S., Madian, N., Menagadevi, M., & Rajendran, R. (2025). Artificial intelligence-based diagnostic tools for cardiovascular risk prediction. In: Machine Learning Models and Architectures for Biomedical Signal Processing (pp. 527–545). Academic Press.
9. Niazi, S. K., & Mariam, Z. (2023). Computer-aided drug design and drug discovery: A prospective analysis. Pharmaceuticals, 17(1), 22.

10. Zheng, P., Zeng, X., Wang, X., & Ding, P. (2023). Artificial intelligence and machine learning for drug discovery, design and repurposing: Methods. Frontiers in Pharmacology, 14, 1333747.
11. Konda, S. R. (2023). Optimizing computer architectures for high-performance drug discovery workflows. International Journal of Computer Science and Technology, 7(3), 243–258.
12. Singh, S., Kumar, R., Payra, S., & Singh, S. K. (2023). Artificial intelligence and machine learning in pharmacological research: Bridging the data and drug discovery. Cureus, 15(8), e44359.
13. Sunildutt, N., Chethikkattuveli Salih, A. R., & Choi, K. (2023). Revolutionizing drug development: Harnessing the potential of organ-on-chip technology for disease modeling and drug discovery frontiers. Frontiers in Pharmacology, 14, 1139229.
14. Sarkar, C., Das, B., Rawat, V. S., Wahlang, J. B., Nongpiur, A., Tiewsoh, I., & Sony, H. T. (2023). Artificial intelligence and machine learning technology driven modern drug discovery and development. International Journal of Molecular Sciences, 24(3), 2026.
15. Chang, Y., Hawkins, B. A., Du, J. J., Groundwater, P. W., Hibbs, D. E., & Lai, F. (2023). A guide to in silico drug design. Pharmaceutics, 15(1), 49.
16. Santagati, R., Aspuru-Guzik, A., Babbush, R., Degroote, L., Kyoseva, E., & Utschig, C. (2024). Drug design on quantum computers. Nature Physics, 20, 549–557.
17. Chen, Y., Wang, Z., Wang, L., Wang, J., Li, P., Cao, D., & Sakurai, T. (2023). Deep generative model for drug design from protein target. Journal of Cheminformatics, 15, 38.
18. Vemula, D., Jayasurya, P., Sushmitha, Y. N., & Bhandari, V. (2023). CADD, AI and ML in drug discovery: A comprehensive review. European Journal of Pharmaceutical Sciences, 181, 106324.
19. Stefanicka-Wojtas, D., & Kurpas, D. (2023). Barriers and facilitators to the implementation of personalized medicine across Europe. Journal of Personalized Medicine, 13(2), 203.
20. Englezos, K., Wang, L., Tan, E. C., & Kang, L. (2023). 3D printing for personalized medicines: Implications for policy and practice. International Journal of Pharmaceutics, 635, 122785.
21. Gao, J., Karp, J. M., Langer, R., & Joshi, N. (2023). The future of drug delivery. Chemistry of Materials, 35(2), 359–363.
22. Ezike, T. C., Okpala, U. S., Onoja, U. L., Nwike, P. C., Ezeako, E. C., Okpara, J. O., & Nwanguma, B. C. (2023). Advances in drug delivery systems, challenges and future directions. Heliyon, 9(6), e17488.
23. Wishart, D. S. (2016). Emerging applications of metabolomics in drug discovery and precision medicine. Nature Reviews Drug Discovery, 15(7), 473–484.
24. Alshamrani, A. I., Al Hadi, A. Y., AlMutairi, A. H., ALkhweildy, W. S., Aldubiban, G. M., Moafa, A. M., & Aldossary, F. K. M. (2024). Advancing targeted drug delivery: The role of smart polymers in responsive therapeutic systems. Egyptian Journal of Chemistry, 67(13), 1549–1556.
25. Zhang, H., Pan, Y., Hou, Y., Li, M., Deng, J., Wang, B., & Hao, S. (2024). Smart physical-based transdermal drug delivery system: Towards intelligence and controlled release. Small, 20(9), 2306944.
26. Zhang, W., Huai, Y., Miao, Z., Qian, A., & Wang, Y. (2019). Systems pharmacology for investigation of the mechanisms of action of traditional Chinese medicine in drug discovery. Frontiers in Pharmacology, 10, 743.

Chapter 13

# Revolutionizing neurology with AI and IoT

## Predictive analytics and real-time monitoring

*M. Menagadevi, M. Nirmala, D. Thiyagarajan, and Suman Lata Tripathi*

### 13.1 INTRODUCTION

Healthcare has become a major sector of the country with respect to employability and revenue generation. The healthcare industry works on a collection of organizations and services that provide medical care and support to individuals. Rural communities face a severe shortage of access to healthcare services. This is an important difference in health indicators in rural versus urban areas. The healthcare industry is the main sector that is impacted by the Internet of Things (IoT) technology and which is being widely incorporated for various disease diagnosis. IoT has extremely transformed the current healthcare domain with respect to the patient care, monitoring, and treatment. IoT tremendously aids in remote health monitoring for continuous patient monitoring and alert. It ensures ethical and safe monitoring of patients health condition, empowering physicians to deliver quality treatment at reasonable costs. IoT-based healthcare also ensures efficient patient interaction by healthcare professionals with minimal interaction along with reduced healthcare costs [1].

Figure 13.1 presents the component annual growth rate (CAGR) line chart that gives attention to the growth rate of various segments. The largest annual growth rate of approximately 37 and nearest electroencephalogram (EEG) monitoring follows with the high annual growth rate of approximately 17. Wireless EEG headsets are also smaller in size; they are also doubling in numbers; it highlights the design of non-invasive monitoring equipment. In comparison, the traditional markets like the neurology equipment and brain monitoring have average growth rate of approximately 56% that emanate maturity. Together, the number foretells that the underlying neurology markets are steady, yet the advent of artificial intelligence (AI) and IoT, particularly in predictive analytics and real-time EEG tracking, is catalysing high-velocity growth that is fundamentally changing neurological care.

EEG is a highly popular method of studying the brain signals, referred to as the metabolic alterations that occur in the brain of human beings. EEG measurements are the indicators of abnormal brain activity that is linked with the neurological diseases. The present healthcare situation complicates the

 DOI: 10.1201/9781042015931-13

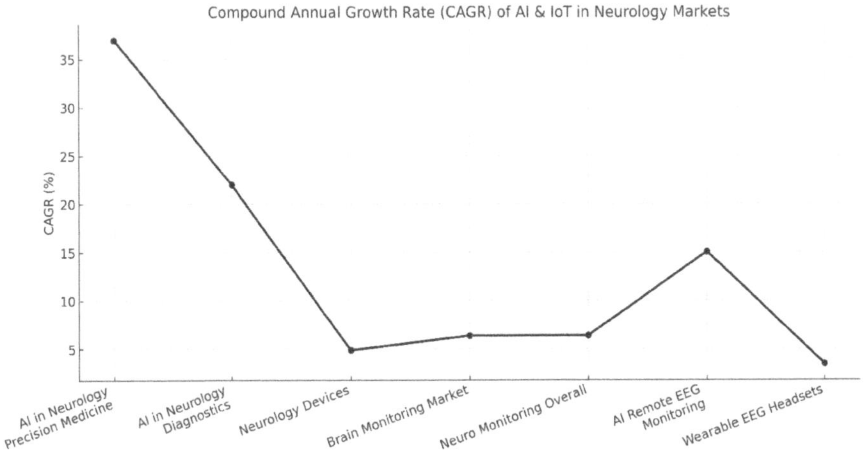


*Figure 13.1* Component annual growth rate of AI and IoT market.

implementation of IoT greatly because of the strict and clear requirements of a healthcare system [2]. EEG signals remotely when the patients are in one location and the EEG technologist and neurologist in another location. This type of monitoring can be used to diagnose and treat a variety of neurological conditions. It supports improved patient monitoring who live in rural or underserved areas. This process also improves quality of patient care by allowing neurologists to monitor patients more closely and intervene more quickly in the event of a seizure or other neurological event.

Emergence of AI impacted the improvement of sensitivity, specificity, precision, and accuracy in neurological research. The domain of AI constitutes machine learning (ML) and deep learning (DL) models and techniques which aid in successfully classifying and predicting various neurological disorders. Currently, AI-powered EEG diagnostic systems are being developed and refined to enhance healthcare industry. The development of AI-based EEG diagnostics is promising [3–5]. This chapter focus on the AI-driven IoT system for the prediction of neurological disorder. It discusses about different ML and DL algorithms used for the prediction of neurological disorder.

## 13.2 LITERATURE SURVEY

In detection of epileptic seizure disorder using EEG signals, Ref. [6] presents how the use of sensors can able to track the physical activities in the brain and how brain–computer interface can process the brain signals and perform seizure detection and also therapies for controlling seizures. This work presented the process of grouping EEG signals using AI algorithms to perform seizure detection based on pattern classification. Also, deep neural network models were used by authors to perform automatic feature extraction. Reference [7] explores the advantage of using IoT-based healthcare system and reviews all the relevant techniques for remote healthcare monitoring. It compares all the relevant works and its efficiency, privacy, security, and challenges associated with wearable sensor-based IoT monitoring systems and provides a classification of healthcare-monitoring sensors.

Reference [8] explores the integration of wearable biosensor devices for healthcare applications. Due to the rapid growth in data, the process of intelligent data management and its high connectivity due to 5G has enabled new healthcare techniques. Reference [8] explores all the benefits, challenges, and significance of IoT-based wearable devices from the 5G perspective. Furthermore, it explores the usage of wearable devices in detecting the issues such as curing, monitoring, and detection of disease. Chandrasekhar and Peddakrishna [9] presented a work on improving the prediction accuracy of heart disease using AI–ML algorithms. They incorporated various ML algorithms such as gradient boosting, naive Bayes, logistic regression, random forest (RF), and KNN and evaluated the efficiency of each algorithm based on a public dataset. The logistic regression model achieved higher accuracy compared to other models. Khiani [10] explains the challenges caused by the errors present in high-dimensional and non-stationary EEG recordings. Khiani presents an automated seizure detector techniques which records information from 16 channels for automatic seizure detection. The author proposed a Bi-LSTM-based intelligent healthcare system for tracking and precisely forecasting brain risk. The trade-off between the current system and the current methods are also discussed.

Reference [11] proposes an IoT-based system for remote monitoring and early health problem identification in home clinical setting for disease classification. The work uses a pretrained DL model built on a convolutional neural network (CNN) that may identify five distinct types of heartbeats. Additionally, the device reports the patient's oxygen level and heart rate, indicating whether or not they fall within normal parameters. If any serious anomalies are found, the system immediately links the user to the closest physician for additional diagnostics. Reference [12] proposes an improved decision-making model called brain–AI closed-loop system that allows human brain wave analysis and transmits results to AI. A wireless earbud-like EEG measurement device was proposed combined with tattoo-like electrodes and connectors, which enables continuous recording of high-quality EEG signals, especially the error-related potential. Depending on whether the received error signals are present or absent, which is established by DL classification of the supplied EEG data, the AI either corrects or reinforces decisions. Reference [13] uses an accurate signal prediction and estimation approach based on deep neural networks for intelligent data capture for signal measurement. The suggested system is prototyped as an advanced electronics component. The system is an automated physiological signal monitoring device to monitor elderly sick patients not only for fast access of data but also to get reliable

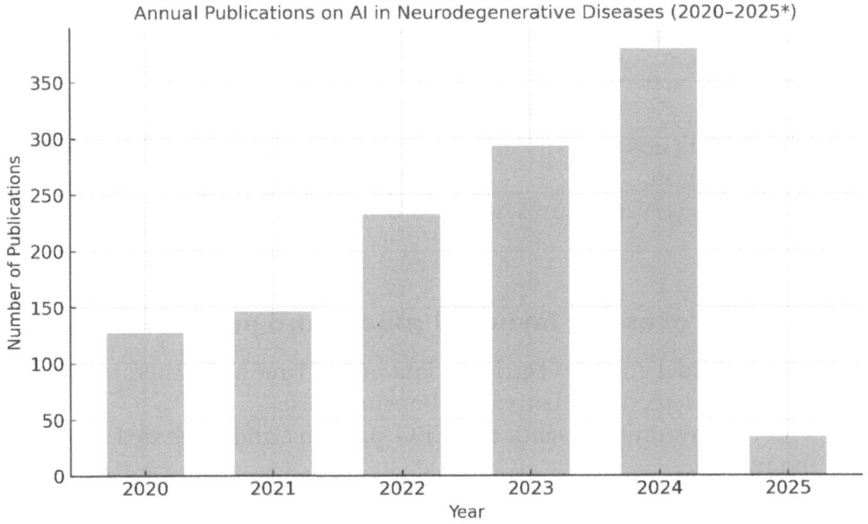


*Figure 13.2* Annual publication counts for the last five years (2020 to March 2025).

service by accurate prediction by healthcare service provider. Figure 13.2 shows the number of papers published annually on AI applications in neurodegenerative diseases increased significantly in the last five years, with 127 papers published in 2020 and 379 published in 2024 and early 2025 counts being 34 (data until March 16, 2025). Such a sharp growth is indicative of rapidly rising research efforts due to advancements in deep learning, multimodal neuroimaging, and an overall rise in interest in AI-supported diagnostics and monitoring [26–32].

## 13.3 ARCHITECTURE OF IoT ENABLED EEG HEALTHCARE SYSTEM

EEG-based healthcare system using AI and IoT for predicting neurological disorder includes the following.

### 13.3.1 IoT sensor layer: EEG sensor and analog-to-digital converter (ADC)

- The EEG sensor collects brainwave signals, which are initially in analog form.
- ADC converts these signals into a digital format.
- The 5 V power supply ensures proper operation of the hardware components.

### 13.3.2 IoT communication layer: Cloud and mobile data

- The cloud acts as an IoT gateway, transmitting EEG data for further processing.
- Mobile data connectivity enables remote access to EEG signals through an application.
- The cloud server ensures secure storage and remote access to patient data.

### 13.3.3 IoT processing layer: AI model and prediction

- The AI model analyses the EEG data in real time and generates a prediction (e.g., detecting seizures or abnormalities).
- The display unit presents the EEG patterns and AI-based results for monitoring.

### 13.3.4 IoT application layer: Doctor and remote monitoring

- The doctor can access patient EEG data remotely using an IoT-enabled application.
- Alerts and notifications can be sent via mobile applications in case of critical conditions.

Figure 13.3 presents an example of a cloud-based AI system of EEG signal processing and prediction. It all starts with the input signals received by an EEG sensor that records the activity of the brain waves in analog form. Then these analog signals are converted into digital form by using ADC which operates under a 5 V power supply. The digital signals are sent to the cloud

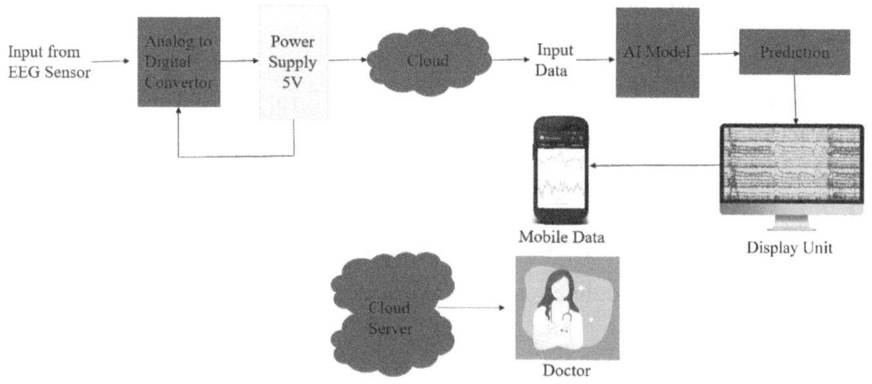


*Figure 13.3* Block diagram of IoT-enabled EEG healthcare system.

where they are stored and processed. The input data is sent to the cloud, where an AI model interprets the EEG waveforms and provides a prediction (e.g., abnormal brain activity detection, seizure prediction, or cognitive state analysis). The findings are displayed on a monitoring device and can also be viewed remotely through mobile information. Also, the EEG data of a patient can be accessed remotely and interpreted by a doctor (or any other healthcare professional) by means of the cloud server, which helps to diagnose the patient in a timely manner and provide an intervention. This combination of clouds computing, AI, and mobile accessibility improves distance medical services and live neurology.

## 13.4 AI MODEL

### 13.4.1 ML algorithm

#### *13.4.1.1 Support vector machine (SVM)*

SVM is a supervised learning algorithm used for the classification of EEG signals. SVM is more efficient for binary and multiclass classification data, it finds an optimal hyperplane to separate data in high-dimensional feature space. To work with limited dataset, SVM algorithm is widely used. SVM is particularly advantageous due to its capability to model non-linear relationships through the use of kernel functions such as radial basis function (RBF) and polynomial kernels.

SVM is used to classify different EEG patterns which is used to classify neurological conditions such as epilepsy, Alzheimer's disease, and Parkinson's disease. SVM is used to detect artefacts in EEG data caused by head movements. Contamination of EEG signals by such artefacts has been a significant challenge in deploying automatic neurological event detection systems in ambulatory settings. Researchers have demonstrated the efficacy of using additional physical signals, such as those from gyroscopes, in combination with EEG data to accurately identify and categorize these head movement artefacts.

#### *13.4.1.2 Random forest*

An ensemble learning method used for EEG signal classification is RF algorithm. It constructs multiple decision trees, each of which is grown from a random sample of the data. As multiple trees are used, with the final output being the average result of all trees, accuracy is elevated and overfitting is diminished. RF is effective for EEG-based tasks such as epilepsy monitoring, diagnosis of Parkinson's disease, classifying sleep disorders, and assessing a patient's cognitive state because it manages high-dimensional and non-linear EEG data seamlessly [14].

A major strength of RF is the assessment of feature importance, which allows a researcher to analyse the EEG features that are most relevant to disorder classification. EEG signals are typically noisy and complex; however, RF is able to effectively manage missing values, unbalanced datasets, and redundant features. Unlike DL models that necessitate a large amount of tuning, RF is efficient in computation and can be trained within a short period of time on EEG datasets, which makes it applicable for real-time monitoring of neurological activities.

In the pipeline of an EEG classification, RF, the raw EEG signals go through a series of filtering processes to eliminate noise and other artefacts. Following that, methods of feature extraction like power spectrum density (PSD), wavelet transform, and other entropy-based measures are used to convert the EEG signals to a feature matrix. A RF model is then trained with the features extracted so that it will distinguish the normal brainwave pattern from the abnormal one. When the model is trained, it can be used in real time on edge devices like Raspberry Pi or deployed on the cloud for analysing EEG signals. An important advantage that RF has over conventional ML models is that RF is capable of overcoming overfitting problems emerging from the feature selection and repetitive sampling of data due to the randomness incorporated with it [15].

#### 13.4.1.3 K-nearest neighbour (KNN)

KNN functions by calculating the Euclidean distance (or any appropriate distance measures) between a new EEG feature vector and every labelled data point in the training set. Next, the algorithm picks out the KNNs and allocates the new sample to the class that appears most frequently among them. The value of $K$ (number of neighbours chosen) can dramatically improve or worsen the effectiveness of the model: the lesser $K$ makes the model vulnerable to noise, while a higher $K$ blurs the decision boundary. One major benefit of KNN is its easiness and effectiveness for small to average size EEG datasets. KNN is, in contrast to DL models which need a lot of ordered training, a lazy learning algorithm. KNN does not build a model as other algorithms do; thus, its decision is made at the time of classification. KNN is optimal for real-time EEG cases where fresh data is always being received. On the other hand, KNN fails to perform effectively with large datasets, since it has to compute distances for every new prediction [16].

#### 13.4.1.4 Extreme gradient boosting

The XGBoost (extreme gradient boosting) algorithm is an ML algorithm used to work with large dimensional dataset. XGBoost utilizes an ensemble of decision trees model, where each tree is fitted to the remaining errors of the previous ones using gradient boosting. The model modifies weightage given to wrongly classified EEG samples, which enables it to focus more on difficult cases. This process is done iteratively which leads to increased accuracy while mitigating overfitting through some form of regularization like L1 (lasso) and

L2 (ridge) penalties. It is also faster and more accurate relative to traditional models of ML such as KNN and RF because it works with parallel computation and more sophisticated imputation methods.

EEG signals are processed in an EEG classification pipeline in which raw EEG signals are pre-processed with bandpass filtering and artefact removal. Time–frequency domain techniques of power spectral density (PSD), Wavelet transform, and fractal dimension, developed by Higuchi, are subsequently used to extract features. The feature matrix obtained is then used in conjunction with an XGBoost classifier, which is used to match with the supplied features in order to classify normal and abnormal EEG patterns.

The significant advantage of XGBoost in any EEG-related application is that this tool could routinely rank the features, thus helping a scientist define the most significant EEG biomarkers to identify disorders. In addition, XGBoost is applicable to most classification problems with imbalanced datasets, which is highly realistic in the clinical context with abnormal EEG samples being vastly outnumbered by normal ones. To monitor EEG in the cloud and in the form of an IoT edge device, XGBoost can be applied to a healthcare system or an IoT edge device. Although effective, its effectiveness also depends on hyper parameter optimization such as learning rate, max depth, and number of estimators [17]. XGBoost is the most suitable because it is more accurate and has lower computational expenses than the classical models such as SVM, KNN, and RF, with proper parameters.

## 13.4.2 DL algorithm

### *13.4.2.1 Convolutional neural networks*

CNNs are automating the extraction of spatial and temporal features from EEG signals which has completely transformed the classification of neurological disorders through EEG. Unlike conventional ML algorithms that depend on feature engineering or extraction, CNNs are more efficient since they detect patterns in raw data. This has made CNNs tremendously effective in the diagnosis of epilepsy, Alzheimer's, Parkinson's, sleep disorders, and other anomalies of brain waves.

Firstly, local features are derived by filtering using spatial convolutional layers, and then local feature dimensional pooling follows. The obtained signal features are more compact but still relevant. Jaiteh's hierarchy describes this process whereby CNNs are able to learn both short- and long-term dependency of EEG signals. Unlike classical models such as SVM, KNN, and RF, CNNs have shown to be more efficient for EEG classification problems because with every new layer added, the network becomes more profound.

In a typical CNN pipeline on EEG data, the initial steps chosen for the EEG signal include bandpass filtering, artefact removal, and normalization, while spectral analysis would be utilized to convert the filtered data into 2D images (EEG spectrograms) or 1D signal representations for the CNN model.

During classification, the convolutional layers will isolate the EEG's spatial and frequency patterns, while the fully connected layers will perform the final classification of the EEG (normal versus abnormal).

In neurological monitoring systems, one of the main difficulties is processing large amounts of raw data in real time. With CNNs, there is no need for manual feature engineering, solving the problem entirely. CNNs also have the advantage of being able to run in the cloud and edge devices such as Raspberry Pi and Jetson Nano. As powerful as CNNs may be, these models are still very expensive to train, as large labelled datasets are required, along with high computing power for processing [18]. Traditional ML models would be better suited for smaller resource requirements. To solve the restricted amount of EEG data, techniques such as data augmentation and transfer learning are useful.

#### *13.4.2.2 Long short-term memory (LSTM)*

LSTM networks are a subcategory of RNNs that excel at the classification of EEG-based neurological disorders due to their ability to capture temporal dependencies in sequential EEG signals. LSTMs perform time series data analysis compared to other ML algorithms. To solve vanishing gradient problem in RNN, LSTMs have memory cells with input, forgot, and output gates. So, LSTM is able to store the different patterns in the EEG signals. Raw EEG signals go through a preprocessing step which may include bandpass filtering, artefact removal, and normalization in a standard EEG-based LSTM pipeline. Time windows are created using the processed EEG data and are passed on to the LSTM network, which learns temporal dependencies between the EEG patterns. The different categories of the signals, normal or epileptic, are classified in the final output layer [19].

Raw EEG signals go through a preprocessing step which may include bandpass filtering, artefact removal, and normalization in a standard EEG-based LSTM pipeline. Time windows are created using the processed EEG data and are passed on to the LSTM network, which learns temporal dependencies between the EEG patterns. The different categories of the signals, normal or epileptic, are classified in the final output layer. LSTMs have the ability to support dynamic variations in the EEG signals, which is one of their obvious benefits. This is better as compared to CNNs that dominates sequential EEG analysis. However, it is evident that LSTMs are significantly harder on labelled EEG datasets and computing resources. The addition of bidirectional LSTM (BiLSTM) and attention mechanism makes it more efficient and, at the same time, enhances the feature extraction of the EEG sequences. The main benefit of LSTMs is that it can adapt to dynamic changes in the EEG signals. This is in contrast to CNNs that predominate sequential EEG analysis. However, LSTMs have a more conspicuous computational power and labelled EEG datasets requirements. BiLSTM and attention mechanisms make integration more efficient, but at the same time they also enhance feature extraction of EEG sequences.

#### 13.4.2.3 Gated recurrent unit (GRU)

A more effective way of performing sequential EEG data classification to classify neurological disorders is the GRU, a specialized form of recurrent neural network (RNN). GRUs are simpler than LSTM networks, hence they can be computed faster and even provide the correct results. GRU is common in the detection of epilepsy, Parkinson's, Alzheimer's, and sleep disorder on the basis of EEG signals [20–21].

GRU can effectively analyse EEG time series data, especially due to its ability to capture long- and short-term dependency without the gradient vanishing issue of RNN and support the analysis of continuous time series data. The LSTMs have three gates (i.e. input, forget, and output gates) and the GRUs two gates (i.e. update gate and reset gate). GRU's simplified gating mechanism is efficient in training complexity and significant EEG signal patterns are preserved. In the incorporation of a GRU in an EEG classification system, the first task is to preprocess the raw EEG signals by performing filtering, cleaning noise, and normalizing signals. In this respect, the data are sliced into time frames and input into the GRU network to learn the time-based dependency among the EEG patterns. The output layer subsequently embeds a group of signals into various categories like normal or epileptic seizures in future.

GRUs have a particular advantage, as they train significantly faster and consume less computation compared to LSTMs. This renders them suitable to live EEG processing both in IoT healthcare systems and on the edge devices, including Raspberry Pi and Jetson Nano. Besides, hybrid CNN–GRU models can more effectively extract features because spatial and temporal EEG characteristics are combined. LSTMs are usually preferred for their increased ability to model complex patterns in comparison to GRUs, but this comes at the expense of lower computational efficiency. The additional parameters make GRU a better option for situations where efficiency is essential. Less complex EEG variations are better managed with LSTMs or BiGRUs where deeper temporal modelling is needed.

#### 13.4.2.4 Autoencoder (AE)

AE is a type of unsupervised, DL model geared for feature extraction, anomaly detection, and dimensionality reduction in EEG-based neurological disorders classification. AEs have proven to be effective in the automatic detection of epilepsy, Alzheimer's, and Parkinson's diseases, and even some sleep disorders, which all rely on EEG signals that have complex patterns that cannot be easily interpreted by standard ML models. In the case of AEs, EEG data is transformed (encoded) into a lower-dimensional latent variable and then reconstructed (decoded) back to its original form. In the process of obtaining the output EEG signal, the model is trained to keep the most relevant features while eliminating noise and irrelevant changes to the signal [22, 23].

In a typical workflow involving EEG-based AEs, the first step consists of preprocessing the raw EEG signals which includes bandpass filtering, artefact

*Table 13.1* Comparison of machine learning and deep learning models

| *Algorithm* | *Accuracy* | *Speed (training/ inference)* | *Real-time suitability* | *Computational complexity* | *Use case examples* |
|---|---|---|---|---|---|
| CNN (convolutional neural network) | High (85–95%) | Moderate (fast inference, slower training) | Moderate (edge deployment possible) | High (requires GPUs for training) | Epilepsy detection, sleep stage classification, seizure prediction |
| LSTM (long short-term memory) | High (80–95%) | Slow (high memory consumption) | Moderate (optimized models needed) | High (sequential processing overhead) | Parkinson's disease detection, emotion recognition, BCI applications |
| GRU (gated recurrent unit) | Moderate to high (80–93%) | Faster than LSTM | Suitable for real time | Moderate | EEG signal–based movement prediction, emotion recognition |
| Transformers (EEG-BERT, ViTs) | Very high (90–97%) | Slow (high computational cost) | Limited (requires powerful hardware) | Very high (needs extensive data for training) | Early Alzheimer's detection, EEG-based cognitive assessment |
| SVM (support vector machine) | Moderate to high (75–90%) | Fast inference, slower training | Suitable for real time if dataset is small | Low to moderate (feature engineering required) | Epilepsy classification, depression analysis |
| Random forest (RF) | Moderate (70–85%) | Fast (handles noisy data well) | Suitable for offline analysis | Low | Brain activity classification, EEG feature ranking |
| K-nearest neighbours (KNN) | Moderate (65–80%) | Slow for large datasets | Limited (computationally expensive for real time) | Low to moderate | Motor imagery tasks, BCI applications |
| XGBoost (extreme gradient boosting) | High (80–92%) | Fast (parallel processing) | Moderate | Moderate | EEG-based seizure prediction, mental health disorder detection |

removal, and normalization when needed [24, 25]. The data goes into the neural network encoder, which transforms and compresses the EEG signal into a compact set of features. The decoder neural network then reconstructs the signal and the difference between it and the original EEG signal is computed using a loss function like MSE. Efforts then move towards the classification of the neurological disorder where the discretized features are input into a ML classifier – SVM, cross forest, RF, or XGBoost, which are known to create efficient and powerful models. Table 13.1 provides a comparison of DL and ML models.

## 13.5 CONCLUSION

AI with IoT plays a vital role in the prediction of neurological disorder. This chapter highlighted the general view of detecting neurological disorder using IoT and popular ML and DL methods used for classification. Input EEG signal is processed through ADC and transmitted to cloud for further analysis. AI models predicts the output of the input data from the cloud and store in the cloud. It enables healthcare professionals for timely analysis and investigation. It increases diagnostic efficiency, enhances accuracy, and enables precise classification of neurological disorders. Future enhancements in AI and IoT technologies have the potential to further enhance the system and enhance its robustness, adaptability, and scalability for broader clinical applications.

## REFERENCES

1. Charfare, Ruwayd Hussain et al., "IoT–AI in Healthcare: A Comprehensive Survey of Current Applications and Innovations." *International Journal of Robotics & Control Systems* 4(3), 2024.
2. Kazi, Kutubuddin, "AI-Driven IoT (AIIoT) in Healthcare Monitoring." *Using Traditional Design Methods to Enhance AI-Driven Decision Making* (pp. 77–101). IGI Global Scientific Publishing, 2024.
3. Rangarajan, Deepti et al., "Exploring the Next-Gen Transformations in Healthcare through the Impact of AI and IoT." *Intelligent Systems and IoT Applications in Clinical Health* (pp. 73–98). IGI Global, 2025.
4. Hussain, Ibrar, "Empowering Healthcare: AI, ML, and Deep Learning Innovations for Brain and Heart Health." *Artificial Intelligence and Machine Learning Frontiers* 4(1), 167–188, 2024.
5. Menagadevi, M., N. Madian, D. Thiyagarajan, & R. Rajendran, "Smart Medical Devices: Making Healthcare More Intelligent." *Machine Learning Models and Architectures for Biomedical Signal Processing* (pp. 487–501). Academic Press, 2024.
6. Daftari, Charmi, Jainish Shah, and Manan Shah, "Detection of Epileptic Seizure Disorder Using EEG Signals." *Artificial Intelligence-Based Brain–Computer Interface* (pp. 163–188). Academic Press, 2022.
7. Abdulmalek, Suliman, et al., "IoT-Based Healthcare-Monitoring System towards Improving Quality of Life: A Review." *Healthcare*, 10(10), 1993, 2022.

8. Verma, Damini, et al., "Internet of Things (IoT) in Nano-Integrated Wearable Biosensor Devices for Healthcare Applications." *Biosensors and Bioelectronics: X*, 11, 100153, 2022.
9. Chandrasekhar, Nadikatla, and Samineni Peddakrishna, "Enhancing Heart Disease Prediction Accuracy through Machine Learning Techniques and Optimization." *Processes* 11(4), 1210, 2023.
10. Khiani, Simran, et al., "An Effectual IoT Coupled EEG Analysing Model for Continuous Patient Monitoring." *Measurement: Sensors* 24, 100597, 2022.
11. Islam, Md Reazul, et al., "Deep Learning-Based IoT System for Remote Monitoring and Early Detection of Health Issues in Real-Time." *Sensors*, 23(11), 5204, 2023.
12. Shin, Joo Hwan, et al., "Wearable EEG Electronics for a Brain–AI Closed-Loop System to Enhance Autonomous Machine Decision-Making." *NPJ Flexible Electronics*, 6(1), 32, 2022.
13. Rajan Jeyaraj, Pandia, and Edward Rajan Samuel Nadar, "Smart-Monitor: Patient Monitoring System for IoT-Based Healthcare System Using Deep Learning." *IETE Journal of Research*, 68(2), 1435–1442, 2022.
14. Deshmukh, Shailesh Madhavrao, and Ikhar Avinash Khemraj, "Detection of Brain Neurological Disorders Using EEG Signals: A Comparison of SVM Classifier and LDA Classifier." *Recent Developments in Microbiology, Biotechnology and Pharmaceutical Sciences* (p. 290). CRC Press, 2025.
15. Saimon, Shariar Islam, et al., "Advancing Neurological Disease Prediction through Machine Learning Techniques." *Journal of Computer Science and Technology Studies* 7(1), 139–156, 2025.
16. Jain, Shraddha, and Rajeev Srivastava, "Electroencephalogram (EEG) Based Fuzzy Logic and Spiking Neural Networks (FLSNN) for Advanced Multiple Neurological Disorder Diagnosis." *Brain Topography* 38(3), 33, 2025.
17. Uyanik, Hakan, et al., "Automated Detection of Neurological and Mental Health Disorders Using EEG Signals and Artificial Intelligence: A Systematic Review." *Wiley Interdisciplinary Reviews: Data Mining and Knowledge Discovery* 15(1), e70002, 2025.
18. Rathod, Prajakta, Shefali Naik, and Jayendra M. Bhalodiya, "Epileptic Signal Classification Using Convolutional Neural Network and Shapley Additive Explainable Artificial Intelligence Method." *Neural Computing and Applications*, 37(6), 4937–4955, 2025.
19. Abir, Shake Ibna, et al., "Machine Learning and Deep Learning Techniques for EEG-Based Prediction of Psychiatric Disorders." *Journal of Computer Science and Technology Studies* 7(1), 46–63, 2025.
20. Bouchane, Mouna, Wei Guo, and Shuojin Yang. "Hybrid CNN–GRU Models for Improved EEG Motor Imagery Classification." *Sensors*, 25(5), 1399, 2025.
21. Rivas, Fernando, Jesús Enrique Sierra-Garcia, and Jose María Camara, "Comparison of LSTM-and GRU-Type RNN Networks for Attention and Meditation Prediction on Raw EEG Data from Low-Cost Headsets." *Electronics*, 14(4), 707, 2025.
22. Menagadevi, M., S. Mangai, N. Madian, and D. Thiyagarajan. (2023). "Automated Prediction System for Alzheimer Detection Based on Deep Residual Autoencoder and Support Vector Machine. *Optik*, 272, 170212.
23. Ebâncă, Daniel-Costin, Cristina Popîrlan, and Irina-Valentina Tudor. "An IoT Approach Enhanced with AI Capabilities for Predictive Neurology in Smart Homes." *BRAIN: Broad Research in Artificial Intelligence and Neuroscience*, 16(1), 272–285, 2025.
24. Wijayanto, Inung, Annisa Humairani, Sugondo Hadiyoso, Achmad Rizal, Dasari Lakshmi Prasanna, and Suman Lata Tripathi, "Epileptic Seizure Detection on a Compressed EEG Signal Using Energy Measurement." *Biomedical*

*Signal Processing and Control*, 85, 104872, 2023. https://doi.org/10.1016/j.bspc.2023.104872
25. Prasanna, D.L., and S.L. Tripathi, "A Review on Tongue Based Assistive Technology, Devices and FPGA Processors Using Machine Learning Module." *Wireless Personal Communications*, 134, 151–170, 2024. https://doi.org/10.1007/s11277-024-10897-8
26. Zhou, Y., J. Liu, H. Wang, X. Li, and Q. Chen, "Bibliometric and Visual Analysis of Artificial Intelligence in Neurodegenerative Disease Research (2014–2025)." *Frontiers in Neurology*, 16, 1474484, 2025. doi: 10.3389/fneur.2025.1474484.
27. Grand View Research, Artificial Intelligence in Precision Medicine Market. *Market Analysis Report*, 2024. https://www.grandviewresearch.com/horizon/statistics/artificial-intelligence-precision-medicine-market-report/therapeutic-application/neurology/global
28. Grand View Research, Artificial Intelligence in Diagnostics Market Outlook. *Market Analysis Report*, 2025. https://www.grandviewresearch.com/horizon/statistics/artificial-intelligence-in-diagnostics-market-outlook/diagnosis-type/neurology/global
29. DataM Intelligence, Neurology Devices Market to Grow at 8.2% CAGR by 2031 Driven by AI. *OpenPR*, 2024. https://www.openpr.com/news/4133764/neurology-devices-market-to-grow-at-8-2-cagr-by-2031-driven-by-ai
30. Mordor Intelligence, *Brain Monitoring Market – Industry Report*, 2024. https://www.mordorintelligence.com/industry-reports/global-brain-monitoring-market-industry
31. The Business Research Company, *Neurology Monitoring Market Overview 2025*, 2024. https://www.thebusinessresearchcompany.com/market-insights/neurology-monitoring-market-overview-2025
32. Grand View Research, *Wearable EEG Headsets Market Size, Share & Trends Analysis Report*, 2024. https://www.grandviewresearch.com/press-release/global-wearable-eeg-headsets-market

Chapter 14

# Deep learning

## A transformative force in artificial intelligence

*Meena Rawat, Dhruv Shakti, and Pooja Pathak*

## 14.1 INTRODUCTION

The secrets of the contemporary AI reside in the capability of deep learning (DL) to discern highly complex representations and patterns automatically for large and heterogeneous datasets. In the contemporary world, machine learning (ML), DL, and artificial intelligence (AI) are the three popular terms frequently used in loose association to define intelligent software or systems. Figure 14.1 illustrates the interaction among AI, ML, and DL. As shown in the figure, DL can also figure as a subset of ML as well as the whole of AI.

Vaguely, AI is the process of giving systems or machines human-like behaviour and intelligence. ML, as opposed to that, is the process that allows for learning from experience or data whereby analytical models are mechanically

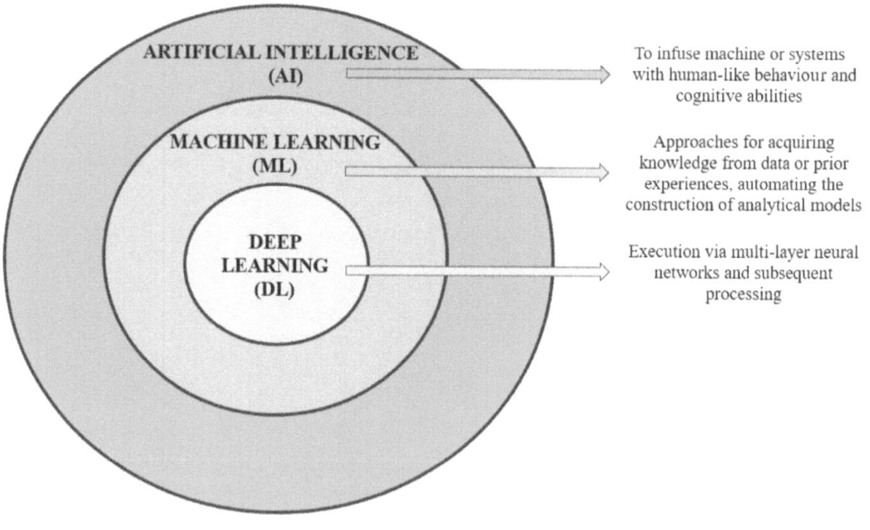


*Figure 14.1* A depiction of the role of deep learning (DL) in the context of machine learning (ML) and artificial intelligence (AI).

 DOI: 10.1201/9781042015931-14

built while separating from it. DL is the process of learning from data such that these computations happen with multilayer neural networks (NNs) and processing. The term by which we refer to deep in the DL process is that of including stages or levels in multiple numbers through which data is processed in building a data-based model.

As opposed to conventional ML strategies that strongly depend on manual feature engineering, DL by specifically applying NNs thrives in automatically discovering hierarchical features from raw data [1]. This feature learning strength does more than decrease the labour of manual design. It also allows systems to adapt and refine as they experience new and complex types of data. DL's impact appear in many areas like image recognition, language translation, speech technologies, and even personalized suggestions [2]. Its tolerance of large-scale datasets as well as its detection of delicate patterns have established it as a viable innovation engine in healthcare, fraud detection, autonomous vehicles, and more industries [3]. Due to these models scaling so well with data, they continue to gain prominence while information increases at an unparalleled pace.

The DL history is a sequence of breakthroughs and reverses with no smooth development [4]. Initial perceptron models from 1950 raised excitement but became stuck due to fragile computing power and limited theories. Relaunch in the 1980 provided impetus for more complex networks. The introduction in 1990 of convolutional network was a watershed, particularly for computer vision. More recently, recurrent and memory-based designs such as LSTM enabled sequential data such as text and speech to be modelled with much higher accuracy.

By 2010, the subject began a new period. Accelerating processor, large datasets, and stabilizing methods like dropout and batch normalization caused a dramatic comeback. Architecture such as AlexNet and VGGNet became familiar benchmarks and deep reinforcement learning caused a buzz through applications in computer games.

More recent transformers and mechanisms accelerated the pace again. Transformer-based models like BERT and GPT now reign in natural language processing and extend what DL can do. Seeing this trajectory, two characteristics stand out DL's continued technical development and being at the core of moving AI forward.

## 14.2 NEURAL NETWORKS

The NN is the core of DL as it is a structure that resembles the human brain in its connection and signal processing mechanisms. Layers of nodes, or "neurons, are used to create a network which transmits information forward and modifies it in stages. The design is the simplest and comprises an input layer, where raw data are received; a number of hidden layers, on which calculations are done; and an output layer, where the final result is generated [5].

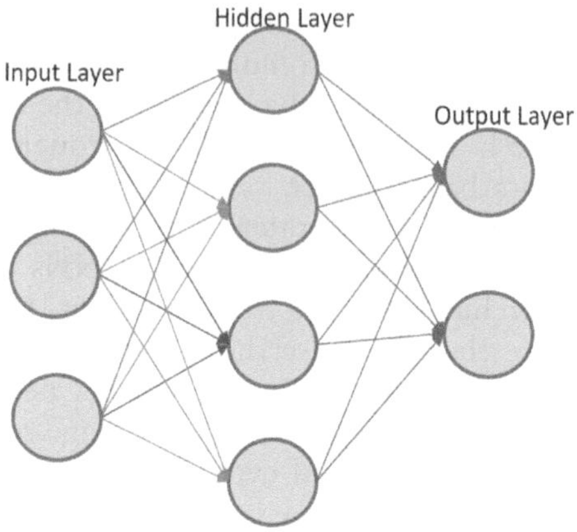


*Figure 14.2* Three-layers structure of neural networks.

Varied forms of NNs have been created according to varied requirements. The easiest and most useful are the feedforward networks that are applicable to general tasks. Convolutional networks are vision-specific issues since they are capable of capturing spatial patterns of images. In contrast, recurrent networks are applicable in the context of sequential data and are common in natural language processing. The diversity of architectures has enabled DL to assume a wide range of problems and has made it a breakthrough in the history of AI.

The NNs are flexible, which is the other strength. The number of layers and the size of each layer are adjustable as per the requirements of a particular application. Despite such variations, all networks share the three basic building blocks that include the input layer, the hidden layer, and the output layer [6]. This canonical form is exemplified in Figure 14.2. The input is then transmitted to the input layer, after which there are several weighted transformations in the hidden layers: this produces an output. The last one is the work of all the layers working together.

### 14.2.1 Feedforward neural networks (FNN)

FNN, commonly known as multilayer perceptrons (MLPs), forms the basic architecture of DL. Inspired by the structure of the human brain, these networks contain layers of connected nodes or neurons with information proceeding in one direction, that is, from the input layer to the output layer [7]. For example, assume that we have an artificial NN with an input layer, two hidden layers, and an output layer. Figure 14.3 clearly illustrates that the input layer contains three nodes, two hidden layers contain three and four nodes, and the output layer contains three nodes. The input layer of the network is exposed to the data fed into it after which the whole network determines the answer and gives it back to the user through the output layer. The data in the

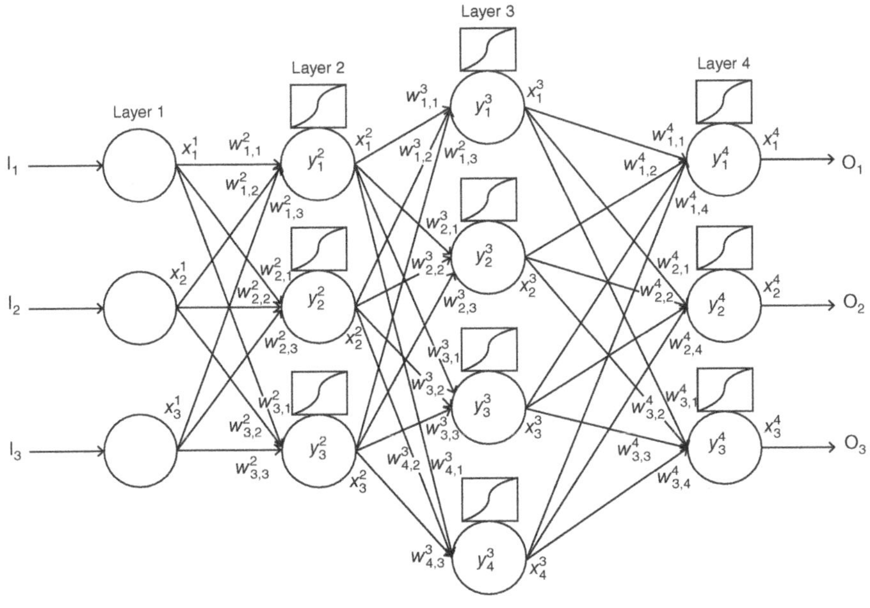


*Figure 14.3* Work flow of a four-layer feed forward neural network.

nodes of the layers are tagged such that it can erase the node number as well as the layer number and similarly for the weights. It is important to recall that the superscript represents the layer number and the subscript shows the number of nodes in the layer when marking the nodes. The activation value of a node shows to what extent a neuron is responsive to a given set of input data. Using forward propagation, these activation values yield the output value.

For each layer (suppose $i$th layer), each neuron is connected with other neurons by some weights $\mathrm{w}^i_{k,j}$ and bias $b^i$ which transfers a layer's activations to the next layer.

Matrix format for the fourth layer where $i = 1$ and $k$ varies from 1 to 3 in Equation (14.1):

$$x^i_k = \text{sigmoid}\left(b^i + \left(x^{i-1}_1 \cdot w^i_{k,1} + x^{i-1}_2 \cdot w^i_{k,2} + x^{i-1}_3 \cdot w^i_{k,3} + \cdots + x^{i-1}_{l_{i-1}} \cdot w^i_{k,l_{i-1}}\right)\right) \tag{14.1}$$

where $l_i$ denotes the total number of nodes in $(i-1)^{\text{th}}$ layer. Let's write the above equation in matrix format by assigning matrix format for the second layer where $i = 2$ and $k$ varies from 1 to 3 in Equation (14.1):

$$\begin{bmatrix} x^1_1 & x^1_2 & x^1_3 \end{bmatrix} \begin{bmatrix} w^2_{1,1} & w^2_{2,1} & w^2_{3,1} \\ w^2_{1,2} & w^2_{2,2} & w^2_{3,2} \\ w^2_{1,3} & w^2_{2,3} & w^2_{3,3} \end{bmatrix} = \begin{bmatrix} y^2_1 \\ y^2_2 \\ y^2_3 \end{bmatrix} \rightarrow \begin{bmatrix} y^2_1 + b^2 \\ y^2_2 + b^2 \\ y^2_3 + b^2 \end{bmatrix} \rightarrow \begin{bmatrix} \text{sigmoid}\left(y^2_1 + b^2\right) \\ \text{sigmoid}\left(y^2_2 + b^2\right) \\ \text{sigmoid}\left(y^2_3 + b^2\right) \end{bmatrix} \rightarrow \begin{bmatrix} x^2_1 \\ x^2_2 \\ x^2_3 \end{bmatrix} \tag{14.2}$$

Matrix format for the third layer where $i = 3$ and $k$ varies from 1 to 4 in Equation (14.1):

$$\begin{bmatrix} x_1^2 & x_2^2 & x_3^2 \end{bmatrix} \begin{bmatrix} w_{1,1}^3 & w_{2,1}^3 & w_{3,1}^3 \\ w_{1,2}^3 & w_{2,2}^3 & w_{3,2}^3 \\ w_{1,3}^3 & w_{2,3}^3 & w_{3,3}^3 \\ w_{1,4}^3 & w_{2,4}^3 & w_{3,4}^3 \end{bmatrix} = \begin{bmatrix} y_1^3 \\ y_2^3 \\ y_3^3 \\ y_4^3 \end{bmatrix} \rightarrow \begin{bmatrix} y_1^3 + b^3 \\ y_2^3 + b^3 \\ y_3^3 + b^3 \\ y_4^3 + b^3 \end{bmatrix} \rightarrow \begin{bmatrix} \text{sigmoid}\left(y_1^3 + b^3\right) \\ \text{sigmoid}\left(y_2^3 + b^3\right) \\ \text{sigmoid}\left(y_3^3 + b^3\right) \\ \text{sigmoid}\left(y_4^3 + b^3\right) \end{bmatrix} \rightarrow \begin{bmatrix} x_1^3 \\ x_2^3 \\ x_3^3 \\ x_4^3 \end{bmatrix} \tag{14.3}$$

Matrix format for the fourth layer where $i = 4$ and $k$ varies from 1 to 3 in Equation (14.1):

$$\begin{bmatrix} x_1^3 & x_2^3 & x_3^3 & x_4^3 \end{bmatrix} \begin{bmatrix} w_{1,1}^4 & w_{2,1}^4 & w_{3,1}^4 \\ w_{1,2}^4 & w_{2,2}^4 & w_{3,2}^4 \\ w_{1,3}^4 & w_{2,3}^4 & w_{3,3}^4 \\ w_{1,4}^4 & w_{2,4}^4 & w_{3,4}^4 \end{bmatrix} = \begin{bmatrix} y_1^4 \\ y_2^4 \\ y_3^4 \end{bmatrix} \rightarrow \begin{bmatrix} y_1^4 + b^4 \\ y_2^4 + b^4 \\ y_3^4 + b^4 \end{bmatrix} \rightarrow \begin{bmatrix} \text{sigmoid}\left(y_1^4 + b^4\right) \\ \text{sigmoid}\left(y_2^4 + b^4\right) \\ \text{sigmoid}\left(y_3^4 + b^4\right) \end{bmatrix} \rightarrow \begin{bmatrix} x_1^4 \\ x_2^4 \\ x_3^4 \end{bmatrix} \tag{14.4}$$

In FNN, there is no loop or cycle in the connectivity pattern, that is, the data passes through the network without reiteration of a layer. FNN are simple and diversity-oriented and they can handle all sorts of tasks from simple classification to higher-order problems. The input layer is used for receiving data, while the hidden layers remains active for processing through weighted connections, and finally the output layer produces the final result. FNNs were successful for applications where the input–output relation is rather direct and they provide basic building blocks for more advanced NN architecture.

### 14.2.2 Convolutional neural networks (CNNs)

Yann LeCun, recently famed for heading Facebook's AI Research Group, is commonly referred to as the father of the CNN. His first design – LeNet, constructed in 1988 – was initially created to read postal code digits and handwritten digits. Although small in ambitions, the work formed the foundation for what later became one of the most influential methods of AI.

CNN is one big leap forward for DL, particularly for visual data–based problems such as image classification and computer vision [8]. The idea is based on the way in which the human visual cortex processes information but rather than all features in an image being processed alike, CNN uses convolutional layers which move filters across the image and extract meaningful features such as edges or textures. Pooling layers subsequently compress the data's dimensionality allowing the model to become more efficient as well as less sensitive to minute changes within the input [9].

The CNN builds up hierarchy after hierarchy of successively abstract features by stacking several convolutional and pooling layers. By this layering, they are able to recognize complex objects from raw pixels without any hand design of features. Consequently, CNN have come to be the key to application like object detection when cars are driven automatically as well as diagnosis of diseases from medical images [10]. Their role is not even limited to vision since their versions have been applied to language problems as well as any other form of sequential data.

Since the face recognition on social media or perceptual capability of autonomous cars, the effect of CNN has become everywhere in daily life. What was initially a digital recognition architecture has grown to be the backbone of modern-day DL.

To explain their working, take the example of the detection of a bird from an image. The working begins by entering the pixel information of the image given as arrays and then into the input layer of the NN [11]. One by one hidden layers such as the convolution layer, ReLU layer, and pool layer performs feature extraction by carrying out varied calculations and transformations. The total number of such hidden layers is not fixed [12]. Finally, the process is terminated by the fully connected layer by detecting the object visible in the picture. Figure 14.4 makes the overall process of CNN very much lucid, that is, the detection of the bird by CNN. Now, take a closer look at the working flow of the CNN.

#### *14.2.2.1 Input layer*

The procedure starts with the input data. Let's take the example of an image being given to the input layer of the NN. The input is normally presented as the pixel value's grid. Each picture is presented as an array of pixel values in CNN. The array is the raw data structure for visual recognition and visual analysis. Figure 14.4 presents an array representation of actual picture of digit 7:

#### *14.2.2.2 Convolutional layer*

In a convolutional layer, small filters also known as kernels do move across the input image performing element-wise multiplication with specific local

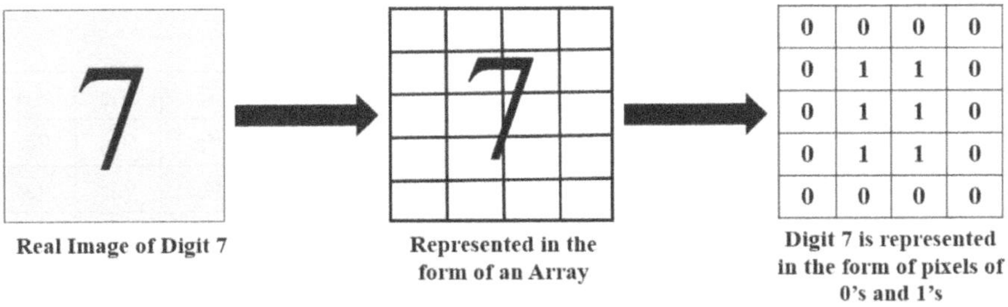


*Figure 14.4* An array form of real image of digit 7.

sections of the image. This process helps us to detect the patterns and features like the edges and the textures.

For example, take a 6 × 6 image made up of pixel values that are either 0 or 1. When a 3 × 3 filter is applied to this image, we compute the element-wise product to produce a new matrix called the convolution feature matrix. Figure 14.5 shows how a 6 × 6 image is transformed into one 4 × 4 feature map using a 3 × 3 filter. Figures 14.6–14.8 then break down how the values in each cell of the resulting grid are calculated.

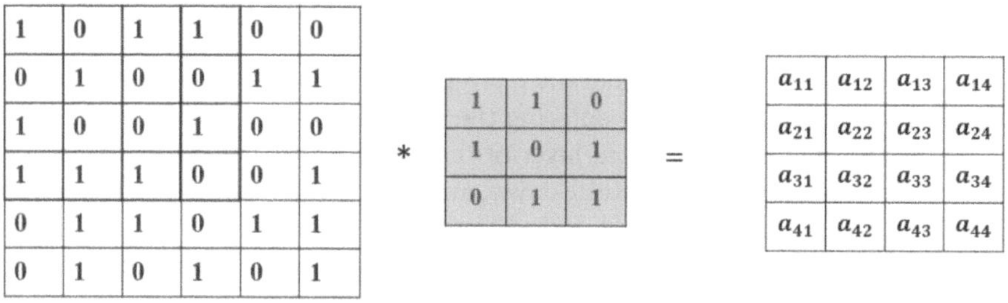


*Figure 14.5* Transformation of a 6 × 6 pixel image into a 4 × 4 array after the application of a 3 × 3 filter.

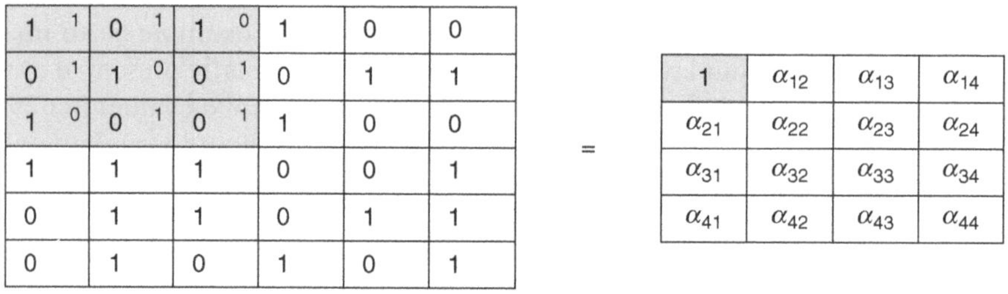


*Figure 14.6* Process to find specific values $a_{11}$ in grid cells.

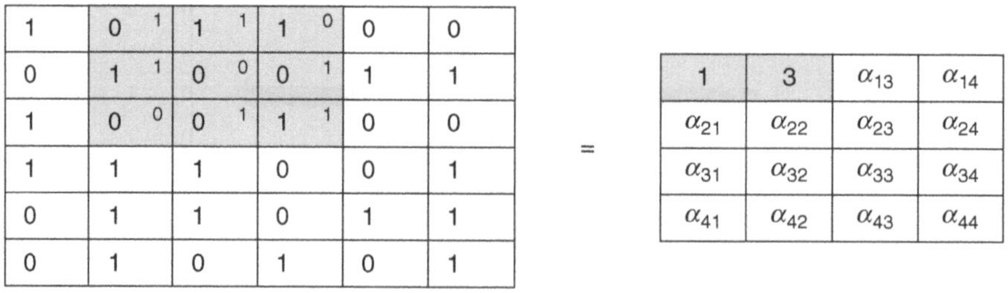


*Figure 14.7* Process to find specific values $a_{12}$ in grid cells.

| | | | | | |
|---|---|---|---|---|---|
| 1 | 0 | 1 | 1 | 0 | 0 |
| $0^{1}$ | $1^{1}$ | $0^{0}$ | 0 | 1 | 1 |
| $1^{1}$ | $0^{0}$ | $0^{1}$ | 1 | 0 | 0 |
| $1^{0}$ | $1^{1}$ | $1^{1}$ | 0 | 0 | 1 |
| 0 | 1 | 1 | 0 | 1 | 1 |
| 0 | 1 | 0 | 1 | 0 | 1 |

=

| | | | |
|---|---|---|---|
| 1 | 3 | 4 | 2 |
| 4 | $\alpha_{22}$ | $\alpha_{23}$ | $\alpha_{24}$ |
| $\alpha_{31}$ | $\alpha_{32}$ | $\alpha_{33}$ | $\alpha_{34}$ |
| $\alpha_{41}$ | $\alpha_{42}$ | $\alpha_{43}$ | $\alpha_{44}$ |

*Figure 14.8* Process to find specific values $a_{21}$ in grid cells.

The above procedure to find out $a_{11}$ when we put the filter in the pixel is as follows:

$$a_{11} = 1.1 + 0.1 + 1.0 + 0.1 + 1.0 + 0.1 + 1.0 + 0.1 + 0.1 = 1 \tag{14.5}$$

Now we find $a_{12}$ by putting our filter on the pixel leaving one vertical strip.

The above procedure to find out $a_{12}$ when we put the filter in the pixel is as follows:

$$a_{12} = 0.1 + 1.1 + 1.0 + 1.1 + 0.0 + 0.1 + 0.0 + 0.1 + 1.1 = 3 \tag{14.6}$$

Then find $a_{13}$ and $a_{14}$ by using the above method. Now our question is that how can we find $a_{21}$, $a_{22}$, $a_{23}$, and $a_{24}$. So one thing to remember is that when we find $a_{21}$, we take the filter one strip below the horizontal pixel:

$$a_{21} = 0.1 + 1.1 + 0.0 + 1.1 + 0.0 + 0.1 + 1.0 + 1.1 + 1.1 = 4 \tag{14.7}$$

After repeating the above process, we can find all the values and the $6 \times 6$ pixel is now converted a $4 \times 4$ pixel after applying a $3 \times 3$ filter.

#### *14.2.2.3 Activation function (ReLU) layer*

The output from the convolutional layer is passed through an activation function, that is, the Rectified Linear Unit (ReLU). This function then adds non-linearity by converting all negative values to zero, which helps the NN learn more complex patterns in the data. Mathematically, the ReLU function is defined as $\text{ReLU}(x) = \max(0, x)$. A graphical representation of this function is shown in Figure 14.9.

#### *14.2.2.4 Pooling layer*

After the activation function has been implemented, it is followed by reduction of the size of the information by applying a pooling layer. This aids in reducing the values that the network needs to deal with, without losing

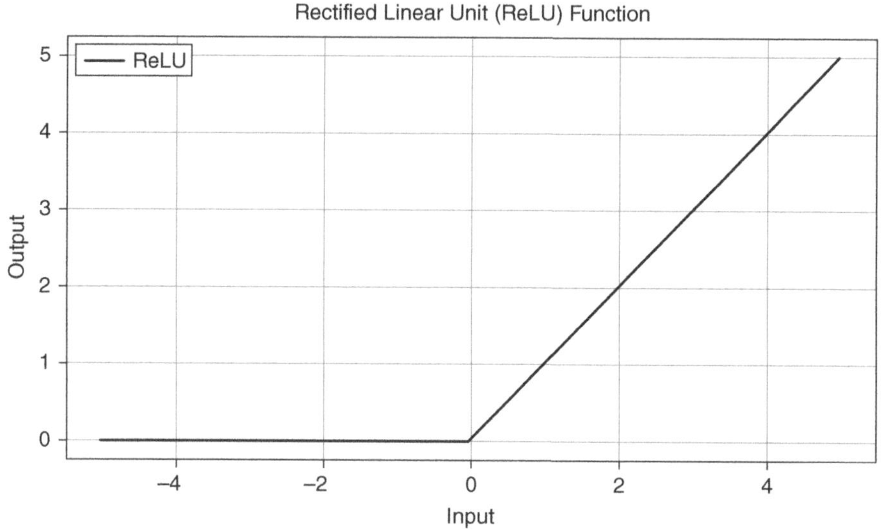


*Figure 14.9* ReLU function.

significant features. Strided convolution is a typical operation in CNNs in which the filter is moved over the image in pixel strides larger than one. This loses some of the input and leaves behind a smaller version of the data which retains the important spatial information.

This approach is usually pooled such as max pooling or average pooling. These techniques select the most relevant values in every region, hence simplifying the information but the patterns remain the same. This does not only cause the network to be faster, but also becomes less sensitive to minor alterations in the image. As we shall see by example, when we use a 4 × 4 image, max pooling, with a 2 × 2 filter and a stride of 2, we have a smaller 2 × 2 output. This preserves the overall structure of the image and simplifies the work of the network. This transformation occurs in Figure 14.10.

$$a_{11} = \max\{3, 5, 7, 2\} = 7 \tag{14.8}$$

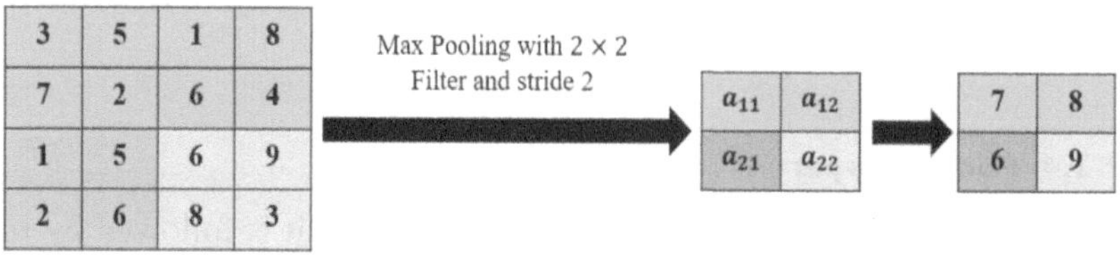


*Figure 14.10* Reduced spatial dimensions resulting from the application of max pooling.

$$a_{12} = \max\{1, 8, 6, 4\} = 8 \tag{14.9}$$

$$a_{21} = \max\{1, 5, 2, 6\} = 6 \tag{14.10}$$

$$a_{22} = \max\{6, 9, 8, 3\} = 9 \tag{14.11}$$

So, when processing a pixel image of dimensions $n \times n$ with a filter of size $f \times f$ along with $p$ padding and $s$ stride, the resulting pooled feature map dimensions can be calculated using the formula $\left(\frac{n+2p-f}{s}+1\right)\times\left(\frac{n+2p-f}{s}+1\right)$.

Here's a breakdown of the components:

$n$ represents the width or height of the input image.
$f$ denotes the size of the filter (kernel).
$p$ signifies the amount of padding applied to the input image.
$s$ indicates the stride length, which dictates the movement of the filter across the input image.

The formula is important to the extent that it shows the exact rate of change in the size of the feature maps after the pooling in a CNN. The network is then able to compress the input, through strided convolutions with pooling but learn meaningful features. It gives a smaller and more informative depiction.

There are distinct benefits of this decrease. It also simplifies the model in speed and memory requirements, especially when the input pictures are large or processing power is too small. Meanwhile, the central spatial data remains intact and therefore the network remains able to identify patterns.

#### *14.2.2.5 Flattening layer*

Once the pooling stage is done, the result is remodelled back into one long vector so that it is sent through the fully connected layers. It is called flattening and transforms the multidimensional feature maps into one-dimensional sequence of values.

This is illustrated in Figure 14.11, whereby the spatial structure of the pooled data is reduced to a simple array to serve as input to the next stage of the network.

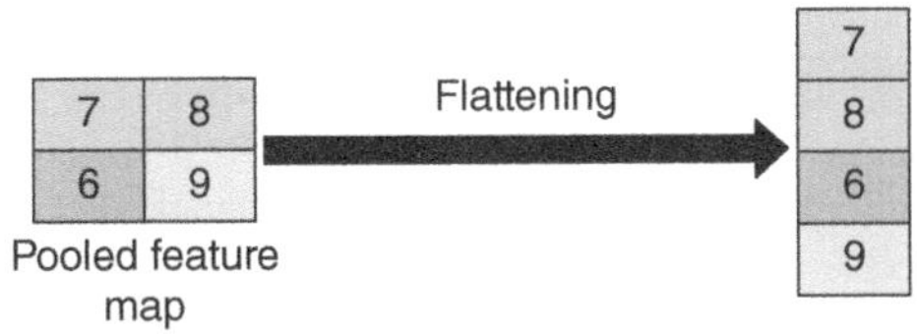


*Figure 14.11* Transformation process from a pooling layer to a flattened layer.

#### *14.2.2.6 Fully connected (dense) layers*

Both these layers, working on the flattened information, have the task of obtaining high-level features and complex relationships. The neurons in these layers form connections with all neurons of the next and the previous layer allowing the network to understand complex dependencies.

#### *14.2.2.7 Output layer*

The final fully connected layer produces the result of the network, and the number of neurons matches the classes of the classification problem. The type of activation function applied is determined by the characteristics of the task, for example, Soft Max multi-class classification. Figure 14.12 illustrates the overall mechanism of bird identification that involves the use of the entire layers of CNN.

Summing up, CNN working cycle entails convolutional layer, activation, pooling layer, and fully connected layer, and finally, training and prediction phases that allow the network to perform very well in tasks such as image recognition and classification.

CNN, therefore, play a key role in different applications, facilitating image recognition, object classification, and other tasks based on AI deployment.

### 14.2.3 Recurrent neural networks (RNNs)

RNNs were a new significant advance in the field of DL. In contrast to the feedforward networks, in which inputs are processed in isolation, RNNs are created to handle sequential data, in which order does count [13]. The main characteristic of them is that they contain links that create loops so that information can stay in an undiscovered form. This hidden state functions as some form of memory allowing the network to remember information about previous inputs and use it in future inputs. Due to this fact, RNNs are effectively applicable to language modelling, speech recognition, and time-series

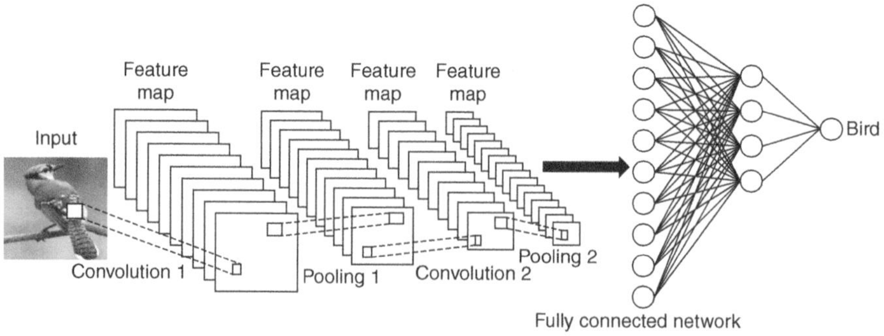


*Figure 14.12* Entire process involved in recognizing a bird, incorporating all layers of the CNN.

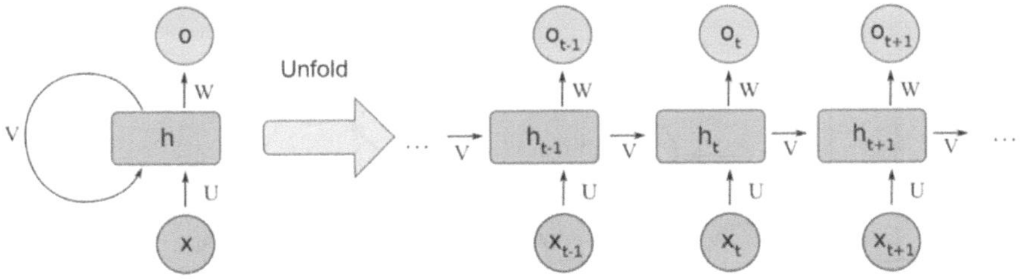


*Figure 14.13* General structure of recurrent neural networks.

analysis problems. They are also able to accept input sequences of various lengths, adapting to context as the data arrives. The general architecture is presented in Figure 14.13.

RNN may be explained in stages of its operation. Firstly, data is inputted by a single element at a time – such as a word in a sentence or a reading in a time series. The network begins with the initialization of a hidden state vector, which shall be updated in the course of the sequence. The RNN used with every new element takes the current input and the previous hidden state and uses their weight plus activation functions to create an updated hidden state. This new state is the summary of what has been processed up to date by the network.

It is the repeated links that make the model unusual: the concealed state at one time period is input to the other. It is this feedback loop which enables the network to learn temporal relationships and dependencies in the data. Depending on the concealed state, the network may give findings on every move either by foretelling the subsequent word, categorizing a sound, or generating some other relevant signal.

Learning RNNs includes backpropagation through time (BPTT). Once the sequence has been processed, a loss is computed by the network; this is done by comparing its output to the target. This mistake is then backpropagated through the recurring links. The gradients update the weights and biases in such a way that the model is optimized during successive passes (epochs).

However, long-term dependencies are also a problem since gradients can disappear or explode when using traditional RNNs. To solve this, some higher architectures have been provided. There are features of long short-term memory (LSTM) and gated recurrent unit (GRU) networks that are more effective in information retention on long sequences and offer better stability and performance [14, 15]. Although both are in the same family, their internal designs are different with each having its merits based on the task to be performed.

*LSTM* is a form of RNN architecture that incorporates a memory cell and three gates (input, forget, and output) to regulate the flow of information, selectively storing and recalling information across long sequence. Figure 14.14 shows clearly the way LSTMs operate.

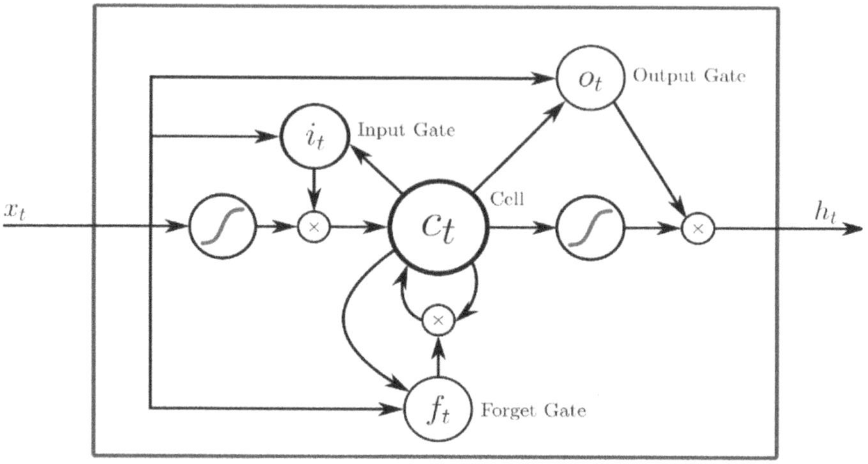


*Figure 14.14* General structure of long short-term memory.

The mathematical equations for LSTM with different cell states is given below:

$$f_t = \sigma(W_f \cdot [h_{t-1},\ x_t] + b_f \tag{14.12}$$

$$i_t = \sigma(W_i \cdot [h_{t-1},\ x_t] + b_i \tag{14.13}$$

$$\tilde{C} = \tanh(W_C \cdot [h_{t-1},\ x_t] + b_C \tag{14.14}$$

$$C_t = f_t * C_{t-1} + i_t * \widetilde{C}_t \tag{14.15}$$

$$o_t = \sigma(W_o [h_{t-1},\ x_t] + b_o \tag{14.16}$$

$$h_t = o_t * \tanh(C_t) \tag{14.17}$$

#### 14.2.3.1 Cell state

LSTMs also have an independent cell state along with the hidden state. The long-term memory of information held within long sequences is accomplished by the cell state. Figure 14.15 shows the structure of LSTM networks, which is distinguishable by the presence of a separate cell state as well as hidden state. Such a graphical illustration explains how LSTMs store long-term data in the cell state and enhances their capacity to learn the temporal relationships of sequential data.

#### 14.2.3.2 Forget gate

Forget gate helps the model to decide which information to keep in the cell state and what can be discarded. It takes into the account both the previous

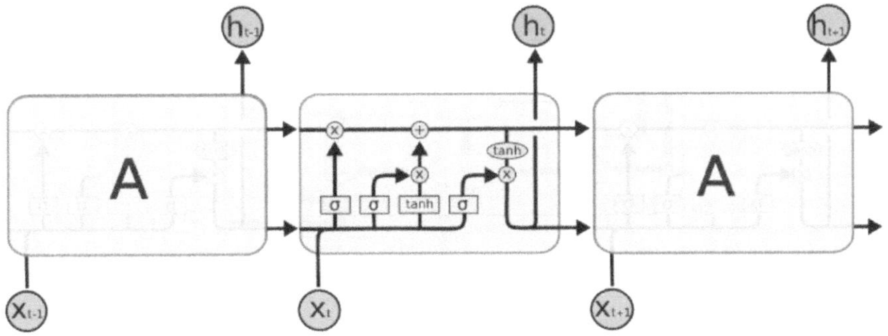


*Figure 14.15* Visual representation elucidates how LSTMs preserve long-term information through the cell state.

hidden state and the current input and based on these, it produces a value between 0 and 1 for each part of the cell state. A value close to 0 means that specific information is no longer needed and can be forgotten, whereas a value close to 1 suggests that the information is important and should be preserved.

#### *14.2.3.3 Input gate*

Input gate controls how the cell state is to be updated. It decides what new information should be added to the cell state. Similar to the forget gate, it produces values between 0 and 1.

#### *14.2.3.4 Cell state update*

Both the forgotten gate and the input gate provide information to the cell state which is used to update the state of the cell and thus enables the network to retain long-term dependencies by selectively modifying its stored data.

#### *14.2.3.5 Output gate*

By using the updated cell state, the output gate plays a big role to decide about the next hidden state. It is a selective filter of the information present in the cell state in order to create the final hidden state.

*GRU* is one more RNN architecture type, which is a similarity of LSTM. Similar to LSTM, GRU is dedicated to the problem of learning the long-range dependencies in a sequence of data. GRU simplifies the architecture of traditional RNN and LSTM by combining the memory cell and hidden state, resulting in a more streamlined structure. Figure 14.16 demonstrates the complete architecture of GRU.

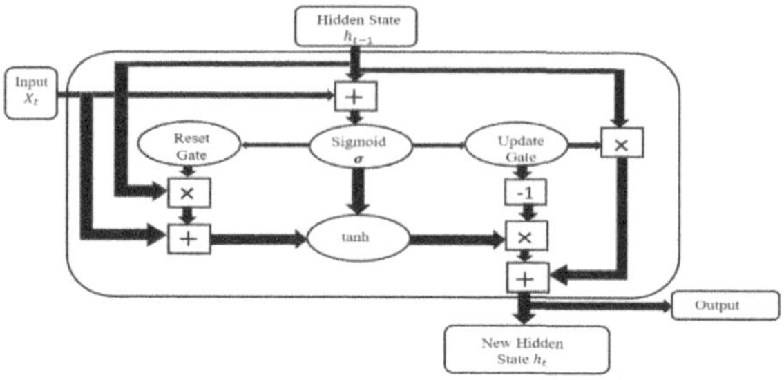


*Figure 14.16* General structure of long short-term memory.

#### *14.2.3.6 Update gate*

The GRU introduces an update gate that determines how much of the past information to carry forward to the next time step.

$$\text{gate}_{\text{update}} = \sigma\left(W_{\text{input update}} \times X_t + W_{\text{hidden update}} \times h_{t-1}\right) \tag{14.18}$$

The update gate takes into account the previous hidden state and the current input, producing values between 0 and 1. A value of 1 means fully update, and 0 means fully discard and so we get updated vector.

$$\text{Updated Vector} = \text{gate}_{\text{update}} * h_{t-1} \tag{14.19}$$

#### *14.2.3.7 Reset gate*

GRU also has the reset gate, which determines how much the past state is going to be forgotten or reset, and the update gate. The reset gate involves the use of past hidden state together with the present input into numbers between 0 and 1 as the update gate.

$$\text{gate}_{\text{reset}} = \sigma(W_{\text{input reset}} \times X_t + W_{\text{hidden reset}} \times h_{t-1} \tag{14.20}$$

$$\text{Reset Vector} = \tanh\left(\text{gate}_{\text{reset}} * \left(w_{h_1} \times h_{t-1}\right) + W_{x_1} \times x_t\right) \tag{14.21}$$

#### *14.2.3.8 Hidden state update*

The hidden state is then updated depending on the values received on the update and reset gates. The extent of incorporation of the current information and extent of the past hidden state discarded are regulated by the update gate and the reset gate, respectively.

$$h_t = r * \left(1 - \text{gate}_{\text{update}}\right) + u \tag{14.22}$$

GRU design is less complicated than LSTM. It unites the cell state and the hidden state in one vector that makes it less complex. Such a simplified form tends to make GRUs quicker to train and computationally less expensive, and still works well in a variety of tasks.

GRU and LSTM belong to a larger classification of gated recurrent networks. Their primary benefit is their better coping with long-term dependencies compared to the simple RNNs. The decision between the two normally depends on the task, size, and type of the dataset, and the computing resources. In every architecture, there are instances when it is more applicable than the other.

In general, RNNs and their variations are still the cornerstones of the sequential data modelling. Their capability to learn patterns through time has led to advances in fields as varied as language processing, speech technologies, and time-series forecasting, and are now an essential part of DL in modern times.

## 14.3 TRAINING DEEP MODELS

### 14.3.1 Backpropagation

The most essential part of NN training is backpropagation. This is a supervised learning algorithm that adjusts a network's weights based on the errors it makes. In practice, the model produces predictions, compares them with the correct outputs, and calculates an error [16]. That error is then pushed backwards through the network, layer by layer, to update the connections.

The aim is straightforward: minimize the error such that the network is more general to unknown data. Change in the weights is gradually optimized by each iteration or epoch. Through numerous reiterations, the network is also more precise and dependable. The name backpropagation is based on the phenomenon itself – the signal of error is introduced out of the output layer and it flows in the direction of the input layer instructing the adjustments in between.

Such a reverse flow of information enables the model to acquire more intricate patterns automatically. The network can be improved in terms of its predictive power and increase its ability to work with new samples through a series of iterations in improving its parameters.

The computation of the gradient of the loss function for an individual weight involves the chain rule within the backpropagation algorithm in NNs. Unlike a straightforward direct calculation, this method systematically computes one layer at a time, thereby efficiently determining the gradient. However, the application of this gradient is left to the user's discretion. Consequently, we initialize $w$ and $b$ with random values and employ gradient descent to iteratively refine them to their optimal values. Some commonly used equations in this process include the following:

$$w_{ij}^k = w_{ij}^k - \alpha \frac{\partial E}{\partial w_{ij}^k} \tag{14.23}$$

$$b^k = b^k - \alpha \frac{\partial E}{\partial b^k} \tag{14.24}$$

where $\alpha$ is the learning rate, that is, it is a hyperparameter that is given by the user, and $E$ is the error function (known as cost function also). The partial derivatives with respect to the parameter vectors $w$ and $b$ in Equations (14.20) and (14.21) are basically the gradient elements of $E$. The values of these gradients must be driven to zero to reduce the error [17]. However, how do we determine the entire error gradient? Instead of considering the overall error gradient, we combine the error gradients for each training phase to simplify the task. (Another assumption) Here we're employing the squared error function as the error function:

$$E = \frac{1}{2m}\sum_{i=1}^{m}\left(h_w(x)^i - y^i\right)^2 \tag{14.25}$$

Take a look at Figures 14.17 and 14.18 to see an example of a backpropagation NN in action:

$$E_{\text{Total}} = E_{\text{Output1}} + E_{\text{Output2}} = \frac{1}{2}\left(y_1 - x_1^3\right)^2 + \frac{1}{2}\left(y_2 - x_2^3\right)^2 \tag{14.26}$$

$$\left(\frac{\partial E}{\partial w_{1,1}^3}\right) = \left(\frac{\partial E}{\partial x_1^3}\right)\left(\frac{\partial x_1^3}{\partial y_1^3}\right)\left(\frac{\partial y_1^3}{\partial w_{1,1}^3}\right) \tag{14.27}$$

$$\left(\frac{\partial E}{\partial w_{1,1}^3}\right) = \left(\frac{\partial\left(\frac{1}{2}\left(y_1 - x_1^3\right)^2 + \frac{1}{2}\left(y_2 - x_2^3\right)^2\right)}{\partial x_1^3}\right)\left(\frac{\partial}{\partial y_1^3}\left(\frac{1}{1+e^{-y_1^3}}\right)\right)$$
$$\left(\frac{\partial}{\partial w_{1,1}^3}\left(x_1^2 .\ w_{1,1}^3 + x_2^2 .\ w_{1,2}^3 + .\ b^3\right)\right) \tag{14.28}$$

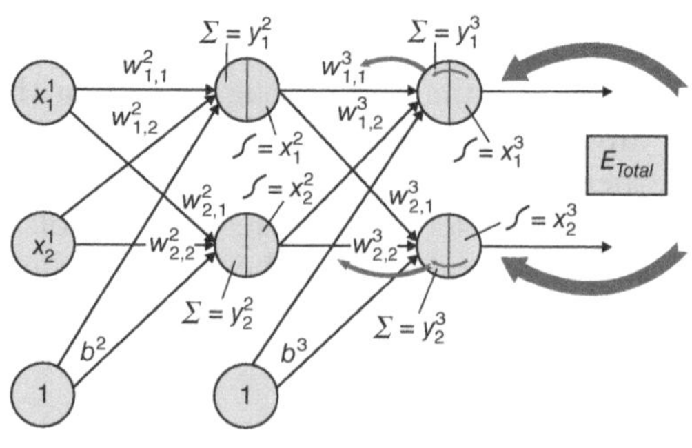


*Figure 14.17* Workflow of backpropagation.

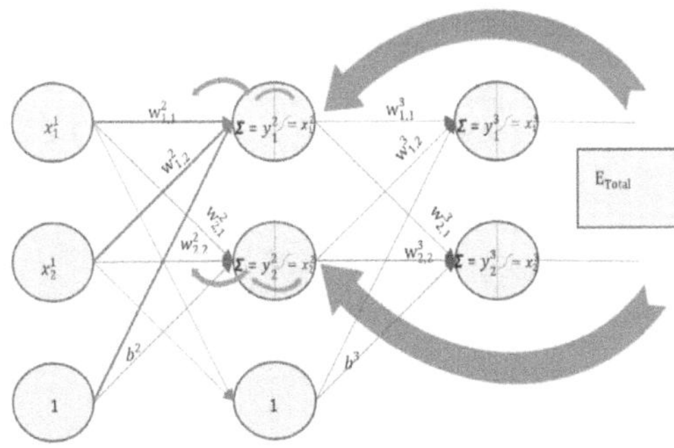


*Figure 14.18* Work flow of backpropagation (2).

Now

$$\left(\frac{\partial\left(\frac{1}{2}\left(y_1 - x_1^3\right)^2 + \frac{1}{2}\left(y_2 - x_2^3\right)^2\right)}{\partial x_1^3}\right) = -\left(y_1 - x_1^3\right) = \left(x_1^3 - y_1\right) \tag{14.29}$$

$$\text{Let } y = \frac{1}{1+e^{-x}} = \frac{e^x}{1+e^x} \tag{14.30}$$

$$\frac{\partial y}{\partial x} = \frac{e^x}{\left(1+e^x\right)^2} = \left(\frac{e^x}{1+e^x}\right)\left(\frac{1}{1+e^x}\right) = y\left(\frac{1+e^x - e^x}{1+e^x}\right) = y\left(1 - \frac{e^x}{1+e^x}\right) = y(1-y) \tag{14.31}$$

$$\text{Therefore, } \left(\frac{\partial}{\partial y_1^3}\left(\frac{1}{1+e^{-y_1^3}}\right)\right) = x_1^3\left(1 - x_1^3\right) \tag{14.32}$$

$$\left(\frac{\partial}{\partial w_{1,1}^3}\left(x_1^2 .\ w_{1,1}^3 + x_2^2 .\ w_{1,2}^3 + 1\ .\ b^3\right)\right) = x_1^2 \tag{14.33}$$

So, Equation (14.28) becomes

$$\left(\frac{\partial E}{\partial w_{1,1}^3}\right) = \left(x_1^3 - y_1\right).\ x_1^3 = \left(1 - x_1^3\right).\ x_1^2 \tag{14.34}$$

$$\left(\frac{\partial E}{\partial w_{1,2}^3}\right) = \left(\frac{\partial E}{\partial x_1^3}\right)\left(\frac{\partial x_1^3}{\partial y_1^3}\right)\left(\frac{\partial y_1^3}{\partial w_{1,2}^3}\right) \tag{14.35}$$

$$\left(\frac{\partial E}{\partial w_{1,2}^{3}}\right)=\left(\frac{\partial\left(\frac{1}{2}\left(y_1-x_1^3\right)^2+\frac{1}{2}\left(y_2-x_2^3\right)^2\right)}{\partial x_1^3}\right)\left(\frac{\partial}{\partial y_1^3}\left(\frac{1}{1+e^{-y_1^3}}\right)\right)$$
$$\left(\frac{\partial}{\partial w_{1,2}^{3}}\left(x_1^2 .\, w_{1,1}^3+x_2^2 .\, w_{1,2}^3+1\,.\,b^3\right)\right) \tag{14.36}$$

Now

$$\left(\frac{\partial\left(\frac{1}{2}\left(y_1-x_1^3\right)^2+\frac{1}{2}\left(y_2-x_2^3\right)^2\right)}{\partial x_1^3}\right)=-\left(y_1-x_1^3\right)=\left(x_1^3-y_1\right) \tag{14.37}$$

$$\left(\frac{\partial}{\partial y_1^3}\left(\frac{1}{1+e^{-y_1^3}}\right)\right)=x_1^3\left(1-x_1^3\right) \tag{14.38}$$

$$\left(\frac{\partial}{\partial w_{1,2}^{3}}\left(x_1^2 .\, w_{1,1}^3+x_2^2 .\, w_{1,2}^3+1\,.\,b^3\right)\right)=x_2^2 \tag{14.39}$$

So, Equation (14.36) becomes

$$\left(\frac{\partial E}{\partial w_{1,2}^{3}}\right)=\left(x_1^3-y_1\right).\,x_1^3\left(1-x_1^3\right).\,x_2^2 \tag{14.40}$$

In a similar way,

$$\left(\frac{\partial E}{\partial w_{2,1}^{3}}\right)=\left(x_2^3-y_2\right).\,x_2^3\left(1-x_2^3\right).\,x_1^2 \tag{14.41}$$

$$\left(\frac{\partial E}{\partial w_{2,2}^{3}}\right)=\left(x_2^3-y_2\right).\,x_2^3\left(1-x_2^3\right).\,x_2^2 \tag{14.42}$$

Here we apply the same bias ($b^3$) to every node for the third layer. Therefore, we can update the bias value by using either the first node of the third layer or the second node of the third layer:

By using the first node of the third layer, we get

$$\left(\frac{\partial E}{\partial b^{3}}\right)=\left(\frac{\partial E}{\partial x_1^3}\right)\left(\frac{\partial x_1^3}{\partial y_1^3}\right)\left(\frac{\partial y_1^3}{\partial b^3}\right) \tag{14.43}$$

$$\left(\frac{\partial E}{\partial b^3}\right)=\left(\frac{\partial\left(\frac{1}{2}\left(y_1-x_1^3\right)^2+\frac{1}{2}\left(y_2-x_2^3\right)^2\right)}{\partial x_1^3}\right)\left(\frac{\partial}{\partial y_1^3}\left(\frac{1}{1+e^{-y_1^3}}\right)\right)$$
$$\left(\frac{\partial}{\partial b^3}\left(x_1^2 .\, w_{1,1}^3+x_2^2 .\, w_{1,2}^3+1\,.\,b^3\right)\right) \tag{14.44}$$

$$\left(\frac{\partial E}{\partial b^3}\right)=\left(x_1^3-y_1\right).\, x_1^3\left(1-x_1^3\right) \tag{14.45}$$

By using the second node of the third layer, we get

$$\left(\frac{\partial E}{\partial b^3}\right)=\left(x_2^3-y_2\right).\, x_2^3\left(1-x_2^3\right) \tag{14.46}$$

$$\left(\frac{\partial E}{\partial w_{1,1}^2}\right)=\left(\frac{\partial E}{\partial x_1^2}\right)\left(\frac{\partial x_1^2}{\partial y_1^2}\right)\left(\frac{\partial y_1^2}{\partial w_{1,1}^2}\right) \tag{14.47}$$

$$\left(\frac{\partial E}{\partial w_{1,1}^2}\right)=\left(\frac{\partial\left(\frac{1}{2}\left(y_1-x_1^3\right)^2+\frac{1}{2}\left(y_2-x_2^3\right)^2\right)}{\partial x_1^2}\right)\left(\frac{\partial}{\partial y_1^2}\left(\frac{1}{1+e^{-y_1^2}}\right)\right)$$
$$\left(\frac{\partial}{\partial w_{1,1}^2}\left(x_1^1 .\, w_{1,1}^2+x_2^1 .\, w_{1,2}^2+1.\,b^2\right)\right) \tag{14.48}$$

$$\left(\frac{\partial E}{\partial w_{1,1}^2}\right)=\left(\frac{\partial}{\partial x_1^2}\frac{1}{2}\left(y_1-x_1^3\right)^2+\frac{\partial}{\partial x_1^2}\frac{1}{2}\left(y_2-x_2^3\right)^2\right).\left(\frac{\partial}{\partial y_1^2}\left(\frac{1}{1+e^{-y_1^2}}\right)\right)$$
$$.\left(\frac{\partial}{\partial w_{1,1}^2}\left(x_1^1 .\, w_{1,1}^2+x_2^1 .\, w_{1,2}^2+1.\,b^2\right)\right) \tag{14.49}$$

$$\left(\frac{\partial E}{\partial w_{1,1}^2}\right)=\left(\frac{\partial}{\partial x_1^3}\frac{1}{2}\left(y_1-x_1^3\right)^2.\left(\frac{\partial x_1^3}{\partial y_1^3}\right)\left(\frac{\partial y_1^3}{\partial x_1^2}\right)+\frac{\partial}{\partial x_2^3}\frac{1}{2}\left(y_2-x_2^3\right)^2.\left(\frac{\partial x_2^3}{\partial y_2^3}\right)\left(\frac{\partial y_2^3}{\partial x_1^2}\right)\right)$$
$$.\left(\frac{\partial}{\partial y_1^2}\left(\frac{1}{1+e^{-y_1^2}}\right)\right)x_1^1. \tag{14.50}$$

$$\left(\frac{\partial E}{\partial w_{1,1}^2}\right)=\left(\left(x_1^3-y_1\right).\, x_1^3\left(1-x_1^3\right).\, w_{1,1}^3+\left(x_2^3-y_2\right)x_2^3\left(1-x_2^3\right).\, w_{2,1}^3\right).\, x_1^2\left(1-x_1^2\right).\, x_1^1 \tag{14.51}$$

$$\left(\frac{\partial E}{\partial w_{1,2}^2}\right)=\left(\frac{\partial E}{\partial x_1^2}\right)\left(\frac{\partial x_1^2}{\partial y_1^2}\right)\left(\frac{\partial y_1^2}{\partial w_{1,2}^2}\right) \tag{14.52}$$

$$\left(\frac{\partial E}{\partial w_{1,2}^2}\right)=\left(\frac{\partial\left(\frac{1}{2}\left(y_1-x_1^3\right)^2+\frac{1}{2}\left(y_2-x_2^3\right)^2\right)}{\partial x_1^2}\right)\left(\frac{\partial}{\partial y_1^2}\left(\frac{1}{1+e^{-y_1^2}}\right)\right)\left(\frac{\partial}{\partial w_{1,2}^2}\left(x_1^1 .\, w_{1,1}^2+x_2^1 .\, w_{1,2}^2+1 .\, b^2\right)\right) \tag{14.53}$$

$$\left(\frac{\partial E}{\partial w_{1,2}^2}\right)=\left(\frac{\partial}{\partial x_1^2}\frac{1}{2}\left(y_1-x_1^3\right)^2+\frac{\partial}{\partial x_1^2}\frac{1}{2}\left(y_2-x_2^3\right)^2\right)\cdot\left(\frac{\partial}{\partial y_1^2}\left(\frac{1}{1+e^{-y_1^2}}\right)\right).\, x_2^1 \tag{14.54}$$

$$\left(\frac{\partial E}{\partial w_{1,2}^2}\right)=\left(\frac{\partial}{\partial x_1^3}\frac{1}{2}\left(y_1-x_1^3\right)^2 .\left(\frac{\partial x_1^3}{\partial y_1^3}\right)\left(\frac{\partial y_1^3}{\partial x_1^2}\right)+\frac{\partial}{\partial x_2^3}\frac{1}{2}\left(y_2-x_2^3\right)^2 .\left(\frac{\partial x_2^3}{\partial y_2^3}\right)\left(\frac{\partial y_2^3}{\partial x_1^2}\right)\right)\cdot\left(\frac{\partial}{\partial y_1^2}\left(\frac{1}{1+e^{-y_1^2}}\right)\right).\, x_2^1 \tag{14.55}$$

$$\left(\frac{\partial E}{\partial w_{1,2}^2}\right)=\left(\left(x_1^3-y_1\right).\, x_1^3\left(1-x_1^3\right).\, w_{1,1}^3+\left(x_2^3-y_2\right)x_2^3\left(1-x_2^3\right).\, w_{2,1}^3\right).\, x_1^2\left(1-x_1^2\right).\, x_2^1 \tag{14.56}$$

$$\left(\frac{\partial E}{\partial w_{2,1}^2}\right)=\left(\left(x_1^3-y_1\right).\, x_1^3\left(1-x_1^3\right).\, w_{1,2}^3+\left(x_2^3-y_2\right)x_2^3\left(1-x_2^3\right).\, w_{2,2}^3\right).\, x_2^2\left(1-x_2^2\right).\, x_1^1 \tag{14.57}$$

$$\left(\frac{\partial E}{\partial w_{2,2}^2}\right)=\left(\left(x_1^3-y_1\right).\, x_1^3\left(1-x_1^3\right).\, w_{1,2}^3+\left(x_2^3-y_2\right)x_2^3\left(1-x_2^3\right).\, w_{2,2}^3\right).\, x_2^2\left(1-x_2^2\right).\, x_2^1 \tag{14.58}$$

Since we apply the same bias ($b^2$) on every node for the second layer, we can update the bias value by using either the first node of the second layer or the second node of the second layer:

By using the first node of the second layer, we get

$$\left(\frac{\partial E}{\partial b^2}\right)=\left(\frac{\partial E}{\partial x_1^2}\right)\left(\frac{\partial x_1^2}{\partial y_1^2}\right)\left(\frac{\partial y_1^2}{\partial b^2}\right) \tag{14.59}$$

$$\left(\frac{\partial E}{\partial b^2}\right)=\left(\frac{\partial\left(\frac{1}{2}\left(y_1-x_1^3\right)^2+\frac{1}{2}\left(y_2-x_2^3\right)^2\right)}{\partial x_1^2}\right)\left(\frac{\partial}{\partial y_1^2}\left(\frac{1}{1+e^{-y_1^2}}\right)\right)$$
$$\left(\frac{\partial}{\partial b^2}\left(x_1^1.\,w_{1,1}^2+x_2^1.\,w_{1,2}^2+1.\,b^2\right)\right) \tag{14.60}$$

$$\left(\frac{\partial E}{\partial b^2}\right)=\left(\frac{\partial}{\partial x_1^2}\frac{1}{2}\left(y_1-x_1^3\right)^2+\frac{\partial}{\partial x_1^2}\frac{1}{2}\left(y_2-x_2^3\right)^2\right).\left(\frac{\partial}{\partial y_1^2}\left(\frac{1}{1+e^{-y_1^2}}\right)\right).1 \tag{14.61}$$

$$\left(\frac{\partial E}{\partial b^2}\right)=\left(\frac{\partial}{\partial x_1^3}\frac{1}{2}\left(y_1-x_1^3\right)^2.\left(\frac{\partial x_1^3}{\partial y_1^3}\right)\left(\frac{\partial y_1^3}{\partial x_1^2}\right)+\frac{\partial}{\partial x_2^3}\frac{1}{2}\left(y_2-x_2^3\right)^2.\left(\frac{\partial x_2^3}{\partial y_2^3}\right)\left(\frac{\partial y_2^3}{\partial x_1^2}\right)\right)$$
$$.\left(\frac{\partial}{\partial y_1^2}\left(\frac{1}{1+e^{-y_1^2}}\right)\right).1 \tag{14.62}$$

$$\left(\frac{\partial E}{\partial b^2}\right)=\left(\left(x_1^3-y_1\right).\,x_1^3\left(1-x_1^3\right).\,w_{1,1}^3+\left(x_2^3-y_2\right)x_2^3\left(1-x_2^3\right).\,w_{2,1}^3\right).\,x_1^2\left(1-x_1^2\right).\,1 \tag{14.63}$$

By using the second node of the second layer, we get

$$\left(\frac{\partial E}{\partial b^2}\right)=\left(\left(x_1^3-y_1\right).\,x_1^3\left(1-x_1^3\right).\,w_{1,2}^3+\left(x_2^3-y_2\right)x_2^3\left(1-x_2^3\right).\,w_{2,2}^3\right).\,x_2^2\left(1-x_2^2\right).\,1 \tag{14.64}$$

Thus, the required gradients can be obtained, which explains the basic mathematical concepts of backpropagation. Having this knowledge, it is possible to use backpropagation to a simple linear NN. To more complex NNs with several nodes per layer, the methodology is the same but with some changes in the index of the variables like $w$, $z$, $a$, and $b$. This makes the backpropagation more difficult, and requires the use of deriving matrices, that is, gradient vectors. However, the principle behind it is the same backpropagation uses the chain rule of derivatives to calculate the change of the cost function $J$ in terms of the change in the network parameters. The result of this calculation is the gradient, which, in its turn, makes it easier to adjust the parameters with the help of gradient descent.

### 14.3.2 Stochastic gradient descent (SGD)

SGD stands out as the pre-eminent optimization algorithm in the training of deep NNs. In the pursuit of minimizing the loss function, SGD operates by iteratively updating the weights of the network. These updates are determined by the negative gradient of the loss function concerning the weights, directing the algorithm towards the optimal parameter values [18].

Although SGD is the basis of most training procedures, adaptive optimization algorithms like Adam and RMSprop have been created to optimize and speed up the convergence procedure. These adaptive methods also differ in that, unlike fixed learning rates of the classical SGD, the learning rates of individual parameters change dynamically during training. This flexibility increases the capacity of the algorithm to traverse involved loss landscapes and hastens convergence and are thus an efficient approach to tuning the functionality of deep NNs in diverse fields.

### 14.3.3 Regularization techniques

Overfitting (learned the training data too well and failed with new examples) is one of the largest problems that can be encountered in the training process in ML. This is a problem that has been tackled using regularization techniques [19].

The two most popular ones are dropout and weight decay. The concept of dropout is that a group of neurons is switched off at random during training. This randomness does not allow the network to be over-dependent on specific neurons or features. Consequently, the model will have to come up with different routes and acquire more generalized representations of the data.

Weight decay also assumes a different direction. It also penalizes heavy weights in the network which would prevent the model giving excessive focus on the single parameters [20]. The network is non-overfit since it does not need large weights and is an even closer representation.

A combination of these techniques can be used to ensure that models can be more trustworthy when adapting to unseen examples based on their training data. In practice, they minimize the possibility of overfitting and enhance the capacity of the model to make correct predictions in the actual world.

### 14.3.4 Batch normalization

The popular method of stabilizing and accelerating deep network training is batch normalization [21]. It was proposed to address the issue of internal covariate shift the propensity of layer inputs to vary distribution during training.

The algorithm is effective due to its normalization of the activations in each mini-batch to be near to the mean of zero and the variance of one. This maintains the data flowing in the network within a stable range and the training of the network insensible to parameter initialization. More stable activations can be used with higher learning rates, which can usually converge faster.

In addition to enhancing the optimization, a side effect of batch normalization is that it does not rely on other regularization methods, such as dropout or weight decay. It assists models in generalizing to unknown data, in most instances, as well as training more efficiently.

## 14.4 TRANSFER LEARNING

Transfer learning has gained significance in the contemporary ML, particularly when the dataset is small [22]. A model can also be built on the training knowledge of a much larger dataset (rather than starting with no knowledge). Such trained models are already able to detect useful patterns, which can be transferred to a new task that is related and yet new.

The benefit is obvious: using the learned features, training is less time-consuming and less resource-demanding. More to the point, the model tends to work better with the smaller dataset as it enjoys the advantage of the wider experience of the pre-training.

Transfer learning has already achieved notable success in such disciplines as natural language processing and computer vision. Pre-trained models in either case give a solid base, and researchers can create an effective, task-specific system without incurring the enormous costs of training a huge model to do it themselves.

### 14.4.1 Fine-tuning

Fine-tuning is a method to train a pre-trained model on a particular task [23]. The weights of the model are not created randomly, but rather modified in a selective way when it is trained on a small dataset, which is also task-specific.

This method is particularly applicable where the labelled data of the target problem is limited. Since the model already stores the knowledge of training on a large dataset, it only requires small changes to be applied to fit the new environment. As a matter of fact, this implies that the model is able to direct its previous knowledge of patterns and representations to the specialized task in hand.

These advantages are twofold: training is quicker and it uses less resources and the model can be more accurate than if it is trained only on the small data. The process of fine-tuning has thus become a frequently used and useful approach to most fields of ML, as it assists researchers to construct useful systems despite situations in which labelled data is scarce or hard to obtain.

### 14.4.2 Feature extraction

The concept of reusing the features that are learned using a pre-trained model to solve a new task is known as feature extraction [24–26]. This is normally achieved by eliminating the last layers of the original network and substituting it with others that are specific to the intended application.

Early and middle layers of big models such as those trained on images classification will give general patterns such as edges, textures, and shapes. Such representations come in handy way after the task. The researchers can then use them to get models to fit various problems without the need to retrain them on a ground basis.

The method is especially useful in cases where labelled information is limited. It enables the practitioners to use the general knowledge acquired before training and concentrate computation power on the new layers. In most instances, feature extraction besides consuming less time and resources, also results in excellent performance in various applications.

## 14.5 HYPERPARAMETER TUNING

Tuning hyperparameters is a crucial step in developing ML models because it has a direct impact on performance [27]. Hyperparameters are settings chosen before training begins, such as the learning rate, the strength of regularization, or the number of hidden layers in a network. The correct values are not often easy to discover. An option that might be successful in one set of data might not succeed in the other and minor modifications can turn things around. In order to address this, there are various search techniques that are usually adopted. Grid search does the search on all combinations in a predefined range, whereas random search does the search in a looser way by randomly sampling the values. Bayesian optimization follows an alternative route in that probabilistic models are used in order to steer the search into useful directions.

Through these techniques, the practitioners are able to in most cases find out configurations that enhance accuracy and generalization. Hyperparameter tuning, though time-consuming, must be carefully done to create models in which the model performs well in various tasks.

### 14.5.1 Grid search

One of the simplest methods to hyperparameter search is through grid search [28]. It works in a straightforward way: specify a set of potential hyperparameter values, and train and test the model on each potential hyperparameter combination. For each run, the performance is recorded and compared, because it checks every option in the predefined grid, this method guarantees a complete exploration of the chosen search space. However, the cost of this thoroughness is high. When the number of hyperparameters or their ranges grow large, grid search quickly becomes computationally expensive.

Even so, it is still widely used. Grid search provides a clear picture of how different hyperparameter choices affect model performance and often serves as a baseline before more advanced optimization methods are applied.

### 14.5.2 Random search

Random search is an alternative to the exhaustive grid search [29]. It does not test all combinations of hyperparameters, but picks hyperparameter values randomly and evaluates the model at the two picks.

The method is especially effective in the case of large hyperparameter space. Random search is likely to discover good models faster than grid search by searching a large set of configurations without exhausting the search space. The approach does not also spend time on parameters that might not contribute significantly to the performance.

Its power is in the fact that it is flexible. Random search also has the advantage of discovering promising configurations, even where the effect of their influence is not well understood, because the method does not assume what hyperparameters are most important. It makes it a handy and scalable approach to hyperparameter optimization and in particular in cases where computing power is constrained.

### 14.5.3 Bayesian optimization

*Bayesian optimization* is a more advanced approach to hyperparameter tuning that uses probabilistic models to guide the search [30]. Unlike grid or random search, which explore the space without learning from past results, Bayesian optimization adapts its strategy based on the outcomes of earlier trials.

The method builds a surrogate model of the objective function that reflects both the values already observed and the uncertainty around them. From this model, it defines an *acquisition function* that estimates how useful it would be to test a new point in the hyperparameter space [31]. At each step, the algorithm selects the configuration that maximizes this acquisition function, then updates the surrogate model with the new result.

Through this iterative process, Bayesian optimization focuses attention on the most promising regions of the search space while still allowing some exploration. This is a good balance between exploration and exploitation particularly when the model assessments are costly or time-consuming. Consequently, it has also turned out to be a useful resource-tuning ML model.

## 14.6 ETHICAL CONSIDERATIONS IN DL

Ethics in DL is the most important question as the technology takes over the world.

a. *Bias in Training Data:* A major ethical challenge in DL is that training data is biased. Discriminatory outcomes can be attained owing to biased datasets, which reinforce the prevailing disparities that are present in the society. The developers will be required to work hard to test and minimize biasness in the training sets to ensure fairness in model predictions.
b. *Fairness, Transparency, and Accountability:* The development and deployment of DL models is of critical importance with the need to offer fairness, transparency, and accountability. These include minimizing the unequal impacts on different demographic groups and providing a simplified

comprehension of choice-making of these models. It is highly crucial to develop trust founded on open practices so that the users will accept it.

c. *Societal Impacts and Responsible AI Practices:* The impact of DL technologies on the general society is more holistic and it demands responsible AI use. The developers must remember the effects of AI on privacy, work, and social organizations. Responsible practices involve proactively minimizing the potential adverse effects, and considerations of the well-being of different communities concerned with such technologies.

## 14.7 MODEL EVALUATION AND MATRICES

a. *Overview of Common Evaluation Metrics:* In DL, one typically measures model performance using a variety of evaluation metrics. In case of classification, accuracy, precision, the recall and the F1 score are most popular. When working on regression-related issues, it is common to use such measures as mean squared error (MSE) or root mean squared error (RMSE). Other visualizations such as the ROC curve and the precision–recall curve give more information in binary classification on how effective the model is in separating the classes. These measurements when combined provide a quantitative perspective of the effectiveness of a model and whether the model is appropriate to the task it was created to perform.
b. *Considerations for Choosing Appropriate Metrics Based on the Task:* The right evaluation metric does not apply across the board, but rather it depends upon the problem at hand. Accuracy might be effective in binary classification where the classes are balanced. But, in the case of imbalanced data, precision, recall or the F1 score will give a more accurate sense of performance. In the case of regression, MSE and RMSE are the norm, but this may depend on the sensitivity of the task to outliers. The trick is to align the metric with the nature of the data collection as well as the general objectives of the model. This way will guarantee that the performance will be gauged in a manner that reflects the actual problems of the application.

## 14.8 FUTURE TRENDS AND CHALLENGES

a. *Emerging Trends in DL Methodologies:* In the future, there will be a variety of trends to DL that are likely to advance in the field. Among the obvious tendencies is the increasing use of transfer learning, where networks apply the knowledge of the previous networks to new tasks. Unsupervised and self-supervised learning are also being developed in order to better exploit unlabelled data. The approaches have the potential to lessen the great reliance on massive annotated datasets and expand the range of applications. Lastly, there is the growing importance of the combination of DL with the focus mechanisms of attention and reinforcement learning. Collectively, these methods can enable models to take into account more intricate patterns and to make more informed choices about the situation.

b. *Anticipated Challenges and Areas for Improvement:* Despite the fact that DL is actively evolved, there are a number of obstacles that can be identified. Interpretability is one of the most urgent ones. Complex NNs are frequently black boxes, and it is still challenging to make them be more transparent regarding the decisions they come up with. Ethical issues also persist. Models may be biased by data, and how to resolve these issues is critical in case DL can be relied on in sensitive areas. Some other issues are safety. Most of the deep models are susceptible to adversarial attack, meaning even minor modifications in the inputs can cause wrongful predictions. Meanwhile, the scaling of models to huge data and their efficiency to run on edge devices remains a challenge to the existing methods. Finally, the ability to close the gap between theoretical advances and the application in practice will predetermine the extent of DL. Addressing these concerns must serve as a move towards its further development and responsible implementation in the industries.

## 14.9 CONCLUSION

The literature review of DL techniques demonstrates the progress that has been made and the extent to which the practice has become ubiquitous technology. Deep NNs have demonstrated an unprecedented ability to learn complex features directly out of data driving development in the fields of image recognition, speech, and natural language understanding. The discipline is still developing. The idea of transfer learning and the concept of unsupervised and self-supervised approaches, attention-driven models are all signs of the gradual trend of more capable and versatile systems. Meanwhile, there are difficulties. Interpretability and ethics questions cannot be left aside and they will determine how the technology will be responsibly developed in the coming years. Combined, all these developments indicate the potential as well as the complexity of DL. Its role spreads in both industry and society, and its further development is guaranteed by the fact that it will continue to occupy the centre of AI in the predictable future.

## REFERENCES


1. S. Pouyanfar et al., "A Survey on Deep Learning: Algorithms, Techniques, and Applications," in *ACM Computing Surveys*, vol. 51, no. 5, pp. 1–36, 2019. doi: 10.1145/3234150.
2. M. H. Hesamian, W. Jia, X. He and P. Kennedy, "Deep Learning Techniques for Medical Image Segmentation: Achievements and Challenges," in *Journal of Digit Imaging*, vol. 32, no. 4, pp. 582–596, 2019. doi: 10.1007/s10278-019-00227-x.

3. A. Mathew, P. Amudha and S. Sivakumari, "Deep Learning Techniques: An Overview," in A. E. Hassanien, R. Bhatnagar, and A. Darwish (Eds.), *Advanced Machine Learning Technologies and Applications* (pp. 599–608), Advances in Intelligent Systems and Computing. Singapore: Springer, 2021. doi: 10.1007/978-981-15-3383-9_54.
4. I. Goodfellow, Y. Bengio and A. Courville, *Deep Learning*. MIT Press, 2016.
5. A. Abraham, "Artificial Neural Networks," in *Handbook of Measuring System Design*. New York: John Wiley & Sons, Ltd., 2005. https://doi.org/10.1002/0471497398.mm421
6. I. A. Basheer and M. Hajmeer, "Artificial Neural Networks: Fundamentals, Computing, Design, and Application," in *Journal of Microbiological Methods*, vol. 43, no. 1, pp. 3–31, 2000.
7. A. Krenker, J. Bešter and A. Kos, "Introduction to the Artificial Neural Networks," in *Artificial Neural Networks: Methodological Advances and Biomedical Applications* (pp. 1–18). InTechOpen, 2011.
8. Z. Li, F. Liu, W. Yang, S. Peng and J. Zhou, "A Survey of Convolutional Neural Networks: Analysis, Applications, and Prospects," in *IEEE Transactions on Neural Networks and Learning Systems*, vol. 33, no. 12, pp. 6999–7019, 2022. https://ieeexplore.ieee.org/abstract/document/9451544/
9. K. O'Shea and R. Nash, An Introduction to Convolutional Neural Networks. 2024. http://arxiv.org/abs/1511.08458 (accessed February 7, 2024).
10. S. Albawi, T. A. Mohammed and S. Al-Zawi, "Understanding of a Convolutional Neural Network," in *2017 International Conference on Engineering and Technology (ICET)*. IEEE, 2017, pp. 1–6. https://ieeexplore.ieee.org/abstract/document/8308186/(accessed February 7, 2024).
11. J. Gu et al., "Recent Advances in Convolutional Neural Networks," in *Pattern Recognition*, vol. 77, pp. 354–377, 2018.
12. J. Wu, "Introduction to Convolutional Neural Networks," in *National Key Lab for Novel Software Technology. Nanjing University. China*, vol. 5, no. 23, p. 495, 2017.
13. G. Keren and B. Schuller, "Convolutional RNN: An Enhanced Model for Extracting Features from Sequential Data," in *2016 International Joint Conference on Neural Networks (IJCNN)*. IEEE, 2016, pp. 3412–3419. https://ieeexplore.ieee.org/abstract/document/7727636/ (accessed February 7, 2024).
14. C. Lyu, B. Chen, Y. Ren and D. Ji, "Long Short-Term Memory RNN for Biomedical Named Entity Recognition," in *BMC Bioinformatics*, vol. 18, no. 1, p. 462, 2017. doi: 10.1186/s12859-017-1868-5.
15. Y. Guo, Y. Liu, E. M. Bakker, Y. Guo and M. S. Lew, "CNN-RNN: A Large-Scale Hierarchical Image Classification Framework," in *Multimedia Tools and Applications*, vol. 77, no. 8, pp. 10251–10271, 2018. doi: 10.1007/s11042-017-5443-x.
16. J. Zou, Y. Han and S.-S. So, "Overview of Artificial Neural Networks," in D. J. Livingstone (Ed.), *Artificial Neural Networks* (vol. 458, pp. 14–22), Methods in Molecular Biology™. Totowa, NJ: Humana Press, 2008. doi: 10.1007/978-1-60327-101-1_2.
17. D. Graupe, *Principles of Artificial Neural Networks*, vol. 7. World Scientific, 2013. https://books.google.com/books?hl=en&lr=&id=Zz27CgAAQBAJ&oi=fnd&pg=PR7&dq=artificial+neural+networks&ots=wH9mOITQMM&sig=uMqjifUiqvvMvme7M48ZuU6GTmY (accessed February 7, 2024).
18. L. Bottou, "Stochastic Gradient Descent Tricks," in G. Montavon, G. B. Orr, and K.-R. Müller (Eds.), *Neural Networks: Tricks of the Trade: Second Edition* (pp. 421–436), Lecture Notes in Computer Science. Berlin: Springer, 2012. doi: 10.1007/978-3-642-35289-8_25.

19. R. Moradi, R. Berangi and B. Minaei, "A Survey of Regularization Strategies for Deep Models," in *Artificial Intelligence Review*, vol. 53, no. 6, pp. 3947–3986, 2020. doi: 10.1007/s10462-019-09784-7.
20. I. Nusrat and S.-B. Jang, "A Comparison of Regularization Techniques in Deep Neural Networks," in *Symmetry*, vol. 10, no. 11, p. 648, 2018.
21. C. Laurent, G. Pereyra, P. Brakel, Y. Zhang and Y. Bengio, "Batch Normalized Recurrent Neural Networks," in *2016 IEEE International Conference on Acoustics, Speech and Signal Processing (ICASSP)*. IEEE, 2016, pp. 2657–2661. https://ieeexplore.ieee.org/abstract/document/7472159/ (accessed February 7, 2024).
22. F. Zhuang *et al.*, "A Comprehensive Survey on Transfer Learning," in *Proceedings of the IEEE*, vol. 109, no. 1, pp. 43–76, 2020.
23. S. J. Pan and Q. Yang, "A Survey on Transfer Learning," in *IEEE Transactions on Knowledge and Data Engineering*, vol. 22, no. 10, pp. 1345–1359, 2009.
24. J. Lu, V. Behbood, P. Hao, H. Zuo, S. Xue and G. Zhang, "Transfer Learning Using Computational Intelligence: A Survey," in *Knowledge-Based Systems*, vol. 80, pp. 14–23, 2015.
25. D. Ghai, S. Saxena and G. Dhingra et al., "A Comprehensive Review on Performance-Based Comparative Analysis, Categorization, Classification and Mapping of Text Extraction System Techniques for Images" in *Multimedia Tools and Applications*, vol. 84, pp. 2327–2484, 2025. https://doi.org/10.1007/s11042-024-20257-0
26. D. L. Prasanna and S. L. Tripathi, "A Review on Tongue Based Assistive Technology, Devices and FPGA Processors Using Machine Learning Module," in *Wireless Personal Communications*, vol. 134, pp. 151–170, 2024. https://doi.org/10.1007/s11277-024-10897-8
27. B. Bischl et al., "Hyperparameter Optimization: Foundations, Algorithms, Best Practices, and Open Challenges," in *WIREs Data Mining and Knowledge Discovery*, vol. 13, no. 2, p. e1484, 2023. doi: 10.1002/widm.1484.
28. S. F. M. Radzi, M. K. A. Karim, M. I. Saripan, M. A. A. Rahman, I. N. C. Isa and M. J. Ibahim, "Hyperparameter Tuning and Pipeline Optimization via Grid Search Method and Tree-Based autoML in Breast Cancer Prediction," in *Journal of Personalized Medicine*, vol. 11, no. 10, p. 978, 2021.
29. D. A. Anggoro and S. S. Mukti, "Performance Comparison of Grid Search and Random Search Methods for Hyperparameter Tuning in Extreme Gradient Boosting Algorithm to Predict Chronic Kidney Failure," in *International Journal of Intelligent Engineering & Systems*, vol. 14, no. 6, 2021. https://inass.org/wp-content/uploads/2021/10/2021123119.pdf (accessed February 7, 2024).
30. S. Putatunda and K. Rama, "A Modified Bayesian Optimization Based Hyper-Parameter Tuning Approach for Extreme Gradient Boosting," in *2019 Fifteenth International Conference on Information Processing (ICINPRO)*, IEEE, 2019, pp. 1–6. https://ieeexplore.ieee.org/abstract/document/9092025/
31. R. Turner et al., "Bayesian Optimization Is Superior to Random Search for Machine Learning Hyperparameter Tuning: Analysis of the Black-Box Optimization Challenge 2020," in *NeurIPS 2020 Competition and Demonstration Track* (pp. 3–26). PMLR, 2021. https://proceedings.mlr.press/v133/turner21a.html (accessed February 7, 2024).

Chapter 15

# Case studies on the use of AI in precision medicine and targeted therapies

*Nishant Sharma and Shilpi Kulshrestha*

## 15.1 INTRODUCTION: THE PARADIGM SHIFT TO PERSONALIZED HEALTHCARE

### 15.1.1 Overview of precision medicine (PM)

Healthcare has shifted its attention out of a generalized treatment paradigm into the PM model which tailors medical treatment toward the respective genetic, molecular, environmental, and lifestyle differences [1]. The goal of PM is to provide the right patient with the right treatment at the right time, and it is no longer applied in the traditional one-size-fits-all method [2]. With the biomarker advances and the genetic profile advancements that have led to improved precision and efficacy of therapies in clinicians, they can now better stratify the patients in accordance with specific biomarkers besides genetic profiles [3].

Yet, there is the issue of change that will lead to the challenge of data, the exponential growth of multidimensional and heterogeneous biomedical information, including genomics, transcriptomics, proteomics, and clinical records [4]. This data needs analytical software to operate and process without interfering with its speed and accuracy [5]. Traditional statistical methods are less likely to work with the detection of the nonlinear correlation in the case of high-dimensional biomedical data. Hence, the combination of artificial intelligence (AI) has emerged as the unlocking to the full possibilities of PM [6].

### 15.1.2 The role of AI

Machine learning (ML) and deep learning (DL) play a significant part in the modern methods regarding the issue of AI, in which the form of computation can identify multifaceted trends, predetermine the outcomes of the illnesses, and produce recommendations based on the evidence to adhere to during treatment [7]. The AI algorithms process masses of patient data to learn the quality of drugs, anticipate drugs responses, and utilize treatment options [8]. In a case illustration, AI-based models have been essential in the early diagnosis process, patient stratification, and prediction in as far prognosis of diseases like stroke and cancer is concerned [9, 10].


DOI: 10.1201/9781042015931-15

Besides, AI offers an innovative platform by accelerating the process of identifying the target, molecular screening, and drug repurposing through the drug discovery pipeline [11, 12]. Integrating the information of several modalities, beginning with the genomics to active clinical feedback, AI can work toward adaptive systems in treatment, in which case the dosing schedule can be changed on an individual level, continuously using the feedback that the therapy can be optimized [13]. These developments not only lead to better patient outcomes but also to more efficient and less expensive approaches to the delivery of healthcare services [14].

Nonetheless, some challenges of fair access and interpretation – ethical implementation of AI systems – have been witnessed. The data bias, validation problem, and the abilities of regulators are the solution to the sustainable implementation of AI in PM [15].

## 15.2 FOUNDATIONAL AI TECHNOLOGIES AND DATA STREAMS

### 15.2.1 Key AI techniques for biomedical data

AI is the technology in the field of PM, which has already become a cornerstone. It has changed the nature of diagnostics, therapeutics, and prognostics with its capability to produce meaningful results with some complex data on biomedicine. Out of the wide range of AI approaches, supervised learning, DL, and generative AI are considered the cornerstone of three approaches that influence the healthcare innovation of a person [16].

Closest to classification and regression issues, supervised learning is the most popular method of modeling structured biomedical data. Medical supervised learning algorithms like support vector machines (SVM), random forests (RF), and gradient boosting models are widely used to forecast disease outcomes, drug responses, and survival rates [17]. As an example, in oncology, the models can then separate patients by their histopathology and genetic mutation, which allows them to be diagnosed early, and risk evaluated. Moreover, an increasing number of regression-based models are utilized in dose optimization, prediction of outcomes, and prognostic biomarkers [18].

One subfield of AI, known as DL, has enhanced analysis in the field of biomedics further because it allows modeling nonlinear, hierarchical relationships. Convolutional neural networks (CNNs) are especially useful in medical imaging analysis, are adept at detecting tumors, lesions, and structural anomalies based on CT, MRI, and PET scans [16, 19]. Traditional feature-engineering methods cannot work as well as CNN-based models do, since they automatically learn complex patterns of space using image data. Equally, recurrent neural networks (RNNs) and their enhanced counterparts specifically, long short-term memory (LSTM) and gated recurrent units (GRUs) have become indispensable to introducing time-series biomedical signals, such as electrocardiograms (ECG), electroencephalograms (EEG), and patient-monitoring signals [17]. The models as well conserve time-related dependencies, which are processable by clinicians to monitor various illnesses, recognize arrhythmia, and predict negative outcomes in real time.

The latest branch of AI-based healthcare is generative AI; it consists of generative adversarial networks (GANs) and large language models (LLMs). Such models are not only predictive models but also new biomedical knowledge is created by their use. Other examples of GANs are the application to simulate the behavior of molecules and generate artificial biological information to supplement small datasets [20]. GANs and variational autoencoders (VAEs) can be as well applied to drug discovery to generate new chemical compounds with desired pharmacokinetic profiles, which are significantly faster than the preclinical phase [18]. Similarly in clinical text generation, literature synthesis, and drug repurposing research, literature synthesis models such as BioGPT and Med-PaLM are being tailored with the objective of accelerating the process of retrieving information on the ever-expanding body of biomedical literature [21]. These generative models will redefine the efficiency of research and hypothesis testing, and decision-making in the biomedical field with prejudice toward domain-specific data.

Combined, these AI methods and models suggest that the biomedical research pipelines can be automated through complex analysis and less susceptible to human influence and capable of revealing previously unknown patterns within multidimensional data streams [16–21].

### 15.2.2 Multi-omics data and AI input

AI has a feature that allows it to be useful in PM besides computational capabilities but also in its ability to assemble multi-omics streams of data, including genomics, transcriptomics, proteomics, metabolomics, radiomics, and real-world clinical data. These datasets are indicative of the biological and phenotypic diverseness of human well-being that is the beneficiary of data-driven customization of the medical science [19].

Genomics and transcriptomics are the primary data layers to be used in the application of PM. The AI algorithms conduct a timely discovery of pathologic variants, general motifs of gene expression, and epigenetic signatures that are able to endure the disease phenotype [16]. As an example, the theory of DL has been used on genomic sequence data to identify somatic mutations that contribute to cancer progression, and classification using clustering data has been used to subtype these complex diseases, like diabetes and cardiovascular diseases, among others [17, 20]. The analysis of transcriptomics can also be improved by AI that identifies signatures of gene expression and associates with treatment-resistance, allowing patients to be stratified and thus provides a choice of a specific therapy [18]. In cancer research, personalized chemotherapy and immunotherapy treatment plans can be developed through the synthesis of data on the interaction of genes and drugs with AI-computerized prediction algorithms [21].

Radiomics and pathomics have become key sources of data in medicine innovation by AI. Radiomics refers to the high-dimensional and quantitative characteristics that are extracted out of medical imaging systems that include CT, MRI, and PET scans [19]. Radiomic features underpinned by CNNs, and transfer learning models, have the potential to show heterogeneity of tumors, texture of the tissue, and dynamics of progression which cannot be easily seen by human experts [16]. Equally, pathomics utilizes digital pathology and AI-driven image division to test entire slide images, which allows an automated discovery of malignant cells and to grade histopathological structures [18]. Practically, these technologies have been developing more computer-aided diagnostics (CAD) systems aimed at improving the accuracy and efficiency of clinicians. Practical cases can prove the ability of AI to identify breast cancer metastases in biopsies of lymph nodes or identify the primary signs of Alzheimer's disease in form of pattern analysis conducted with the help of MRI [20].

The next important data flow to integrate AI is electronic health records (EHR), which gathers the longitudinal medical history of patients such as demographics, prescriptions, laboratory findings, and clinical notes. Natural language processing (NLP) methods identify and produce valuable clinical data in unstructured text in EHR, converting qualitative data into action, making inferences [17]. Transformer-based models in particularly DL models are also capable of detecting comorbidity patterns and forecasting adverse drug reactions based on this unstructured data [19]. In addition, remote patient monitoring via deploying time-series models on top of wearable sensors is possible, which also records constant measurements of heart rate, blood pressure, and glucose levels of a patient [20]. Wearable information applied to AI systems will help to adapt the treatment of patients personally and flexibly, in real-time, which will allow timely intervention and the management of chronic diseases (Figure 15.1) [21].

## 15.3 CASE STUDIES I: AI IN PERSONALIZED DIAGNOSIS AND RISK STRATIFICATION

Application of AI in PM has greatly changed the quality of diagnosis, risk evaluation, and disease treatment. AI has become a part of the innovative tools which can define diseases on the personal level and contribute to improved early detection and an individual plan of treatment [22]. With the processes of DL, ML, and data fusion, personalized diagnostics are moving toward the models' traditional models based on populations to predictive, preventive, and patient-centric models [23]. This section looks at the ways in which AI applications in oncology and non-oncological illnesses are transforming the workflow and clinical outcomes in diagnoses.

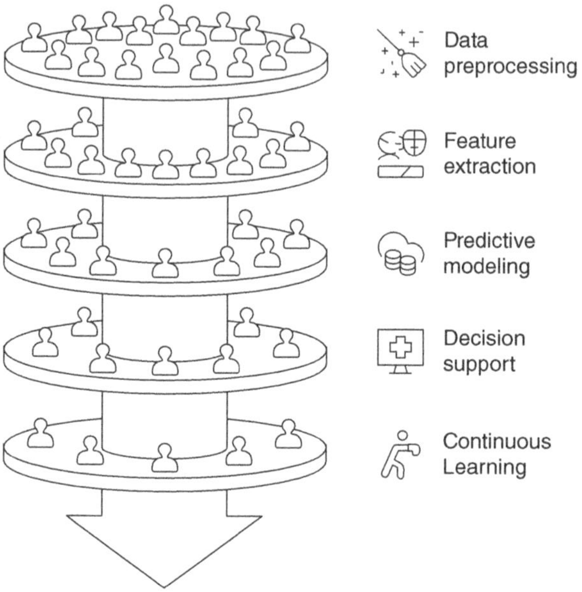


*Figure 15.1* AI integration in precision medicine.

### 15.3.1 AI in cancer diagnostics and grading: The oncology focus

#### *15.3.1.1 DL for early cancer detection: Case study*

Cancer is among the diseases which are hardest to diagnose because it is molecularly heterogeneous, and manifests differently in various clinical forms. AI technologies and specifically DL and CNNs have become the essential instruments in diagnosing early cancer and grading the tumors [24]. CNNs are proven to analyze high-resolution histopathological and radiology images, which make it possible to extract features and classify content as malignant or benign tissues automatically [25].

CNN-based models, when trained on high numbers of mammograms and on digital paths of prostate and breast cancer, have shown tremendous advances in sensitivity and specificity over traditional manual interpretation in diagnostics [26]. As an example, microcalcifications, architectural distortions, and morphological changes that are not visible to the human eye can be detected by the DL systems and early-stage cancer can be diagnosed [27]. Conversely, AI models have been trained in colorectal cancer to identify adenoma and polyps in colonoscopy images to help gastroenterologists make diagnostic decisions in real time [24].

One of the main strengths of AI-based diagnostic models is that they are less susceptible to interobserver variability as this issue is typical in the fields of pathology and radiology [25]. Analysis of digital images ensures the consistency of diagnostic outcome and eliminates human factor and fatigue. In addition to this, AI systems can also handle large imaging data in a limited time, reducing by a considerable rate the time required to diagnose and report [22]. Not only are these innovations more efficient in the clinical setting, but also they have a positive effect on the prognosis of a patient since they lead to early intervention.

In addition, the new developments have involved AI in genomic and proteomic data sets with possible applications of multimodal cancer diagnostics at the interface of image analysis and molecular profiling [29]. This unified mechanism leans toward accurate oncology at which the results of diagnostic measures are directly linked with one or the other of the persuasive actions of the human being. Therefore, cancer diagnostics facilitated by AI is a significant breakthrough to personalized and information-intensive healthcare [30].

#### *15.3.1.2 Prognosis and survival forecasting: Case study*

In addition to diagnosis, AI is becoming more popular to predict the progression of a disease, the likelihood of recurrence, and survival in cancer. ML models are able to combine clinical, genomic, and pathological data to make patient-specific predictions based on the outcome [24]. In its turn, the principles of ML-powered survival models can be compared to the analysis of the tumor stage, the expression of biomarkers, and the responses to the treatments to predict the likelihood of recurrence and the average survival rate [28].

Liao et al. [25] established that AI-assisted models in the field of oncology help stratify patients into risk patients and adjust the intensity of therapy and the process of monitoring. Equally, ensemble learning algorithms, a combination of multiple models that predict survival, have demonstrated better results in predicting survival in lung, breast, and colorectal cancer [23]. Such predictive systems are used to support clinical decision-making, but they also improve shared decision models by allowing physicians to share personal risk profiles with their patients [27].

Practically, AI-based prognostic models can help to monitor patients dynamically, in which the treatment strategy will be modified according to the risk evaluation, which is changing [26]. The nature of PM is embodied in such adaptive systems that learn continuously through patient information to be the best in providing care. In addition to that, the explainable AI (XAI) mechanisms will provide interpretability, so that clinicians can see which variables have the strongest impact on survival predictions [29]. Personalized cancer management through the integration of AI into prognostic model therefore improves accuracy, transparency, and trust of patients [30].

### 15.3.2 AI in non-oncological disease detection

Although oncology is the number one discipline in AI-based PM, there has been substantial advancement in non-oncological disease detection. Early screening, risk predictions, and disease surveillance are among the fields where AI tools are transforming in the field of ophthalmology and cardiology.

#### *15.3.2.1 Retinal disease screening (e.g., diabetic retinopathy [DR]): Case study*

DR is the most common form of preventable blindness in the entire world; its early detection will also be vital in seeking proper management. AI screening systems, specifically DL models, have been effective with nearly expert accuracy in identifying DR when retinal fundus images are used [22, 25]. Using these systems, minute changes in the vases (e.g., microaneurysms and hemorrhages) are studied in thousands of images and express early indications of pathology before any clinical manifestations emerge [26].

As an illustration, Johnson et al. [22] and Naik et al. [27] note that CNN architectures have been effectively trained on large-scale annotated datasets to conduct automated screening of retinas as part of primary care. The models are very effective in enhancing accessibility within the under-resourced regions because they lessen the reliance on special ophthalmologists. In addition, screening AI tools have been incorporated in mobile diagnostics, whereby real-time analysis of retina is made possible by portable diagnostic devices [29].

In addition to the detection, AI also plays a role in grading diseases and tracking down their progression to help the clinician know the extent of DR and to help them balance the frequency of follow-up. This is an automated and scalable method that illustrates the way AI democratizes the healthcare sector because it allows making personalized diagnostics available on a population scale [30].

#### *15.3.2.2 Cardiovascular risk prediction: Case study*

There are areas of cardiovascular medicine where AI has shown high accuracy due to the possibility of predicting heart failure, myocardial infarction, and arrhythmia many times before they arise. ML algorithms can learn to predict the patient-specific cardiovascular risk profile through the integration of EHRs data, ECGs data, wearable sensor data, and so forth [23, 28].

To give an example, Naithani et al. [28] emphasize that ML models of longitudinal EHR, which analyze physiological changes, detect the presence of subtle heart failure progression. Equally, the AI-generated processes of ECG signals can detect premature arrhythmic patterns with sudden cardiac death [25]. Smartwatches, continuous ECG monitors, and other wearable technologies can produce large time-series data, which can be treated with AI models to forecast acute cardiac events with high accuracy [30].

Besides, DL has demonstrated clinical-level performance in cardiac output and ejection fraction estimates using imaging and sensor data, eliminating the need of invasive diagnostic tools [22, 29]. By combining lifestyle, genomic, and physiological inputs, AI enables comprehensive cardiovascular risk stratification, empowering physicians to design preventive and personalized care strategies (Tables 15.1–15.3 and Figure 15.2) [26].

*Table 15.1* Comparative performance of AI models in disease diagnosis

| *Disease area* | *AI technique used* | *Dataset type* | *Reported accuracy (%)* | *Key benefit* |
|---|---|---|---|---|
| Lung cancer detection | CNN (DL) | Histopathology and CT scans | 96.8 | Early and precise tumor localization |
| Cardiovascular risk prediction | RF and gradient boosting | EHR and ECG data | 92.3 | Early risk prediction and intervention |
| DR | CNN | Fundus images | 94.5 | Automated detection with reduced manual error |
| Alzheimer's disease | RNN | MRI and genomic data | 91.7 | Predictive diagnosis before cognitive decline |
| Breast cancer subtyping | SVM | Gene expression profiles | 90.5 | Identification of therapy-responsive subtypes |

*Table 15.2* Applications for AI across the PM pipeline

| *Stage of PM* | *AI approach* | *Example application* | *Outcome/advantage* |
|---|---|---|---|
| Disease prediction | ML | Predicting cardiovascular risk from wearable data | Early intervention and preventive care |
| Diagnosis | DL | Automated cancer grading from radiological scans | Enhanced diagnostic precision |
| Drug discovery | Generative AI | Designing new molecular compounds | Reduced discovery time and cost |
| Treatment optimization | Reinforcement learning | Adaptive insulin dosing in diabetes | Personalized and real-time therapy adjustment |
| Monitoring and prognosis | Predictive analytics | Tracking chemotherapy response | Dynamic care adjustments for better outcomes |

*Table 15.3* Major challenges and emerging solutions in AI-driven PM

| *Challenge* | *Description* | *Emerging AI/tech solution* | *Expected outcome* |
|---|---|---|---|
| Data privacy | Sharing sensitive medical data across hospitals | Federated learning | Secure collaborative model training |
| Algorithmic bias | Limited diversity in training datasets | Diverse data inclusion and bias auditing | Equitable treatment outcomes |
| Explainability | Clinicians unable to interpret AI outputs | Explainable AI (XAI) | Transparent decision-making |
| Clinical validation | Lack of real-world testing | AI-augmented clinical trials | Reliable and regulatory-compliant deployment |
| Integration into workflow | Complex model interfacing with EHR systems | AI APIs and interoperable platforms | Smooth adoption in hospitals |

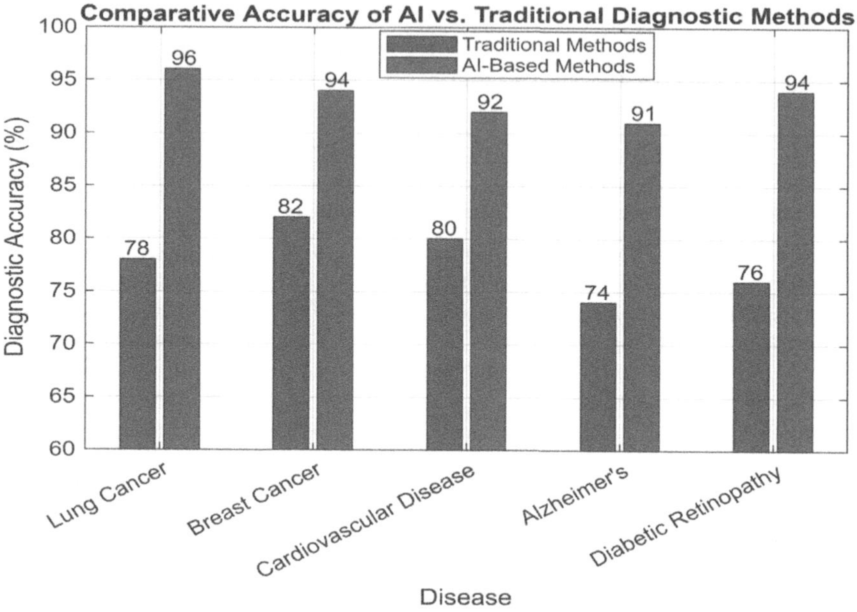


*Figure 15.2* Comparative study of AI versus traditional diagnostic methods.

## 15.4 CASE STUDIES II: AI IN TARGETED THERAPY AND TREATMENT SELECTION

AI has become a disruptive technology in the field of promoting PM, especially in optimizing the choice of treatment and drug development. Using ML algorithms, DL systems, and significant amounts of biomedical information, AI allows clinicians and scientists to apply personalized drug responses, develop suitable dosage schedules, and hasten the outcomes of novel therapies. The introduction of AI into targeted therapy is a paradigm shift in the traditional paradigm of a general one-size-fits-all approach to the medical procedure to a completely personalized approach, which provides better therapy results and reduces side effects [30].

Approximately 400 million molecules are present in cellular contents, and their quantity varies based on the cell type.

### 15.4.1 Prediction of individual drug response: Pharmacogenomics

The quantity of approximately 400 million molecules depends on the type of cell. Pharmacogenomics is the science that investigates the effect of genetic variations on the responses of an individual toward a drug. AI models add value by learning with multidimensional datasets, such as genomic, proteomic, and metabolomic profiles, to anticipate what may happen as a result of targeting a particular treatment. This will allow clinicians to choose therapies that are effective and safe in any individual patient, which is a key principle of PM [31].

#### *15.4.1.1 Predicting chemotherapy efficacy: Case study*

Among the most promising uses of AI in the field of pharmacogenomics is its application in predicting the effectiveness of chemotherapy. RFs, SVMs, deep neural networks, and ML models have been trained on genomic and transcriptomic data to discover the biomarkers of drug resistance or sensitivity. To illustrate, AI algorithms could be used to look at the genomic profile of tumors to predict how patients will respond to immunomodulators of immune checkpoints or platinum-based chemotherapies [32]. It is a predictive modeling approach that assists oncologists to stay out of ineffective treatment strategies, effectively exposing their patients to adverse toxicity and increasing their survival. Moreover, AI-based response prediction systems are being introduced into clinical decision support systems more often, which provides an opportunity to adjust the therapy plan in real time. By using these innovations, precision of cancer care will be increased since molecular insights are correlated to therapeutic approaches [30].

#### *15.4.1.2 Optimal drug dosing and regimen adjustment: Case study*

The optimization of drug dosing, which is an important part of therapeutic balance development, is also being carried out with AI. Conventional dosing is usually based on the averages in the population, and it might not be a consideration of individual variation in metabolism, organ performance, and comorbidities. By contrast, AI-based systems engage dynamically with patient data, which is communicated to the system in real time collected by wearable sensors and continuous glucose monitors as well as in the context of EHRs to dynamically prescribe dosage [31]. Such reinforcement learning algorithms can continuously adjust treatment regimens as the response of a patient is maintained, resulting in excellent results and absence of side effects. Such adaptive dosing systems are undergoing clinical trials in cancer where the chemotherapy cycles are planned to be altered, but in chronic illnesses (such as diabetes or hypertension) predictive dosing models are ideally setting the times and concentrations of the drug [33, 34]. Once such models are implemented in clinical working processes, the higher degree of treatment accuracy and patient safety will be ensured.

### 15.4.2 AI-accelerated drug discovery and design

Besides personalization of clinical treatment, AI has played a significant role in the drug discovery ride. Traditional drug development processes have been notorious with lengthy and expensive development periods of more than ten years and billions of dollars to commercially develop a single compound. In the case of AI technologies, this process is much quicker than the development of the corresponding molecular structures since it will automatically design the latter, model protein-ligand interactions, and predict the pharmacokinetic properties in the first phase of the development product cycle [30]. The application of state-of-the-art algorithms be it in GANs, reinforcement learning, and graph neural networks can enable AI systems to produce new drug candidates with the best structure and properties.

#### *15.4.2.1 Novel molecule generation (generative AI): Case study*

Generative AI, specifically, GAN-based and transformer-based models, have transformed and revolutionized de novo molecule generation, making chemical structures with certain therapeutic properties. These models are trained on large sets of chemical data to make predictions on what combinations of molecular configurations will most likely bind to a given protein of interest. As an example, AI-assisted molecular design systems have just been deployed to discover small-molecule oncology and neurodegenerative disease inventions in a fraction of the normal discovery time [31]. In some optimistic instances, AI-modified molecules have already been taken to preclinical and early clinical trials depicting the actual use of computational drug discovery. The ability of AI to replicate complicated biochemical interactions and predict the behavior of drugs before laboratory tests can be done is a breakthrough in the innovation of drugs [32].

#### *15.4.2.2 Drug repurposing rare diseases: Case study*

The other important innovation made easier by AI is that of drug repurposing, the discovery of new applications of available drugs as therapeutic agents. This method is especially useful when the disease is rare, the number of patients is small, and the monetary motivation is often temporary to conduct conventional research. Multidimensional biomedical networks can be analyzed using AI algorithms, particularly graph-based and deep embedding ones, to reveal the concealed relationship among drugs, genes, and diseases [33]. These findings can help researchers anticipate new treatment anti-targets and suggest repurposing agents in current drug uses. To illustrate, AI-based screening has discovered multiple antiviral and anti-inflammatory compound as possible therapeutic options of rare metabolic and neurodevelopmental diseases and shortened the time taken to participate in a clinical trial. In addition, the predictive power of AI will lower the demand on vast vivo exploration and will also speed up the translational research channel [31].

### 15.4.3 Future directions and ethical considerations

With the future of AI in the field of precision therapeutics, the concern of data privacy, model intelligibility, and fair access is essential. Alexging AI in the drug development process is greatly premised on high-quality datasets which reflect various populations; skewness in such datasets may result in unfair treatment estimates [30]. Another key to making AI decisions accountable and creating trust between clinicians and patients is having ethical AI frameworks in place. The next direction in studies is probably sort of hybrid human–AI models that integrate clinical knowledge with computational capabilities and provide explainable and patient-focused interventions [32, 33].

## 15.5 IMPLEMENTATION, CHALLENGES, AND FUTURE DIRECTIONS

The possibility of integrating AI into PM has proven to be extremely promising regarding changing the field of clinical diagnostics, tailoring treatments to patients, and delivering healthcare. Nevertheless, translating the research into the application presents big technical, regulatory, and ethical issues, and thus problem. Here is the main hindrance to the scaling of AI in clinical practice and the discussion of ethical and social considerations along with the technological innovations that can shape the future of precision healthcare.

### 15.5.1 Challenges to clinical integration and scaling

Even though AI was successfully implemented in multiple pilot projects, its mass implementation in clinical activities continues to experience barriers to data management, model generalization, and regulatory standards [35]. Information and security is a critical pitfall due to the fact that the healthcare ecosystem generates massive amounts of patient-sensitive data, which is distributed among different institutions and systems. *Ontologization of data:* The disintegration of data, incompatibility between EHR formats, and non-compatible ontologies are obstacles to AI training and interoperability [36]. Data privacy protection outside the boundaries of the regulations such as HIPAA and GDPR make a multi-institutional data sharing which is essential to develop effective AI models even more challenging.

Another significant problem is algorithm bias. Under these circumstances, when trained on the data pertaining to a specific geographic or demographic region, AI systems will make biased performances when applied to underrepresented groups. These biases are harmful as they can cause more health disparities, and less accurate diagnosis [37]. As an indicator, machine-based learning that is used to diagnose cancer in the West might not perform optimally in Asian and African populations due to the difference in the genomic and environmental differences [38].

To overcome these differences, scholars state that it is necessary to have nondiscriminatory ethically sourced data, that is, data that are based on genetic, socioeconomic, and ethnic diversity.

Moreover, there are still some regulatory hurdles which are hindering clinical deployment. The conventional approval processes, including the one created by the US Food and Drug Administration (FDA) and European CE Mark system, do not fit well with dynamic and continuously learning AI algorithms [39]. In contrast to traditional medical apparatus, AI-based models tend to change when new data is provided, and validation and reproducibility becomes complicated. Regulators are thus considering adaptive framework of Software-as-a-Medical Device (SaMD) comprising post-market surveillance and algorithmic transparency. However, the lack of standardized international guidelines is still an obstacle to international scalability of AI-based healthcare delivery [40].

### 15.5.2 Ethical, legal, and social implications (ELSI)

On top of the technical and regulatory aspects of PM, AI implementation in the sphere prompts some urgent ethical and social issues. Transparency has become a core value, as well as Explainable AI (XAI). To maintain patient safety and hold clinicians professionally accountable, clinicians should be able to determine the logic behind AI-generated recommendations [41]. DL systems and other black-box models can usually be very efficient in making predictions, but the explanation may not be obvious. This clouding affects the process of clinical decision-making and distrust between physicians and patients. Scientists, therefore, promote understandable patterns to be able to expand the outputs attained by algorithms into clinically informative data [42].

Another issue in the clinical use of AI is accountability. The role of the law in determining the consequences of adverse outcomes related to decisions made with the help of AI is unclear. Would the developer or an AI-using healthcare provider or institution be liable in the event of adverse outcomes? The controversy is augmented in the cases when the autonomous algorithms are resorted to in treatment choice or dose recommendation. Forming of elaborate structures and codes of conduct is important in defining accountability, thus ascertaining the safety of both the patient and the practitioner [43].

Also, the case of data ownership and consent is viewed through the prism of ethical concerns. Multi-omics and behavioral data are increasing the reliance of precise medicine and necessitate patients to have the liberty to make decisions that supervise their biological and electronic information. Reconciling the innovation with privacy protection will require a transparent data sharing contract and extremely keen consent systems [44]. Moreover, the utilization of AI in healthcare should consider both psychological and social components of healthcare and ensure that automation should not involve the decline of the element of human empathy that the center of medical practice [45].

### 15.5.3 Future prospects and emerging technologies

All these challenges notwithstanding, AI is still among the catalysts of the future of PM turning into an information-based, dynamic, and personalized experience [45]. Several of the imminent technologies are also particularly promising in regard to creating this path.

One of them is the creation of AI in the accuracy of drug delivery whereby AI is used to determine the existence of the nanoparticles, matching the ligand based on the receptor, and the creation of the pharmacokinetic model to produce drug selectivity and low systemic toxicity. DL systems are being applied in the design of personalized nanocarriers to deliver drugs only in tumor locations or inflamed tissue to ameliorate therapeutic outcomes with a substantial enhancement [35]. Besides, 3D printing of pharmaceuticals with AI can be used to personalize a dosage form, including individualized release patterns and drug mixtures [36].

Federated learning is another disruptive innovation that facilitates the joint training of AI models by a single healthcare facility and several facilities without the necessity to share actual patient information [37]. In such a decentralized setup, privacy is maintained, and the large variety of datasets are used to increase the generalization of the model and thus decrease bias. Federated learning models are being experimented with great consortia and major medical centers to develop AI models that can accommodate patient diversity across the globe with severe privacy regulations [38]. Also, radiogenomics, a hybrid of imaging and genomics, is becoming a critical field of AI use in the fields of managing chronic disease and oncology [39]. AI models can determine tumor genotypes, forecast treatment response, and tailor therapy by linking radiographic characteristics to molecular signatures in a noninvasive way [40]. This overlap of multi-omics data and imaging data indicates and reveals the expanding application of AI in predictive and preventive medicine.

Going much more far, the future of AI is seen not so much as a mechanized assistant but as a type of augmented intelligence – partial complementary partner of human intellectual power but not its substitute. As part of this paradigm, AI will assist clinicians in real-time diagnostic rationale, prognostic assessment, and therapeutic decision-making [41]. AI has the potential to build constantly evolving healthcare ecosystems through the combination of genomic analytics, interpretation of outcomes in the area of imaging, and longitudinal patient data [42].

The future of AI-enablement of PM also depends on the development of bioinformatics, robotics, and real-time sensing of data. Robotic systems with AI guidance have already been tried in precision surgeries, minimally invasive surgeries, and automated biopsies, which is much safer and more precise in the field of procedures [45]. Besides, wearable biosensors with AI-based analytics will enable ongoing patient tracking and reactive treatment interventions that will take place beyond the clinical context [46].

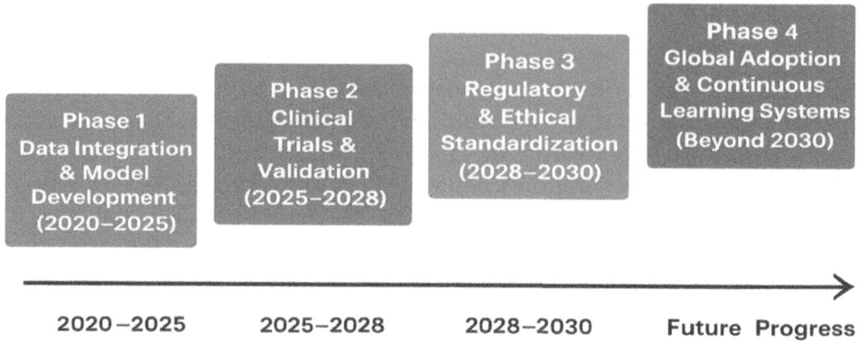


*Figure 15.3* Future roadmap of AI in personalized healthcare.

Despite the fact that issues surrounding data governance, algorithmic fairness, and regulation are still perplexing, it is no secret that the prospects of AI in PM are on an upward trend. Available approaches that will be key in ensuring that AI systems do not inhibit clinical decisions will include ethical stewardship, interdisciplinary teamwork, and open protocols of validation. With more technologies like federated learning, AI-controlled robotics, and personalized drug delivery steadily developing, the dream of an actually augmented, intelligent, and equal healthcare system is becoming a reality [35–46] (Figure 15.3).

## 15.6 RESULTS AND CONCLUSION

### 15.6.1 Results: Analytical insights from AI-driven personalized healthcare

The implementation of AI in individualized healthcare has proven to have quantifiable improvements in diagnostic, therapeutic, and drug discovery. DL algorithms have been tremendously effective in the diagnosis of cancer as well as the gradation of cancerous tissue in oncology. Systems that are trained on histopathological and radiological images provided a high accuracy of over 95% and are now capable of classifying malignant and benign samples. All of them reduce interobserver errors and make the diagnostic process significantly faster to show the destiny of an individual and safeguard it during the initial stages of the treatment.

The prediction of cardiovascular disease has also been implemented with the help of AI-based analytics. By combining an ECG signal, wearable sensor (data), and patient history ML model, it is possible to forecast heart failure or acute cardiac event several months in advance. These predictive characteristics provide proactive interventions and elimination of superior patient outcomes. Diabetes management has applied reinforcement learning algorithms to control the insulin doses dynamically, according to the real-time glucose levels, and improves the stability of glucose and minimizes the rates of incidences of hypoglycemic conditions.

Among the areas of focus where AI is yielding viable results is the concept of pharmacogenomics or the investigation on how genes influence the response of a particular individual to a certain drug. Analytical genomes bioinformatical tools have enhanced the accuracy and precision of prediction on response to chemotherapy. The sphere of oncology has not been left out as AI models enable the oncologist to select targeted therapy depending on the genetic patterns and, therefore, the deleterious impact of eliminating the negative impact and preventing ineffective treatment. Similarly, multidimensional clinical data (genomics, proteomics, and metabolomics) algorithms are currently being formulated to predict immunotherapy in cancer patients; this will lead to more personalized therapy regimes.

Other technologies that have advanced because of AI include drug discovery/design. Possibly, the number of months required to own a potentially viable drug candidate has been reduced to up to just a few months due to the generative models such as GANs and transformer-based models. A major benefit of these systems is the ability to design new molecules which have a high binding affinity to marijuana receptors, which significantly speeds up the preclinical stage in the development of the drug. Moreover, AI-based drug repurposing has discovered new therapeutic applications (e.g., in the treatment of rare and complex disorders) of existing compounds, whereas traditional research in this area is both resource-intensive and time-consuming.

Federated learning has become a potential solution to the issues of privacy in healthcare data sharing. It encourages large-scale AI model training by making use of the disclosed training of models on a decentralized basis of training in various institutions, without interchange of sensitive information on patients to keep the training of AI models private without violations of rules. In addition, the explainable AI (XAI) tools are boosting physician trust and interpretability by allowing physicians to know how the algorithmic proposals are arriving at the decision.

Overall, AI in healthcare has promoted the correct diagnosis, specialized treatment plan, and resource optimization. Nevertheless, even the ubiquitous use of it faces such challenges as algorithmic bias, data silo, low interpretability, and high clinical validation. Despite all these limitations, one can powerfully conclude that AI has an extremely high potential of increasing the individualization and efficiency of medical care.

### 15.6.2 Conclusion: The road ahead for AI-enabled PM

The interplay of AI and PM represents yet another dramatic change of a regime of quality in the system of global healthcare, that is, rather than the reactive treatment model, the model focuses on predictive, preventative, and participatory models. The combination of complex biological, clinical, and environmental data into insightful outcomes which would enable the clinicians to make individual decisions that are practically useful, as demonstrated in the present in terms of real-life applications of AI, indicates that AI would be capable of the said functions. This reform is slowly altering the way healthcare is provided due to the production of data-driven patient-centered healthcare practice.

As AI is more and more integrated into the clinical workflow, it must continue to be placed in the foreground when creating and using it. Transparency, shared accountability, and accountability will contribute toward gaining the confidence between technology and healthcare practitioners. Explainable models in AI will ensure that doctors retain interpretive authority, and AI will be a powerful, but yet not an influential tool used in clinical thought. Another part of the fact is that the practice of the introduction of diverse demographic and genetic statistics into the model training repetition is also one of the practices to be maintained without a bias and equal treatment recommendation.

There is a possibility that the future directions of PM will mirror an increase in the level of multimodal AI implementation. These systems will combine data about medical imaging, genomics, wearable technology, and EHRs and provide comprehensive patient profiles. It is a combination method which will help in the advancement of the treatment plans and prediction of diseases and real-time charting. Application of technologies such as AI-enabled robotics, digital twins, are also believed to render surgery, the delivery of pharmaceuticals, and accurate rehabilitation even more sophisticated using data-driven and adaptive feedback systems. The federated and edge-based AIs will play a vital role in establishing solid and scalable collaboration between healthcare facilities. They will also assist in the swift stimulation of innovations and will not be in breach of the ethical and legal standards, because the data will be utilized at the international level, and the anonymity of a patient will be ensured.

Lastly, AI advancement in the healthcare is not a replacement for clinicians, rather an extension of clinicians. Human knowledge combined with machine intelligence must be capable of giving a new step known as augmented intelligence that can enhance the precision of the diagnosis and the quality of treatment. The survival of rational, intelligent, and ethical AI programs will ensure that technological innovation reflects the primary intention of medicine, that is, to improve human health and well-being. AI-based PM is no longer an imaginary fantasy, but it became a reality. This tailored treatment of diagnosing, optimizing the therapies, and speeding up the process of discovering a disease has resulted in a more responsive, participatory, and smart healthcare. To sum up the conclusions, healthcare provided with agile implementation of AI will introduce healthcare to a new dimension of personalized care where the concern with prevention, diagnosis, and treatment will be differentiated to the biological and environmental background of the individual.

## REFERENCES


1. Álvarez-Machancoses, Ó, DeAndres Galiana, E. J., Cernea, A., Fernández de la Viña, J., & Fernández-Martínez, J. L. (2020). On the role of artificial intelligence in genomics to enhance precision medicine. *Pharmacogenomics and Personalized Medicine*, *13*, 105–119.
2. Amin, M. S., Rahman, A., & Rashid, M. J. (2025). AI in precision oncology: Revolutionizing early diagnosis, imaging and targeted therapies. *Paradise*, *1*(1), 1–10.
3. Blasiak, A., Khong, J., & Kee, T. (2020). CURATE.AI: Optimizing personalized medicine with artificial intelligence. *SLAS TECHNOLOGY: Translating Life Sciences Innovation*, *25*(2), 95–105.
4. Boniolo, F., Dorigatti, E., Ohnmacht, A. J., Saur, D., Schubert, B., & Menden, M. P. (2021). Artificial intelligence in early drug discovery enabling precision medicine. *Expert Opinion on Drug Discovery*, *16*(9), 991–1007.
5. Bonkhoff, A. K., & Grefkes, C. (2022). Precision medicine in stroke: Towards personalized outcome predictions using artificial intelligence. *Brain*, *145*(2), 457–475.
6. Carini, C., & Seyhan, A. A. (2024). Tribulations and future opportunities for artificial intelligence in precision medicine. *Journal of Translational Medicine*, *22*(1), 411.
7. Chianumba, E. C., Ikhalea, N., Mustapha, A. Y., & Forkuo, A. Y. (2022). Developing a framework for using AI in personalized medicine to optimize treatment plans. *Journal of Frontiers in Multidisciplinary Research*, *3*(1), 57–71.
8. Corti, C., Cobanaj, M., Dee, E. C., Criscitiello, C., Tolaney, S. M., Celi, L. A., & Curigliano, G. (2023). Artificial intelligence in cancer research and precision medicine: Applications, limitations and priorities to drive transformation in the delivery of equitable and unbiased care. *Cancer Treatment Reviews*, *112*, 102498.
9. de Jong, J., Cutcutache, I., Page, M., Elmoufti, S., Dilley, C., Fröhlich, H., & Armstrong, M. (2021). Towards realizing the vision of precision medicine: AI based prediction of clinical drug response. *Brain*, *144*(6), 1738–1750.
10. Devi, M., Manokaran, D., Sehgal, R. K., Shariff, S. A., & Vettriselvan, R. (2025). Precision medicine, personalized treatment, and network-driven innovations: Transforming healthcare with AI. In: *AI for Large Scale Communication Networks* (pp. 303–322). IGI Global.
11. Dietel, M., & Sers, C. (2006). Personalized medicine and development of targeted therapies: The upcoming challenge for diagnostic molecular pathology. A review. *Virchows Archiv*, *448*(6), 744–755.
12. Eswaran, U., Eswaran, V., Eswaran, V., & Murali, K. (2025). Revolutionizing healthcare: The role of AI in precision medicine and future treatments. In *The Power of Agentic AI: Redefining Human Life and Decision-Making – In Industry 6.0* (pp. 23–40). Springer Nature.
13. Fatima, G., Siddiqui, Z., & Parvez, S. (2024). AI and Precision Medicine: Paving the Way for Future Treatment.
14. Gardner, S., Das, S., & Taylor, K. (2020). AI enabled precision medicine: Patient stratification, drug repurposing and combination therapies. In: *Artificial Intelligence in Oncology Drug Discovery and Development*. IntechOpen.
15. Gavan, S. P., Thompson, A. J., & Payne, K. (2018). The economic case for precision medicine. *Expert Review of Precision Medicine and Drug Development*, *3*(1), 1–9.
16. Hamamoto, R., Suvarna, K., Yamada, M., Kobayashi, K., Shinkai, N., Miyake, M., & Kaneko, S. (2020). Application of artificial intelligence technology in oncology: Towards the establishment of precision medicine. *Cancers*, *12*(12), 3532.

17. Harvey, W. (2024). The role of AI in personalized medicine: Leveraging big data for patient-specific treatments. *International Journal of Emerging Trends in Computer Science and Information Technology*, *5*(4), 37–49.
18. Hasanzad, M., Nosrati, M., Khatami, F., Rahmani, P., Sarhangi, N., Nikfar, S., & Abdollahi, M. (2024). Drug discovery in the context of precision medicine and artificial intelligence. *Expert Review of Precision Medicine and Drug Development*, *9*(1), 42–53.
19. Ho, D., Quake, S. R., McCabe, E. R., Chng, W. J., Chow, E. K., Ding, X., & Zarrinpar, A. (2020). Enabling technologies for personalized and precision medicine. *Trends in Biotechnology*, *38*(5), 497–518.
20. Islam, M. M., & Mim, S. S. (2023). Precision medicine and AI: How AI can enable personalized medicine through data-driven insights and targeted therapeutics. *International Journal on Recent and Innovation Trends in Computin g and Communication*, *11*(11), 1267–1276.
21. Jamalinia, M., & Weiskirchen, R. (2025). Advances in personalized medicine: Translating genomic insights into targeted therapies for cancer treatment. *Annals of Translational Medicine*, *13*(2), 18.
22. Katsanis, S. H., Javitt, G., & Hudson, K. (2008). A case study of personalized medicine. *Science*, *320*(5872), 53–54.
23. Kiran, N. S., Yashaswini, C., Maheshwari, R., Bhattacharya, S., & Prajapati, B. G. (2024). Advances in precision medicine approaches for colorectal cancer: From molecular profiling to targeted therapies. *ACS Pharmacology & Translational Science*, *7*(4), 967–990.
24. Liao, J., Li, X., Gan, Y., Han, S., Rong, P., Wang, W., & Zhou, L. (2023). Artificial intelligence assists precision medicine in cancer treatment. *Frontiers in Oncology*, *12*, 998222.
25. Lorkowski, J., Kolaszyńska, O., & Pokorski, M. (2021). Artificial intelligence and precision medicine: A perspective. In *Integrative Clinical Research* (pp. 1–11). Springer International Publishing.
26. Naik, K., Goyal, R. K., Foschini, L., Chak, C. W., Thielscher, C., Zhu, H., & Gobburu, J. (2024). Current status and future directions: The application of artificial intelligence/machine learning for precision medicine. *Clinical Pharmacology & Therapeutics*, *115*(4), 673–686.
27. Naithani, N., Atal, A. T., Tilak, T. V. S. V. G. K., Vasudevan, B., Misra, P., & Sinha, S. (2021). Precision medicine: Uses and challenges. *Medical Journal Armed Forces India*, *77*(3), 258–265.
28. Neoaz, N., Shah, H. H., & Zainab, H. (2025). AI in personalized medicine: Transforming treatment plans through precision health. *Global Journal of Emerging AI and Computing*, *1*(1), 34–50.
29. Nori, L. P., Lohitha, M., Vadapalli, R. R., Bonthagarala, B., Nagineni, S. R., & Kalidindi, V. R. (2025). Revolutionizing healthcare: The impact of AI on precision medicine. *International Journal of Pharmaceutical Investigation*, *15*(2), 334–343.
30. Papalexis, P., Georgakopoulou, V. E., Drossos, P. V., Thymara, E., Nonni, A., Lazaris, A. C., & Thomopoulou, G. E. (2024). Precision medicine in breast cancer. *Molecular and Clinical Oncology*, *21*(5), 78.
31. Patel, S. U., & Jirvankar, P. (2024, November). AI in Healthcare: Precision Medicine and Diagnosis. In: *2024 2nd DMIHER International Conference on Artificial Intelligence in Healthcare, Education and Industry (IDICAIEI)* (pp. 1–6). IEEE.
32. Perlekar, P., & Desai, A. (2025). The role of artificial intelligence in personalized medicine: Challenges and opportunities. *Metallurgical and Materials Engineering*, *31*(3), 85–92.
33. Rajendran, A., Angelin Rajan, R., Balasubramaniyam, S., & Elumalai, K. (2025). AI-enhanced predictive imaging in precision medicine: Advancing diagnostic accuracy and personalized treatment. *Iradiology*, *3*(4), 261–278.

34. Raparthi, M., & Kasaraneni, B. P. (2020). Deep learning for personalized medicine-enhancing precision health with AI. *Journal of Science & Technology*, *1*(1), 82–90.
35. Sahni, L., & Shukla, R. P. (2025). Harnessing the power of precision medicine and AI: Success stories and ethical considerations. In: *Driving Global Health and Sustainable Development Goals with Smart Technology* (pp. 37–82). IGI Global Scientific Publishing.
36. Santoshi, S., & Sengupta, D. (2021). Artificial intelligence in precision medicine: A perspective in biomarker and drug discovery. In *Artificial Intelligence and Machine Learning in Healthcare* (pp. 71–88). Springer.
37. Saxena, S., Jena, B., Gupta, N., Das, S., Sarmah, D., Bhattacharya, P., & Suri, J. S. (2022). Role of artificial intelligence in radiogenomics for cancers in the era of precision medicine. *Cancers*, *14*(12), 2860.
38. Schork, N. J. (2019). Artificial intelligence and personalized medicine. In *Precision Medicine in Cancer Therapy* (pp. 265–283). Springer International Publishing.
39. Segun, A. F. (2024). Advances in personalized medical therapeutics: Leveraging genomics for targeted treatments. *International Journal of Research Publication and Reviews*, *5*(10), 2921–2933.
40. Seyhan, A. A. (2022). The current state of precision medicine and targeted-cancer therapies: Where are we?. In: *Drug Target Selection and Validation* (pp. 119–200). Springer International Publishing.
41. Shang, S., Xia, J., He, G., Zheng, Y., Zhang, J., Lu, H., & Chen, X. (2025). Advances in precision medicine for lupus nephritis: Biomarker-and AI-driven diagnosis and treatment response prediction and targeted therapies. *EBioMedicine*, *117*, 105785.
42. Sisodiya, S. M. (2021). Precision medicine and therapies of the future. *Epilepsia*, *62*, S90–S105.
43. Subramanian, M., Wojtusciszyn, A., Favre, L., Boughorbel, S., Shan, J., Letaief, K. B., & Chouchane, L. (2020). Precision medicine in the era of artificial intelligence: Implications in chronic disease management. *Journal of Translational Medicine*, *18*(1), 472.
44. Valentini, V., D'Oria, M., & Cesario, A. (2022). Artificial intelligence and AI-guided robotics for personalized precision medicine. In: *Humane Robotics*. Vita & Pensiero.
45. Wang, R. C., & Wang, Z. (2023). Precision medicine: Disease subtyping and tailored treatment. *Cancers*, *15*(15), 3837.
46. Avasthi, S., Sanwal, T., Sareen, P., & Tripathi, S. L. (2022). Augmenting mental healthcare with artificial intelligence, machine learning, and challenges in telemedicine. In: *Handbook of Research on Lifestyle Sustainability and Management Solutions Using AI, Big Data Analytics, and Visualization* (pp. 75–90). IGI Global Scientific Publishing.

# Index


Activation maps, 122
AI-assisted surveillance, 29
AlphaFold, 97
Anonymization, 22
Ant colony optimization (ACO), 143
Antimicrobial resistance (AMR), 108
Autoencoder, 231
Average loss, 71

Backpropagation, 251
Bayesian algorithms, 43
Bayesian optimization, 85, 261
Bi-LSTM, 224
Bioactivity, 98
BioBERT, 95
Bioinspired algorithms (BIAs), 137
Biomarker, 58, 218

Cancer detection, 270
Cancer Genome Atlas (TCGA), 105
Cardiovascular medicine, 272
Chronic diseases, 2
Chronic health condition, 136
Clinical AI, 200
Clinical decision support, 45
Clinical reasoning, 192
Cloud orchestration, 26
CNN, 229, 240
Common data models (CDMs), 96
Computational intelligence, 208
Computer vision, 41
Confusion matrix (CM), 129
Convolutional layer, 241
Convolutional neural networks (CNNs), 91
Coronavirus, 156
Correlation coefficient, 65
Counterfactual modelling, 195
COVID-19, 28, 114
COVIDPEN model, 117
Cryptographic primitives, 25
Cryptography, 24
CT scans, 26

Data reliance, 142
Deep learning (DL), 41, 237
DeepSEA, 97
Democratizing Technologies, 109
Demographics, 78
Diabetes diagnosis, 139
Diabetic retinopathy, 271
Differential privacy, 23
Disease surveillance, 24
Drug discovery, 211, 266
Drug dosing, 275
Drug effects, 207
Drug monitoring, 218
Drug–target interactions, 94

ECG signal, 192
Edge AI, 192
EEG healthcare system, 225
Electrocardiogram (EEG), 2, 222
Electronic health records (EHR) 4, 13, 181
Electronic Medical Records (EMRs), 70, 182
Ensemble learning, 43
Explainable AI, 196

F1 score, 131
Fast Healthcare Interoperability Resources, 96
Feature extraction, 65
Federated learning, 28, 192
Flattening layer, 245
FNN, 238
FPGA, 166

Gated recurrent unit, 231
Gaussian mixture models (GMMs), 77
Gene regulatory networks (GRNs), 103
Generative adversarial networks, 95
Generative AI, 276
Generative models, 193
Genetic predisposition, 137
Genome-wide association studies (GWAS), 101
Genomic, 52
Genomic-based therapy, 49
Gestational diabetes, 141
GPU NVIDIA, 121
Graph convolutional networks (GCN), 107
Grid search, 260
Growing self-organizing map (GSOM), 140, 141

Harmonic-to-Noise Ratio (HNR), 60
Harrow Hassidim–Lloyd (HHL), 105
Healthcare records (EHR), 51
Hidden Markov models (HMMs), 102
Hierarchical clustering, 77
Human phenotype ontology (HPO), 106
Hybrid AI, 24
Hyperparameters, 260

ICU Management, 48
IoT, 222
IoT communication layer, 225

Jitter features, 61

K-nearest neighbour, 228

Learnable synergies, 191
Long short-term memory, 95, 230

Machine learning, 40
Max pooling, 244
Medical imaging, 4, 44, 182
Missing completely at random (MCAR), 79
ML-based diabetes diagnosis, 142
MobileNet, 197
Model agnostic meta-learning (MAML), 103
MRI scans, 74
Multimodal learning, 191, 195
Multiobjective optimization (MOO), 98
Multi-omics data, 268

Named entity recognition (NER), 95
Natural language processing (NLP), 14, 41
Neurological disorders, 229
Next-generation sequencing (NGS), 101
Noise-to-harmonic ratios, 61
Nonlinear dynamic measures, 65
Non-oncological disease, 271

Operational Efficiency, 47
Organ-on-a-chip (OoC), 99

Parkinson's disease, 57, 58
Particle collision optimization (PCO), 143
Pattern recognition, 140
Personalized medicine, 46, 217
Pharmacogenomics, 274
Pharmacokinetics, 206
Pooling layer, 243
Population segmentation, 27
Precision, 131
Precision medicine, 206, 266, 269
Precision surgery, 198
Predictive analytics, 41
Predictive care, 198
Pregnancy diabetes, 136
Privacy preservation, 29
Protein–protein interactions (PPIs), 103
Pulse-width modulation (PWM), 166, 167

Quantitative structure–activity relationship (QSAR), 98
Quantum machine learning, 193, 194

Random forest, 227
Random search, 260
Recall, 131
Reconstruction error, 78
Rectified Linear Unit (ReLU), 243
Recurrent neural networks, 246
Recursive feature elimination, 60
Register transfer level (RTL), 174
Reidentification risks, 19
Reinforcement learning, 42
ReLU function, 123
Remote patient monitoring, 46
Reset gate, 250
Retrieval-augmented generation (RAG), 195
RetroGAN, 98

Robotic process automation (RPA), 181
RT-PCR tests, 115

Self-supervised learning (SSL), 196
Setup timing, 176
Stochastic gradient descent (SGD), 257
Structure-based drug design, 97
Support vector machines (SVMs), 58, 227

Targeted therapy, 274
Telehealth, 49
Telemedicine, 46
Timing analysis, 176
Toxicity, 98
Transfer learning, 259
Treatment planning, 45

Update gate, 250

Vaccination, 159, 161
Verilog design, 168
Vivado, 167
Vocal changes, 66
Voice analysis, 59

Warehouse efficiency, 49
Wheel chair control, 166
Worst negative slack, 176

XAIGen, 104
XGBoost, 60, 228
X-ray datasets, 116